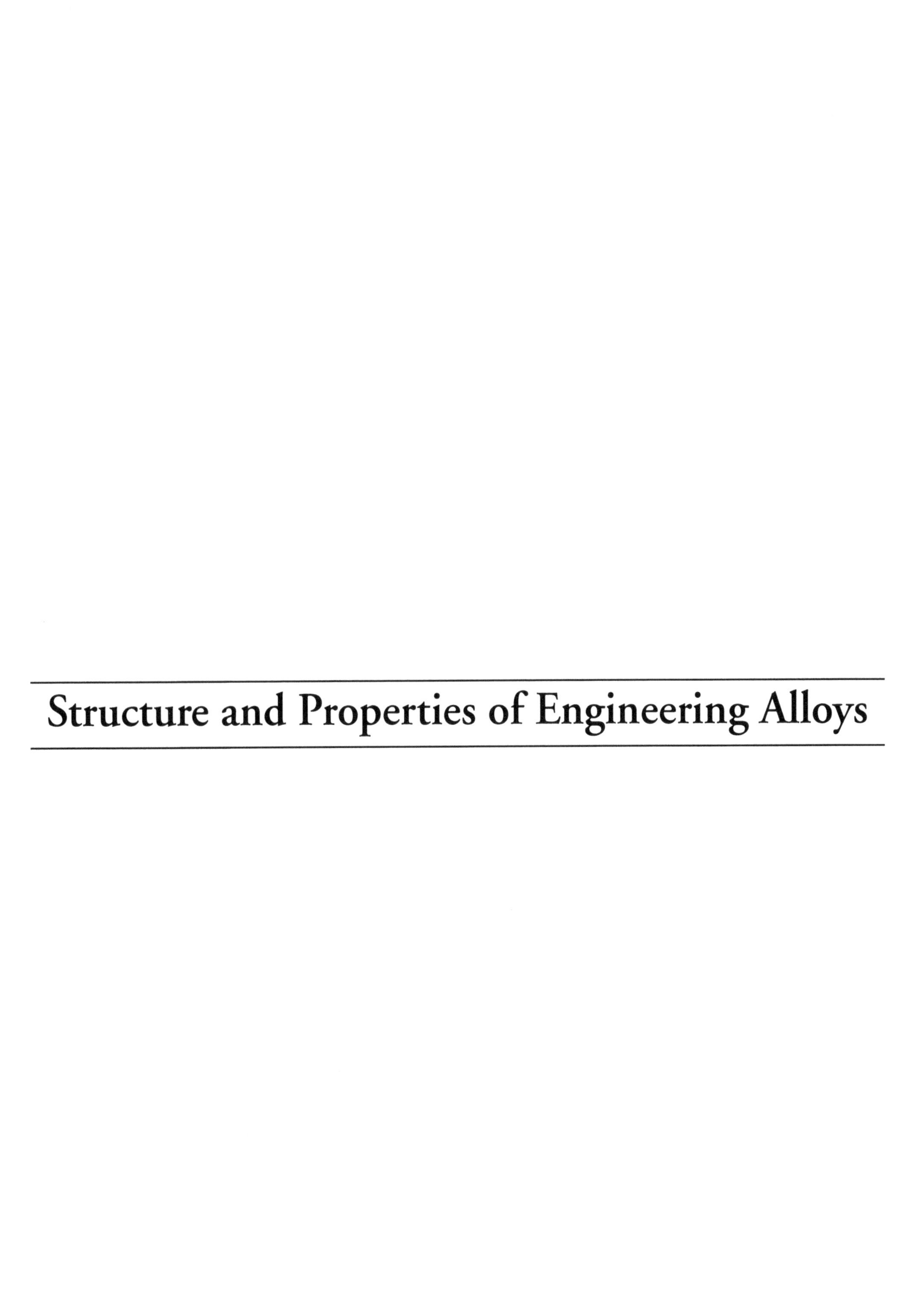

Structure and Properties of Engineering Alloys

Structure and Properties of Engineering Alloys

Editor: Nathaniel Gordon

New York

Published by NY Research Press
118-35 Queens Blvd., Suite 400,
Forest Hills, NY 11375, USA
www.nyresearchpress.com

Structure and Properties of Engineering Alloys
Edited by Nathaniel Gordon

International Standard Book Number: 978-1-64725-436-0 (Hardback)

Cataloging-in-Publication Data

Structure and properties of engineering alloys / edited by Nathaniel Gordon.
p. cm.
Includes bibliographical references and index.
ISBN 978-1-64725-436-0
1. Alloys. 2. Metallurgy. 3. Metallic composites. 4. Metals. 5. Engineering. I. Gordon, Nathaniel.
TA483 .A45 2023
669--dc23

Contents

Preface

It is often said that books are a boon to mankind. They document every progress and pass on the knowledge from one generation to the other. They play a crucial role in our lives. Thus I was both excited and nervous while editing this book. I was pleased by the thought of being able to make a mark but I was also nervous to do it right because the future of students depends upon it. Hence, I took a few months to research further into the discipline, revise my knowledge and also explore some more aspects. Post this process, I begun with the editing of this book.

An alloy refers to a type of mixture made up of chemical elements. It contains at least one metal. Alloys are used widely as construction materials in many industries. The manufacturing technology, the chemical composition, and (micro) structure of the alloy, all influence the functional qualities of these materials. Some of the popular alloys are steel, brass, bronze, and sterling silver. They have a wide range of applications in the making of tools and automobiles. They are also used within the aerospace industry. Research in this field focuses on improving the functional qualities of construction materials in order to lower their weight and boost their safety of usage. Alloying elements are combined with the base metal in order to obtain various desired qualities such as toughness, hardness, and ductility. This book explores all the important aspects of engineering alloys in the present day scenario. It contains a detailed explanation of their structure and properties. Those in search of information to further their knowledge will be greatly assisted by this book.

I thank my publisher with all my heart for considering me worthy of this unparalleled opportunity and for showing unwavering faith in my skills. I would also like to thank the editorial team who worked closely with me at every step and contributed immensely towards the successful completion of this book. Last but not the least, I wish to thank my friends and colleagues for their support.

Editor

Microstructural, Mechanical, Corrosion and Cytotoxicity Characterization of Porous TiSi Alloys with Pore-Forming Agent

Andrea Školáková [1,*], Jana Körberová [1], Jaroslav Málek [2], Dana Rohanová [3], Eva Jablonská [4], Jan Pinc [1], Pavel Salvetr [1], Eva Gregorová [3] and Pavel Novák [1]

[1] Department of Metals and Corrosion Engineering, University of Chemistry and Technology Prague, Technická 5, 166 28 Prague 6, Czech Republic; janca.korberova@seznam.cz (J.K.); pincj@vscht.cz (J.P.); pavel.salvetr@comtesfht.cz (P.S.); Paja.Novak@vscht.cz (P.N.)
[2] UJP Praha a.s., Nad Kamínkou 1345, 156 10 Prague 16, Czech Republic; jardamalek@seznam.cz
[3] Department of Glass and Ceramics, University of Chemistry and Technology Prague, Technická 5, 166 28 Prague 6, Czech Republic; Dana.Rohanova@vscht.cz (D.R.); eva.gregorova@vscht.cz (E.G.)
[4] Department of Biochemistry and Microbiology, University of Chemistry and Technology Prague, Technická 5, 166 28 Prague 6, Czech Republic; Eva.Jablonska@vscht.cz
* Correspondence: skolakoa@vscht.cz

Abstract: Titanium and its alloys belong to the group of materials used in implantology due to their biocompatibility, outstanding corrosion resistance and good mechanical properties. However, the value of Young's modulus is too high in comparison with the human bone, which could result in the failure of implants. This problem can be overcome by creating pores in the materials, which, moreover, improves the osseointegration. Therefore, TiSi2 and TiSi2 with 20 wt.% of the pore-forming agent (PA) were prepared by reactive sintering and compared with pure titanium and titanium with the addition of various PA content in this study. For manufacturing implants (especially augmentation or spinal replacements), titanium with PA seemed to be more suitable than TiSi2 + 20 wt.% PA. In addition, titanium with 30 or 40 wt.% PA contained pores with a size allowing bone tissue ingrowth. Furthermore, Ti + 30 wt.% PA was more suitable material in terms of corrosion resistance; however, its Young's modulus was higher than that of the human bone while Ti + 40 wt.% PA had a Young's modulus close to the human bone.

Keywords: biomaterials; Ti-Si alloy; mechanical properties; reactive sintering

1. Introduction

Even though there has been an extensive research and development of implant materials, the ideal solution has not been found yet. The crucial factors for their development are mechanical properties and corrosion, wear resistance and biocompatibility. However, one of the possible disadvantages is the stress shielding effect which can result in implants failure. This undesirable effect is associated with the different values of Young's modulus of implants and human bone resulting in the non-uniform loading at the bone/implant interface [1]. For this reason, the porous materials have been studied extensively because they could possess Young's modulus close to the humane bone and can solve the problem with the stress-shielding effect. Moreover, materials with porous structure also allow better osseointegration due to the ingrowth of bone tissue into implants. Moreover, the osseointegration takes place for a short time with good quality [1], and is the important property for biomaterial.

The porous biomaterials provide better fixation to the bone in contrast with dense biomaterials. The transport of body fluids through the implants is easier due to pores resulting in the ingrowth of tissue [1]. This ingrowth depends on material, the size and shape of pores, overall porosity, the character of pores system, biocompatibility and Young's modulus. The optimal pore size varies in the range from 100 μm to 500 μm and pores must be interconnected for maintaining the vascular system [2]. The porous biomaterial possesses low Young's modulus allowing to adjust its value to the value corresponding to human bone, which can solve the problem with stress shielding effect. The value of Young's modulus of human cortical bone is stated as 10 to 30 GPa [3].

The porous metallic implants can be classified based on the various criteria. The first of them divides implants to the ones a porous surface and to the implants with volume porosity [2]. The second classification divides the porous implants to the open- and close-pore systems. Implants with close pores have each pore completely separated by a thin metallic wall, while the implants with open pores have the pores interconnected. The interconnected net of pores enables the penetration and the incorporation of the tissue to materials [3]. The porous materials with open pores are preferred for the processing of biomaterials.

Titanium and its alloys [4], mainly TiAl6V4 [2,5], nitinol (NiTi) [6–9], zinc, cobalt and its alloys [10,11], magnesium [12], tantalum [13,14] and stainless steel [15,16] are the most studied biomaterials. Nowadays, most of the implants are usually produced from titanium and its alloy due to their superior corrosion resistance, good mechanical properties and non-toxicity; however, the stress shielding effect occurs when titanium is applied as the implant. It is caused by their stiffness which is much higher than stiffness of bone tissue [1]. Young's modulus of pure titanium is too high (110 GPa) in comparison with the modulus of the human bone (approximately 20 GPa) [17]. Thus, the mentioned presence of pores can hinder this unwanted effect. The titanium also possesses good biocompatibility and its alloys are bioinert [1,18]. The titanium alloys free of toxic elements show even the best biocompatibility [16]. The porous titanium is mainly suitable for the dental and orthopaedic implants although, the process of the surface treatment is still necessary to obtain bioactive surface [5]. On the contrary, the ions of vanadium and aluminum contained in the most used alloy, TiAl6V4, were considered as the components causing health issues such as neurological or enzymatic disorders or disturbers [19]. However, its releasing amount is negligible and does not cause the health issues. Nitinol, with extraordinary properties such as superelasticity and shape memory effect [6], is applied as orthodontic wires and intravascular stents [20]. Cobalt and its alloys excel in terms of fatigue strength, wear resistance and corrosion resistance as well. Hip replacements or dental implants are made of Co-Cr-Mo alloy [11]. Fractures repairs, such as bone plates, bone screws, pins or rods used solely stainless steel [16]. Many of the mentioned and listed biomaterials are designed mainly for permanents, traumatological implants or for replacements of joints. The new porous alloy Ti-Si was chosen as a potential candidate for implants specially determined for application where the ingrowth of bone tissue into implants is required. The silicon belongs to the group of biogenic elements and thus, this one should not cause undesirable reactions, such as toxicological or allergic reactions in the human body. Moreover, silicon is a biocompatible element strengthening the titanium alloys [21]. Another indisputable advantage for processing is a narrow crystallization range and good fluidity [21]. Furthermore, it was found that the number and size of pores increase with increasing content of silicon in titanium alloys [22].

As it was mentioned, one of the possibilities how to improve osseointegration is using of porous biomaterials. One of the ways to obtain porous biomaterials is the preparation of materials containing pore-forming agent (also called as space-holder) by powder metallurgy. This method comprises the mixing of powders, compaction, removing of pore-forming agent and sintering [23]. Firstly, the pore-forming agent in powder form, e.g., NaCl, TiH_2 ZrH_2 or NH_4HCO_3, is mixed with a mixture of metals whose particle sizes must be less than pore-forming agent powder. The second step of the process is producing the green bodies through the uniaxial or isostatical compression at ambient temperatures. Subsequently, the pore-forming agent is removed from the green bodies during low-temperature heat treatment or during dissolution in the solvent which generates pores (the initial stage of neck formation). After removing the pore-forming agent, the sintering at higher temperatures allows us to obtain more porous alloy due to the Kirkendall effect in the case of alloys—mainly due to the developing sinter neck growth [3,17,18,24]. It is believed that the biggest advantage is the possibility to control the mechanical properties by choice of size, shape or the amount of pore-forming agent [3].

In this work, the porous Ti and Ti-Si alloys with the pore-forming agent were prepared by Self-propagating High-temperature Synthesis (SHS). The important characteristics and properties for biomaterials applications, such pore size, Young's modulus, corrosion resistance or cytotoxicity were studied. These alloys were studied as potential candidates for the bone supports and substitutes (e.g., augmentation).

2. Materials and Methods

2.1. Preparation of Samples

The porous titanium samples with 20, 30 and 40 wt.% of the pore-forming agent (PA), TiSi2 alloy with 20 wt.% of PA, pure titanium and TiSi2 alloy without PA were prepared by powder metallurgy. Titanium powder (particle size <45 μm, purity 99.98%, Sigma-Aldrich, St. Louis, MO, USA) and 2 wt.% of silicon (<45 μm, 99.5%, Alfa Aesar, Kandel, Germany) were mixed with the appropriate amount of PA for 10 h at 45 rpm in Turbula 3D blender. Ammonium bicarbonate NH_4HCO_3 was chosen as PA because its decomposition is accompanied by the formation of gaseous NH_3 and CO_2. Subsequently, the obtained mixture was cold-pressed at an isostatic pressure of 400 MPa. The samples with PA were decomposed thermally for 16–18 h at 100 °C. The samples TiSi2 were heated under inert atmosphere (Ar) at a heating rate of 100 °C/min in an induction furnace with a holding time of 10 min at 1100 °C, during which the samples were reactively sintered. Titanium without PA and with various contents of PA was sintered in a vacuum furnace at 800 °C for 1 h with a subsequent increase in temperature to 1300 °C, at which they were sintered for 4 h.

2.2. Microstructure and Phase Composition

All sintered samples were ground by sandpapers P180–P2500. Mechanical-chemical polishing was performed using colloidal suspension Eposil F (ATM GmbH, Mammelzen, Germany) mixed with hydrogen peroxide (volume ratio 1:6). Kroll´s agent (10 mL HNO_3, 20 mL HF, 170 mL H_2O) was used for etching which took for 2 s.

The microstructure was observed by the optical microscope Olympus PME3 with Axio Vision Rel. 4.8 software (Olympus, Tokyo, Japan). The phase composition was studied by PANalytical X´Pert Pro (Cu anode, K_α radiation; PANalytical, Almelo, The Netherlands) and results were analyzed using X´Pert High Score Plus software with PDF2database only in the cases TiSi2 and TiSi2 + 20 wt.% of PA to found out the kind of formed silicides.

2.3. Porosity

The evaluation of resulted porosity (labeled as porosity by image analysis) and pores size were performed using program Lucia 4.80 (Laboratory Imaging, Prague, Czech Republic) and the optical macrographs were used. The volume porosity (labeled as porosity by weight) was calculated according to Equation (1) expressed as:

$$V_{porosity} = \frac{m_0}{m_1}, \quad (1)$$

where m_0 is the weight of the compacted specimen and m_1 is the actual weight of the porous alloy.

2.4. Mechanical Properties

The Vickers microhardness was measured with a load of 100 g (HV 0.1) using device Future-Tech FM-700 (Future-Tech, Kawasaki, Japan). Furthermore, the compressive tests were performed using universal loading machine LabTest5.250SP1-VM (Labortech, Opava, Czech Republic) and three tested samples were prepared by cutting to obtain the dimensions 5 × 5 × 7.5 mm. The three-point flexural tests were performed via the same universal loading machine, but the dimensions were 5 × 5 × 25 mm.

The impulse excitation technique was applied to calculate Poisson's ratio, Young's modulus and shear modulus on the base of measurements of the resonant frequencies. The cylindrical samples with a height of 4 mm and a diameter of 38 mm were tested using the RFDA IMCE machine (IMCE, Genk, Belgium) where the samples were fastened through the nylon fiber. Subsequently, a small projectile tapped on the surface of tested samples and the induced vibration signals were recorded by the microphone. The acquired vibration signals were converted to the resonant frequency through Fourier transformation which enabled us to calculate the Poisson´s ratio, Young's modulus and shear modulus.

2.5. Electrochemical Measurements

The cylindrical samples with a diameter of 15 mm and height of 3 mm were used for corrosion tests. These samples were ground by sandpapers P400–P2500 and, subsequently, ultrasonically cleaned in water and ethanol.

The electrochemical measurement was carried out in physiological solution (9 g/L NaCl) using a potentiostat Gamry 600 with a Gamry Instrument Framework software (Gamry Instruments, Warminster, PA, USA). A standard three-electrode set-up with a Ag/AgCl/sat. KCl electrode (ACLE) as a reference electrode and a graphite electrode as the counter electrode was used. For the determination of the corrosion rate, two methods of measurements were applied—the measurement of polarization resistance and potentiodynamic curves. In the case of polarization resistance measurement, the open-circuit potential (OCP) was stabilized for 3600 s followed by the polarization in the range from −20 to +20 mV/OCP and with scanning rate of 0.125 mV/s. The total exposed area of the sample was 1 cm^2. The potentiodynamic curves were obtained during anodic (from −0.05 V/OCP to 1 V/OCP) polarization with scanning rate 1 mV/s. The exposed area was also 1 cm^2.

2.6. In Vitro "Bioactivity" Tests

The samples were cut into plates with a height of 3.5–4 mm. Furthermore, the samples were ground by sandpapers P80–P800. The hole with a diameter of 2 mm was drilled in the middle of samples for nylon fiber. The polyethene (PE) vials with a volume of 100 mL were filled with ethanol and the samples were hung separately into vials. The vials with samples were inserted to the ultrasonic bath for 10 min. The samples were taken out and dried in air at ambient temperature for a few days.

The in vitro tests started after drying. Studied alloys (pure Ti, Ti + 40 wt.% PA, TiSi2, TiSi2 + 20 wt.% PA) were immersed in simulated body fluid (SBF) for 7 days at 37 °C in biological thermostat. The SBF solution was prepared according to the study [25]. All preparation was performed at 36.5 ± 1.5 °C and tris(hydroxymethyl)aminomethane (TRIS)+HCl was used as a buffer. Samples were hung to PE vials filled by SBF (100 mL) and pH was measured by intoLab 7110 (WTW) with a glass electrode (3 mol/L KCl) after the first, fourth and seventh day of exposure. At the same time, 1 mL of extract was removed and the quantities of ions Ca^{2+} and $(PO_4)^{3-}$ were determined. The same volume of extract was also removed from the SBF solution without sample at the first and last day of the experiment. This enabled us to observe the concentration of these ions without samples. Finally, samples were rinsed with distilled water and dried with flowing air at ambient temperature after the end of *immersion* tests. The evaluation of immersion tests included measurements of many crucial parameters. The first group of parameters consisted of the measurement of pH of extracts, gravimetry of samples and their surface. As it was mentioned, the values of pH were observed after the first, fourth, and seventh day. Furthermore, samples were weighed using analytical balance METTLER TOLEDO AG204 (Mettler Toledo, Amar Hill, India) before and after tests. The surfaces of studied samples were observed using an optical microscope Olympus BX51 (Olympus, Tokyo, Japan) and scanning electron microscope Tescan Vega 3 LMU (Tescan, Brno, Czech Republic) equipped with energy dispersive spectrometry OXFORD Instruments X-max EDS SDD 20 mm^2 (Oxford Instruments, High Wycombe, UK).

The second group of parameters included the determination of Ca^{2+} and $(PO_4)^{3-}$ ions in SBF. The concentration of Ca^{2+} was analyzed by atomic absorption spectrometer GBC 932 plus (GBC Scientific Equipment, Dandenong, Australia) and by flame atomizers (acetylene and N_2O). On the contrary, the concentration of $(PO_4)^{3-}$ ion was determined by ultraviolet-visible spectroscopy (UV-VIS) at a wavelength of 830 nm using spectrophotometer UV-2450 SHIMADZU (Shimadzu, Kyoto, Japan).

2.7. *In Vitro Cytotoxicity Tests*

The samples with the highest amount of pore-forming agent (Ti + 40 wt.% PA and TiSi2 + 20 wt.% PA) were chosen for in vitro cytotoxicity tests and Ti and TiSi2 were used as referent samples. The shape of the studied samples was cylindrical, with a diameter of 15–16 mm and height 3–4 mm. The surface was ground by sandpapers P80–P2500 and rinsed in acetone and ultrasonically cleaned in ethanol for 15 min.

First of all, samples were sterilized in sterilizer Memmert UFP 700 (Memmert GmbH, Schwabach, Germany) at 180 °C for 2 h. Thereafter, sterilized samples were transferred to MEM (Minimal Essential Medium) with 5% FBS (Fetal Bovine Serum). The reduced amount of FBS was used as recommended in the standard ISO 10993-5 [26]. The total amount of MEM + FBS solution was 4–5 mL per sample in order to reach surface-to-volume ratio of 1.25 cm/mL as recommended in the standard ISO 10993-5 [26]. The incubation of samples proceeded at the temperature of 37 °C for 24 h on an orbital shaker Infors HT (Infors HT, Bottmingen, Switzerland) with speed 120 rpm. The extracts were further used for in vitro cytotoxicity tests (elution test).

The elution in vitro cytotoxicity test was performed according to ISO 10993-5 standard [26] with murine fibroblasts L929 (ATCC® CCL-1™, Manassas, VA, USA) which were cultivated in MEM + 10% FBS at standard conditions. The suspension of 1×10^5 cells/mL was prepared. This suspension was seeded into 96-well plates (6 repetitions per each sample). Well plates, located at the edges, were filled by MEM + 10% FBS, but they were not used for the experiment due to the different temperature and moisture. These plates were inserted into incubator Sanyo where the cells were cultivated at 37 °C for 24 h under 5% CO_2.

Thereafter, murine fibroblasts were observed by optical microscope Olympus CKX41 (Olympus, Tokyo, Japan) in order to confirm their uniform growth. The medium was removed from wells and 100 μL of extracts of tested samples was added to semi confluent cell layer. The second and eight columns of plates containing only 100 μL MEM + 5% FBS served as a negative control (Figure 1).

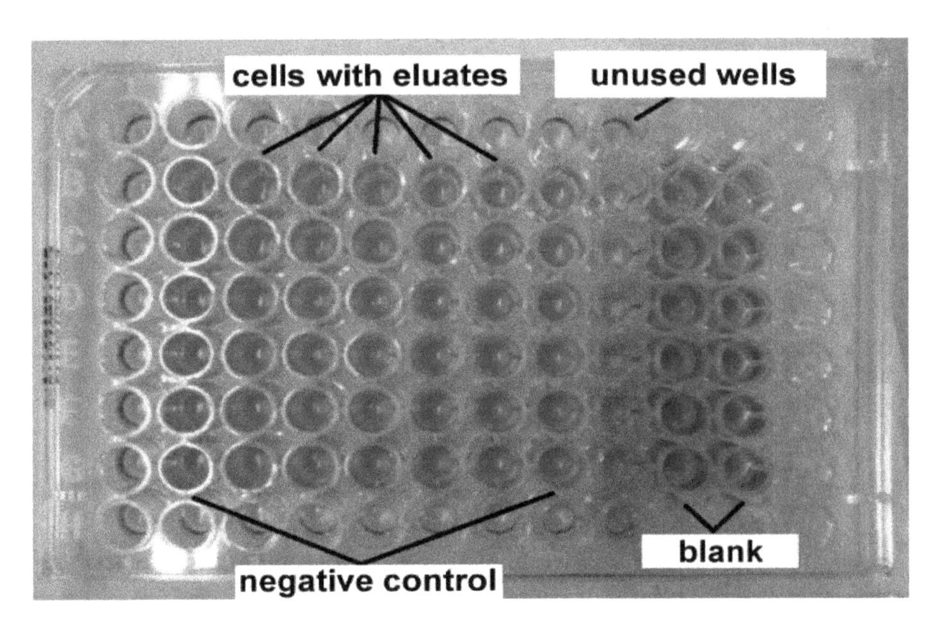

Figure 1. The arrangement of 96-wells plates for cytotoxicity tests.

After 24 h exposition to extracts, the metabolic activity was evaluated. The extracts were removed and wells were filled by 100 μL of MEM + 10% FBS without phenol red + resazurin (25 μg/mL). The wells, labelled as a blank, contained 100 μL of MEM + 10% FBS without phenol red + resazurin and no cells. These plates were incubated for 60 min in the incubator.

The test is based on the ability of viable cells to reduce blue resazurin to fluorescent and violet resorufin. The fluorescence of resorufin was measured using fluorometer Spectramax Minimax i3x platform Molecular devices (Molecular device, San Jose, CA, USA) at input wavelength 560 nm and output wavelength 590 nm and compared to the negative control. The metabolic activity 70% of the control was taken as a minimum value of cytocompatibility.

3. Results and Discussion

3.1. Microstructure and Phase Composition

Optical micrographs are shown in Figure 2a–f. The present pores can be observed as black spots from the smallest ones to the largest ones depending on chemical composition. It seems that pores are open and interconnected which was also observed in [27]. Residual porosity was low and caused by the reactive sintering of pure Ti. Furthermore, the obvious differences in the structure of Ti were affected by the orientation of titanium grains which were etched differently (Figure 2c). It can be seen that the pores were much larger in TiSi2 alloy than in pure Ti (Figure 2a,c), which is caused by the Kirkendall effect and by the different lattice parameters of titanium and silicon influencing the resulted porosity. In the cases of studied alloys with PA, the porosity was caused by both sintering and the decomposition of the pore-forming agent. It is obvious that the porosity evidently increased with the addition of PA. The pore-forming agent was decomposed to gaseous ammonia and carbon dioxide resulting in the formation of the pores. Moreover, in the cases of TiSi2 + 20 wt.% PA alloy (Figure 2b), pores were formed by Kirkendall effect due to unbalanced diffusivities of titanium and silicon during reactive sintering itself and by the decomposition of PA.

Table 1 listed porosity by image analysis and porosity by weight calculated from the theoretical and real weight of samples. If the samples are isotropic (the absence of a directional dependence of porosity), uniform (the absence of porosity gradients) and the location of pores is random (without periodicity of the porosity), the Delesse–Rosiwal law should be valid [28]. According to this law, the porosity by image analysis is equal to porosity by weight. This one was confirmed by the presented values shown in Table 1. Only in the case of TiSi2 alloy, the porosity by weight was significantly higher than porosity by image analysis, suggesting the non-uniform distribution of pores within the alloy.

As was expected, the porosity increased with increasing addition of PA which was already obvious from microstructure (Figure 2a–f). The porosity of TiSi2 + 20 wt.% PA was approximately twice as high as of Ti + 20 wt.% PA alloy. The explanation lies in the mentioned Kirkendall effect and the differences between titanium and silicon lattice parameters. This one explained also the porosity determined for pure Ti and TiSi2 because all processes of sintering elemental powders are susceptible to the formation of Kirkendall pores.

Table 1. Resulted porosity.

Sample	"Porosity by Image Analysis" (%)	"Porosity by Weight" (%)
TiSi2	2 ± 1	15 ± 3
TiSi2 + 20 wt.% PA	37 ± 11	47 ± 1
Ti	1 ± 0	2 ± 1
Ti + 20 wt.% PA	24 ± 5	27 ± 1
Ti + 30 wt.% PA	31 ± 3	35 ± 1
Ti + 40 wt.% PA	46 ± 5	49 ± 1

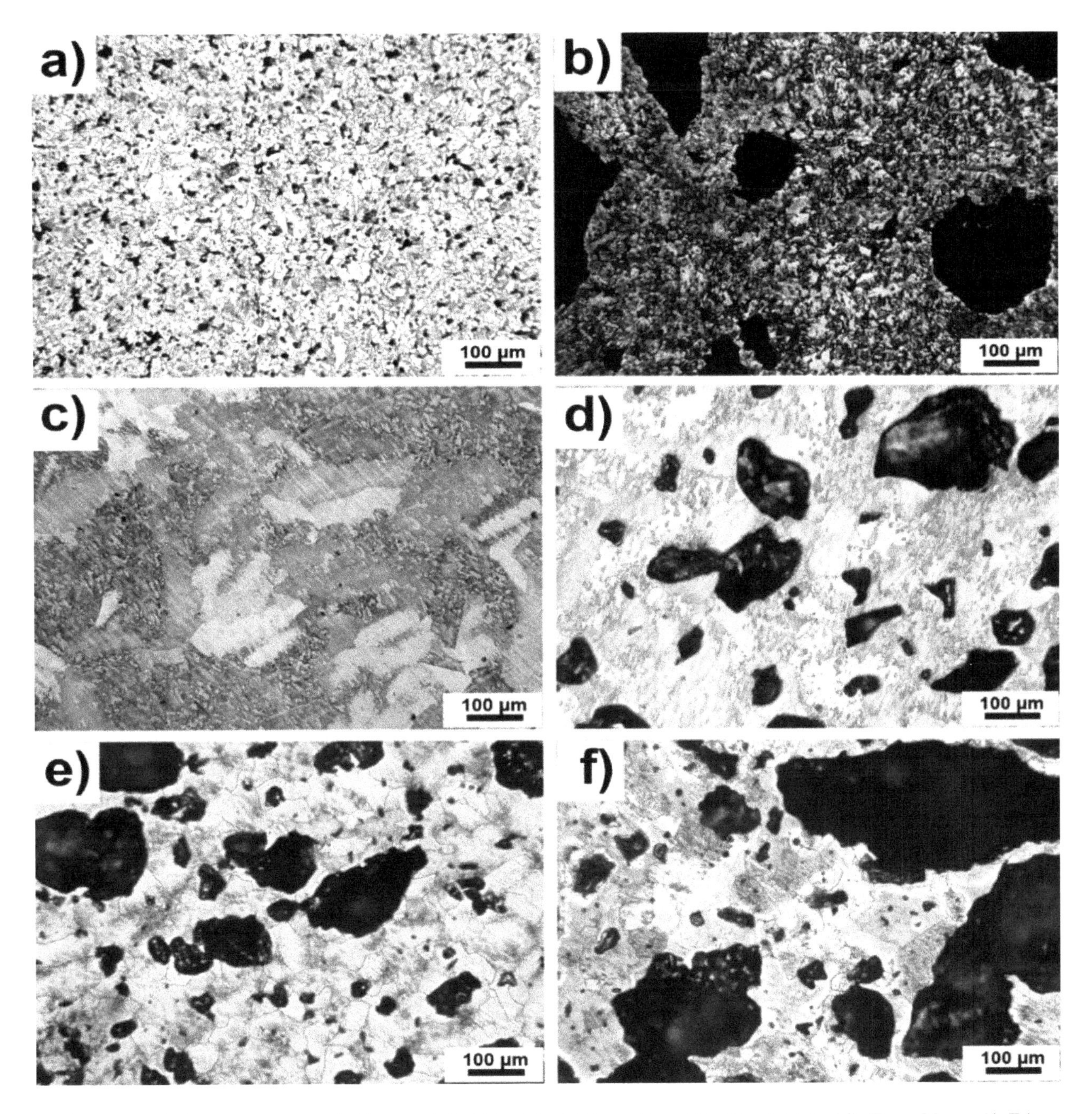

Figure 2. Microstructure of alloys (**a**) TiSi2; (**b**) TiSi2 + 20 wt.% PA; (**c**) Ti; (**d**) Ti + 20 wt.% PA; (**e**) Ti + 30 wt.% PA; (**f**) Ti + 40 wt.% PA.

The histograms shown in Figure 3 illustrate the relative representation of pore size for all studied samples. These results confirmed the increasing pore size with the amount of added PA and the beneficial addition of PA for the synthesis of porous alloys. The observed pores with the size smaller than 100 µm were found in TiSi2 + 20 wt.% PA. Only a few of pores have the size larger than 200 µm. On the contrary, titanium with 20 wt.% PA contained above 90 % of pores with size below 100 µm, 5% of pores smaller than 200 µm and 2 wt.% of pores whose dimensions did not exceed 300 µm. The rest of the observed pores have a size of approximately 500 µm. It is known that the optimal pore size varies in the range of 100–500 µm [2] for the ingrowth of bone tissue into the implants. Only the samples with pore-forming addition contained such large pores; however, almost 80% of them were smaller than 100 µm.

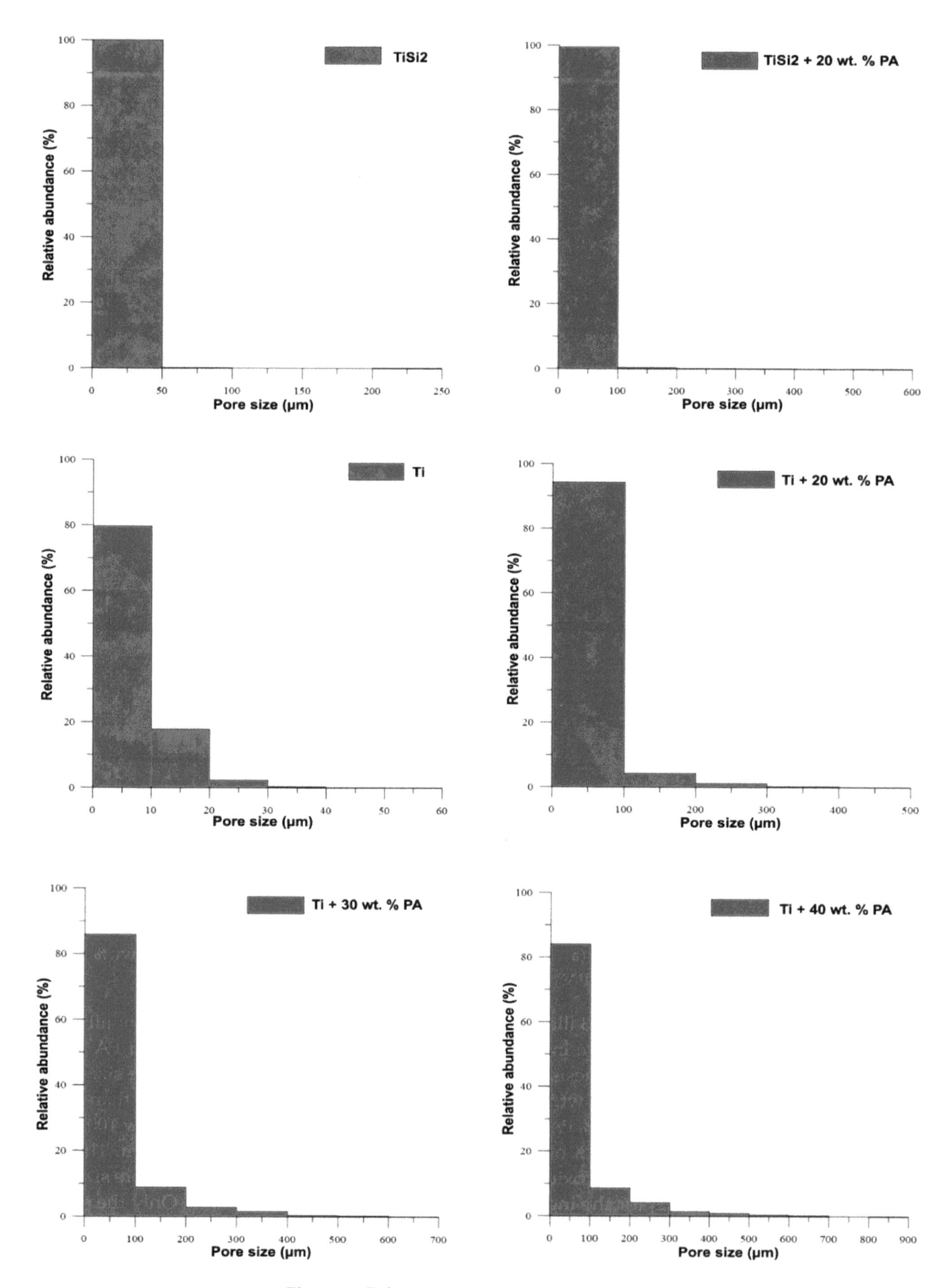

Figure 3. Relative representation of pore size.

The matrix was formed by hexagonal titanium and a solid solution of silicon in α-Ti, which was confirmed by X-ray analysis. X-ray diffraction was performed to identify the types of silicides formed in TiSi2 and TiSi2 + 20 wt.% PA alloys during reactive sintering. X-rays analyses revealed the presence of Ti_5Si_3 and $TiSi_2$ silicides (Figure 4) suggesting the nonequilibrium behavior during reactive sintering. This claim was confirmed by comparison of the expected phase composition with the equilibrium phase diagram [29]. Therefore, it is obvious that Ti_5Si_3 phase formed due to the kinetical factors [30]. Despite the presence of the pore-forming agent, phase composition was not influenced.

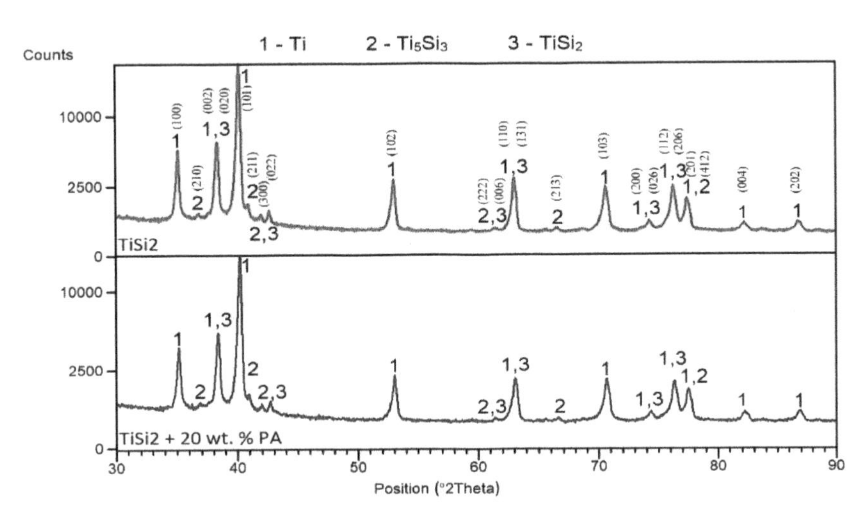

Figure 4. XRD patterns of TiSi2 and TiSi2 + 20 wt.% PA alloys.

Found phases are marked in Figure 5 and was identified according to work [22], which studied the formation of silicides in porous Ti-Si alloy. Titanium formed matrix (white area) while the light grey round particles at the boundaries were Ti_5Si_3 phase (confirmed during EDS analysis). Silicides $TiSi_2$ occurred primarily at the edges of pores. The presence of those phases is not a surprise. Trambukis et al. [31] found the phases sequence of formation during SHS, which is: $TiSi_2 \rightarrow TiSi \rightarrow Ti_5Si_4 \rightarrow Ti_5Si_3$. According to this sequence, it is clear that $TiSi_2$ phase forms preferentially and it is gradually transformed to the more stable Ti_5Si_3 phase. Moreover, it was confirmed that Ti_5Si_3 phase can also form by the direct reaction between titanium and silicon [32] and it is the most thermodynamically stable phase ($\Delta H_f = -579$ kJ/mol) in the Ti-Si system [33]. The high surface-to-volume ratio of small titanium particles enables full contact between titanium and silicon, resulting in the formation of intermediate phase at lower temperatures and the mechanism in the initiation of the SHS reaction was the transformation of α-Ti to β-Ti, as shown in [33].

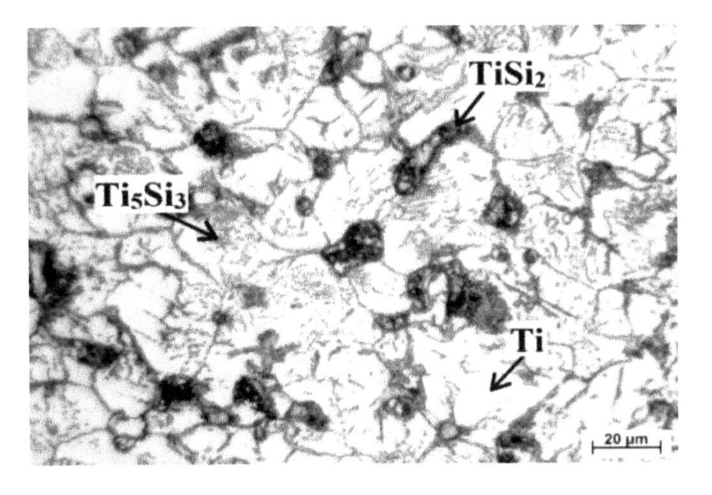

Figure 5. Detail of microstructure of TiSi2 alloy.

3.2. Mechanical Properties

First of all, the microhardness was measured only on areas without pores and results are shown in Figure 6. In the case of TiSi2 and TiSi2 + 20 wt.% PA, the microhardness of the matrix (not silicides) was determined. The microhardness of silicides was impossible to measure because as it was shown, they appeared as thin and elongated particles close to pores (Figure 2a,b). The resulted values of microhardness were approximately similar for all tested alloys.

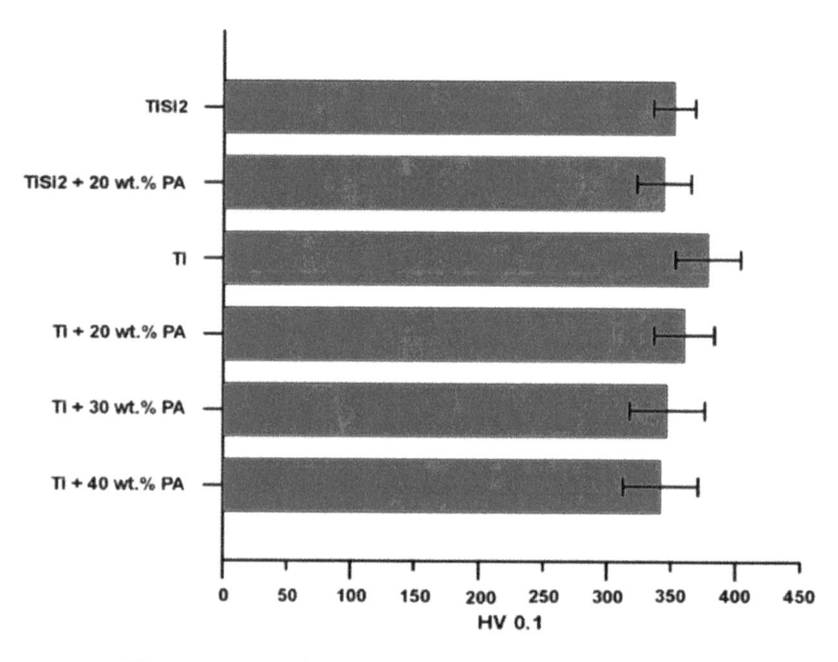

Figure 6. Microhardness HV 0.1 of studied alloys.

The course of compressive stress–strain curves of pure Ti and TiSi2 alloy was similar (Figure 7) and the slight decrease in compressive strength belonging to TiSi2 alloy was mainly affected by higher volume porosity. However, both samples exhibited high compressive strength (Table 2) resulting from their microstructure.

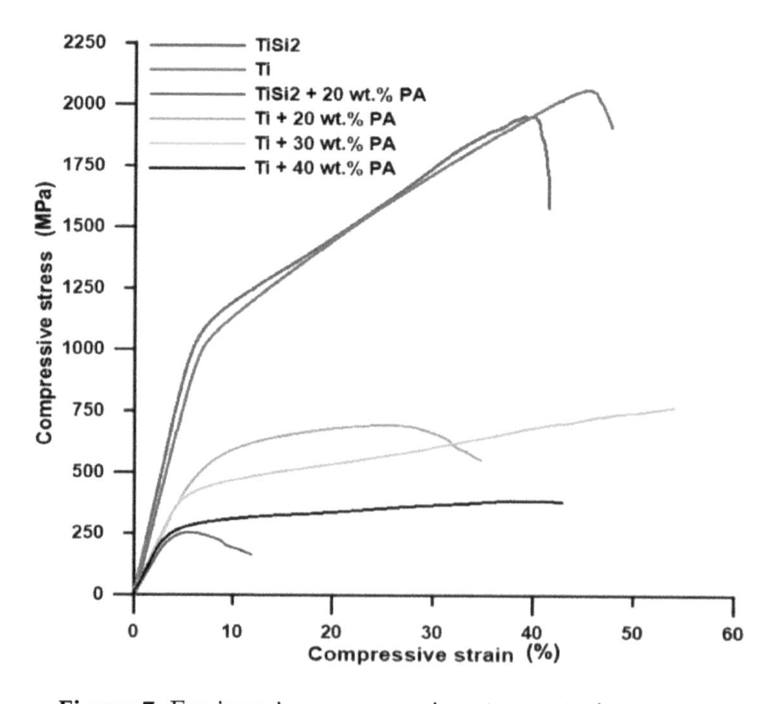

Figure 7. Engineering compressive stress–strain curves.

Table 2. Mechanical properties obtained from compressive stress–strain curves.

Sample	$R_{p0.2}$ (MPa)	R_m (MPa)	E (GPa)
TiSi2	980 ± 18	1846 ± 153	18 ± 0
TiSi2 + 20 wt.% PA	250 ± 52	291 ± 26	8 ± 0
Ti	1039 ± 31	2104 ± 61	19 ± 0
Ti + 20 wt.% PA	508 ± 13	686 ± 9	10 ± 0
Ti + 30 wt.% PA	386 ± 4	-	9 ± 1
Ti + 40 wt.% PA	233 ± 22	-	8 ± 0

The shape of the compressive stress–strain curve of Ti + 40 wt.% PA was typical for highly porous metallic materials. The present pores absorb the applied strain which can be seen as the low increase or almost constant stress accompanied by a significant increase in compressive strain (Figure 7). Tang et al. [34] presented the deformation course of Ti with 40–70% porosity compared to deformation curve observed for Ti + 40 wt.% PA alloy in this work. The course of the deformation curve belonging to Ti + 30 wt.% PA (Figure 7) was the same. The increase in compressive strain is also obvious, together with compressive stress, which is associated with lower porosity and pore size for both alloys. In these two cases, the compressive stress–strain tests were stopped manually when the compressive stress increased significantly. Thus, the values of $R_{p0.2}$ (yield strength) could be deducted while the values of R_m (ultimate compressive strength) were not listed in Table 2. All prepared alloys possessed much higher yield strength, compressive ultimate strength and higher or same the value of Young's modulus than those of porous titanium scaffolds shown in work [1] and TiAl6V4 scaffolds presented in work [5]. Nevertheless, the values were too high for human cancellous bone (E = 0.9 GPa and R_m = 22 MPa) [1], but the obtained modulus generally suits the modulus for cortical bone varying in the range of 10–30 GPa [16].

The sample Ti + 20 wt.% PA exhibited low plasticity in comparison with samples containing more pores (Figure 7). However, the fracture could be observed in Ti, Ti + 20 wt.% PA, TiSi2 + 20 wt.% PA and TiSi2 alloy. The deviations of stress were observed on the stress–strain curve of TiSi2 + 20 wt.% PA during loading. It can be assumed that they are caused by the presence of silicides around pores. These silicides could hinder the absorption of loading by pores which influenced the plastic behavior of the material. The unusual large pores also caused very low R_m of TiSi2 + 20 wt.% PA (Table 2) suggesting the decreasing of $Rp_{0.2}$ and E with increasing the porosity (Table 2).

The fractographic characterizations were performed after compressive stress–strain tests to reveal the mechanism of failure and morphologies are shown in Figure 8a–c. The dimples, meaning the ductile fracture, are obvious only in pure Ti (Figure 8b). The morphology of fractures of TiSi2 and Ti + 20 wt.% PA alloys did not contain dimples (Figure 8a,c) but no facets were also observed, so the cleavage fracture could not be considered. However, the courses of obtained curves can be proof of the ductile fracture of studied materials. The fracture took place diagonally suggesting the shear stress acting at the end of measurement due to dislocation slip. The grain boundaries were weakened by the presence of brittle Ti_5Si_3 phase in the case of TiSi2 alloy.

The results obtained by impulse excitation are shown in Table 3. The value of the Poisson number (μ) of the studied titanium sample is very close to the value of pure titanium (0.32) [35]. On the other hand, the Poisson number of TiSi2 is the lower which is probably caused by the higher volume porosity and by the presence of silicides. Table 3 also revealed that porosity has a significant effect on the value of the Poisson number. Its value decreased with the increment of porosity. The shear modulus was calculated from the Poisson number and Young modulus and results are also shown in Table 3. The calculated value of the shear modulus of studied titanium is close to the value stated for pure titanium, which is approximately 43 GPa [35]. Because the values of shear modulus are dependent on measured Poisson number and Young's modulus, the trends between values are the same as mentioned above. The values of modulus of TiSi2, Ti, Ti + 20 wt.% PA samples are satisfied for the using as the trabecular bone replacement [36]. On the other hand, the values of the modulus of all samples were higher than those of human cancellous bone, although they fulfill the requirements to pore size and porosity (Table 1, Figure 5) for human cancellous bone (pore size of 20–1000 μm and porosity 30–95%) [18]. Obviously, Ti + 30 wt.% PA having volume porosity 35 ± 1%, Young's modulus 47 ± 1 GPa (from compressive tests) and 9 ± 1 GPa (impulse excitation) is suited as the replacement of human cortical bones [37]. The elastic modulus of porous alloys strongly depends on the number of pores, the pore size and also the pore morphology [34] which is reflected by our results in Tables 1 and 3, and confirmed this statement. The different values of Young's modulus obtained by impulse excitation and from the compressive stress–strain curves are affected by the method of measurement. The samples contained the pores acting as stress concentrator and, thus, they make easier the spreading of crack. The value of Young's modulus is deducted from the linear part of obtained compressive stress–strain curve. Meanwhile, the samples are not loaded during impulse excitation and, therefore, the present pores do not affect the mechanical properties.

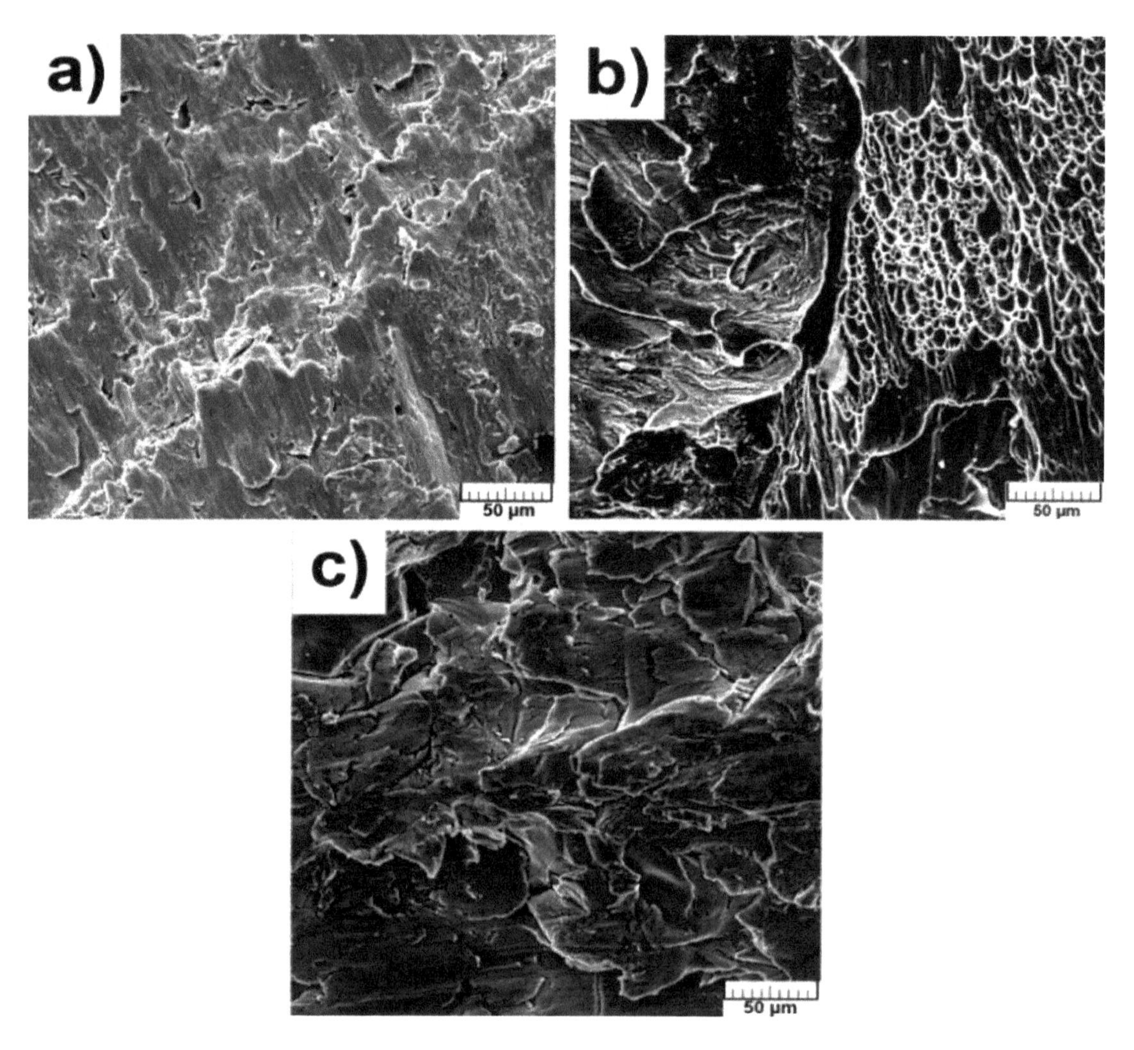

Figure 8. Fracture surface of alloys: (**a**) TiSi2; (**b**) Ti; (**c**) Ti + 20 wt.% PA.

Table 3. Mechanical properties of studied alloys obtained by impulse excitation.

Sample	μ	G (GPa)	E (GPa)
TiSi2	0.28	30 ± 1	77 ± 2
TiSi2 + 20 wt.% PA	0.26	12 ± 1	30 ± 1
Ti	0.33	44 ± 1	116 ± 4
Ti + 20 wt.% PA	0.28	24 ± 1	61 ± 1
Ti + 30 wt.% PA	0.27	18 ± 1	47 ± 1
Ti + 40 wt.% PA	0.26	11 ± 1	28 ± 1

The Young's modulus obtained from impulse excitation was compared with the theoretical one calculated according to the Gibson–Ashby model [18] is expressed by Equation (2):

$$\frac{E}{E_0} = k\,(1-P)^2, \tag{2}$$

where E is Young's modulus for porous titanium, E_0 is Young's modulus for compact titanium, k is the constant (equal to 1 for metals) and P is the porosity of porous titanium. The value 113 GPa [35] belonging to pure titanium without pores was considered for calculation. As can be seen, the values obtained from impulse excitation lies on the curve of theoretically calculated values (Figure 9). This means that the impulse excitation is an accurate method for the determination of the Young's modulus, unlike the compressive stress–strain tests. Furthermore, for both of the experimental (impulse excitation) and calculated results, Young's modulus decreases with the increasing porosity (Figure 9).

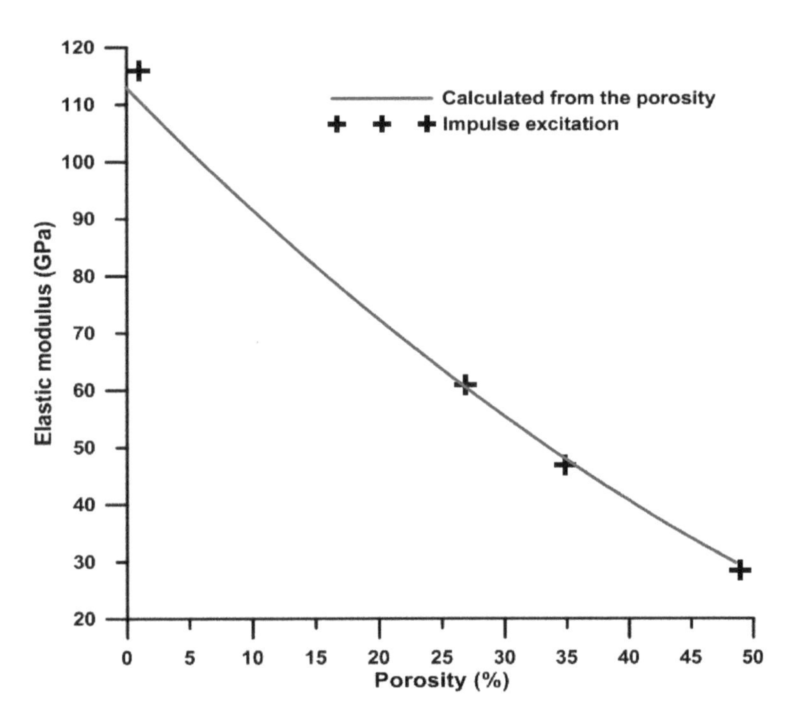

Figure 9. The dependence of Young's modulus (calculated and experimental determined) on porosity of titanium.

The bending strength obtained during three-point flexural tests is listed in Table 4. Besides the results of compressive stress–strain tests, the bending strength during flexural tests varied in a wide range. The reason lies in the presence of silicides and less the volume porosity. When the TiSi2 + 20 wt.% PA is compared to Ti + 40 wt.% PA with the same volume porosity, the different bending strength is affected by silicides, but increasing porosity decreased the bending strength. The bending strengths were also about 2.5–16 times lower than as-cast TiSi1 in [21], which can be explained by the presence of pores.

Table 4. Strength of studied alloys during three-point flexural tests.

Sample	R_m (MPa)
TiSi2	454 ± 36
TiSi2 + 20 wt.% PA	111 ± 15
Ti	652 ± 31
Ti + 20 wt.% PA	412 ± 0
Ti + 30 wt.% PA	328 ± 21
Ti + 40 wt.% PA	252 ± 35

3.3. Corrosion Resistance

3.3.1. Electrochemical Measurements

The anodic curves of potentiodynamic measurement are shown in Figure 10 and only these curves are presented because they illustrated the oxidation of metals. The oxidation of titanium was already described systemically in [38,39]. In our work, the sample of Ti exhibited an increase in the current density in range 0.2–1 V, suggesting the passivation of the sample. In addition, the passivation could be observed in the case of Ti + 20 wt.% PA and Ti + 30 wt.% PA as well. On the contrary, the instability of the passivation layer can be found in Ti + 40 wt.% PA which resulted in the local maxima and impossibility of stabilization of current density from 0.4 V/ACLE. The alloys with Si addition were not able to passivate (Figure 10).

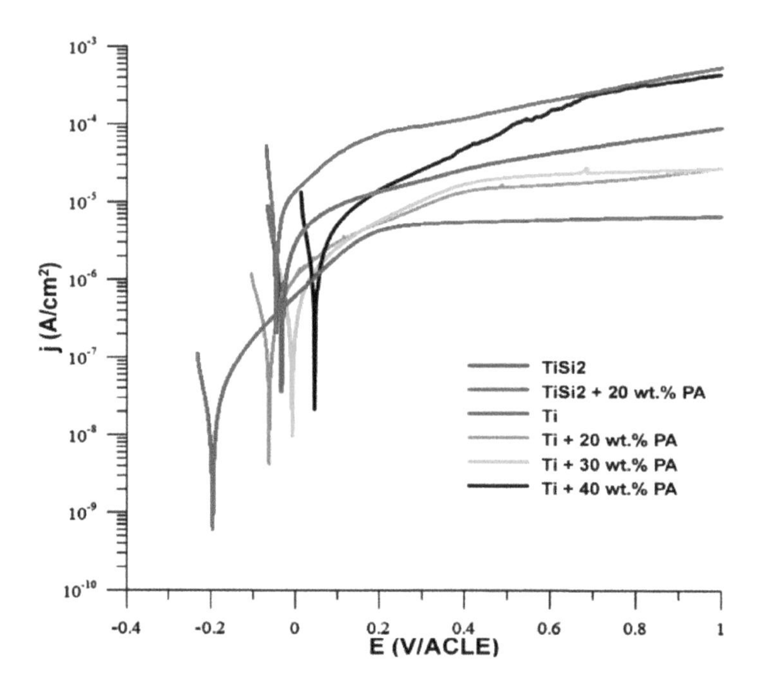

Figure 10. Anodic polarization in simulated body fluid (SBF).

The measured values obtained from the anodic and cathodic polarization are listed in Table 5. Tafel slopes (b_a and b_c) were determined from the anodic and cathodic parts of potentiodynamic curves. Both quantities were used for the calculation of corrosion rate which was related to geometric area 1 cm^2. However, the real area of samples depends on the size and the number of pores contacted with electrolytes and, therefore, the real area is much higher than the geometric area in the case of porous materials. It is known that the pores decrease the corrosion resistance of titanium because they increase the specific surface area its contact with the medium [34]. The present pores could also induce the localized corrosion influencing the values shown in Table 5. The corrosion rate increased with increasing porosity. The R_p (polarization resistance) value of studied porous alloys is significantly smaller than those obtained for the cast titanium [1] which confirmed the important effect of pores in structure. The silicon addition also affected the corrosion rate. As can be seen, TiSi2 alloy had significantly higher corrosion rate in comparison with the sample with the same volume porosity Ti + 20 wt.% PA. For this reason, this sample was studied in detail after corrosion experiments. The study of the exposed structure revealed the intergranular corrosion meaning the weakening of grain boundaries (Figure 11) due to the present of silicides. The higher corrosion rate in Ti + 40 wt.% PA than in TiSi2 + 20 wt.% PA with similar volume porosity was probably caused by the high current density obtained from cathodic curves where the reduction in environmental components took place.

Table 5. Results of anodic and cathodic polarization.

Sample	Anodic Polarization			Cathodic Polarization			R_p	v_{corr}
	b_a	E_{corr}	j_{corr}	b_c	E_{corr}	j_{corr}		
			$\times 10^{-6}$			$\times 10^{-6}$	$\times 10^3$	$\times 10^{-3}$
	(V/dec)	(V/ACLE)	(A/cm^2)	(V/dec)	(V/ACLE)	(A/cm^2)	($\Omega \times$ cm^2)	(g/cm^2 $\times$ a)
TiSi2	0.68	−0.033	6.1	0.41	−0.020	5.4	12	36.8
TiSi2 + 20 wt.% PA	0.90	−0.046	38.1	0.41	−0.073	35.0	2	272.1
Ti	0.19	−0.196	0.1	0.17	−0.253	0.1	330	0.5
Ti + 20 wt.% PA	0.43	−0.062	1.3	0.34	−0.069	0.8	48	6.8
Ti + 30 wt.% PA	0.35	−0.008	1.4	0.54	0.093	4.0	11	33.7
Ti + 40 wt.% PA	0.37	0.045	5.3	-	0.133	114.3	1	573.5

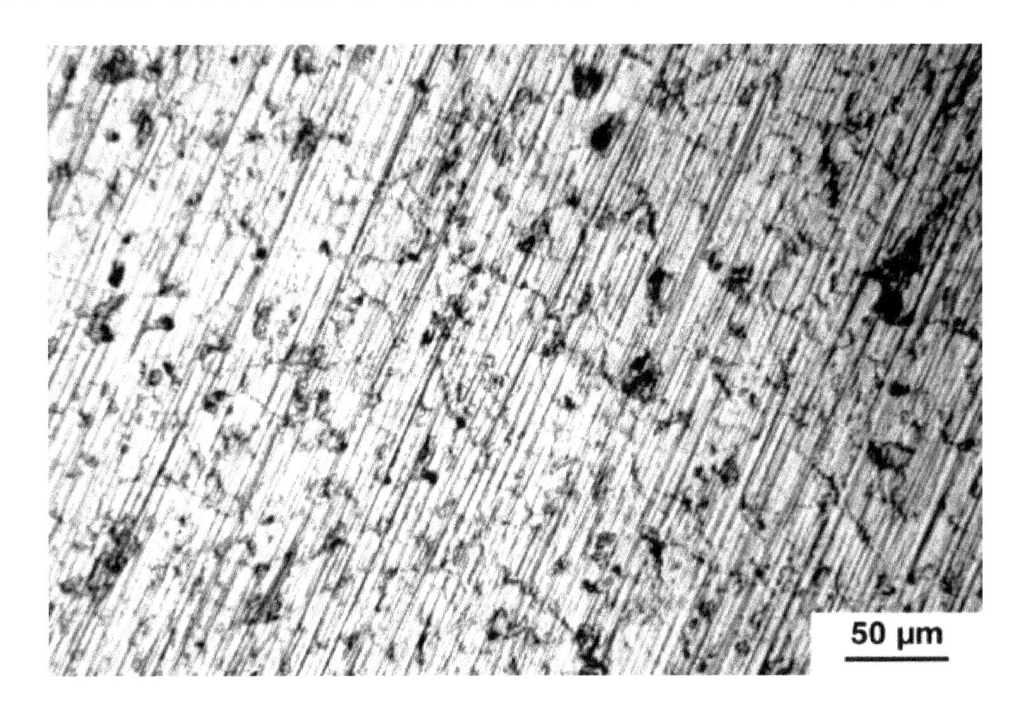

Figure 11. Intergranular corrosion of TiSi2 alloy.

3.3.2. In Vitro "Bioactivity" Tests

No changes of pH measured for extracts were observed during exposure in comparison with the SBF without samples. The value of pH was approximately 7.42 at 36.5 °C. In addition, no significant changes in weight were observed (Table 6). The negative weight loss was observed in the case of TiSi2 and TiSi2 + 20 wt.% PA suggesting the dominance of corrosion of those samples over the precipitation of Ca-P phase. The slight increase in weight (Ti and Ti + 40 wt.% PA) could be caused by the precipitation of low amounts of sodium chloride on the surface whose presence was confirmed during EDS analysis.

Table 6. The changes in weight after 7 days in SBF solution.

Sample	Weight Change (g)
TiSi2	−0.023 ± 0.004
TiSi2 + 20 wt.% PA	−0.050 ± 0.016
Ti	0.001 ± 0.000
Ti + 40 wt.% PA	0.003 ± 0.000

The surface of samples did not change after exposure in SBF and no hydroxyapatite, $Ca_{10}(PO_4)_6(OH)_2$, was found which could point to bioactivity. The hydroxyapatite has exceptional biocompatibility and bioactivity properties [12]. SEM analysis did not reveal the presence of Ca-P phase. The sodium chloride occurred mainly at the edges of the pores (Figure 12a–d).

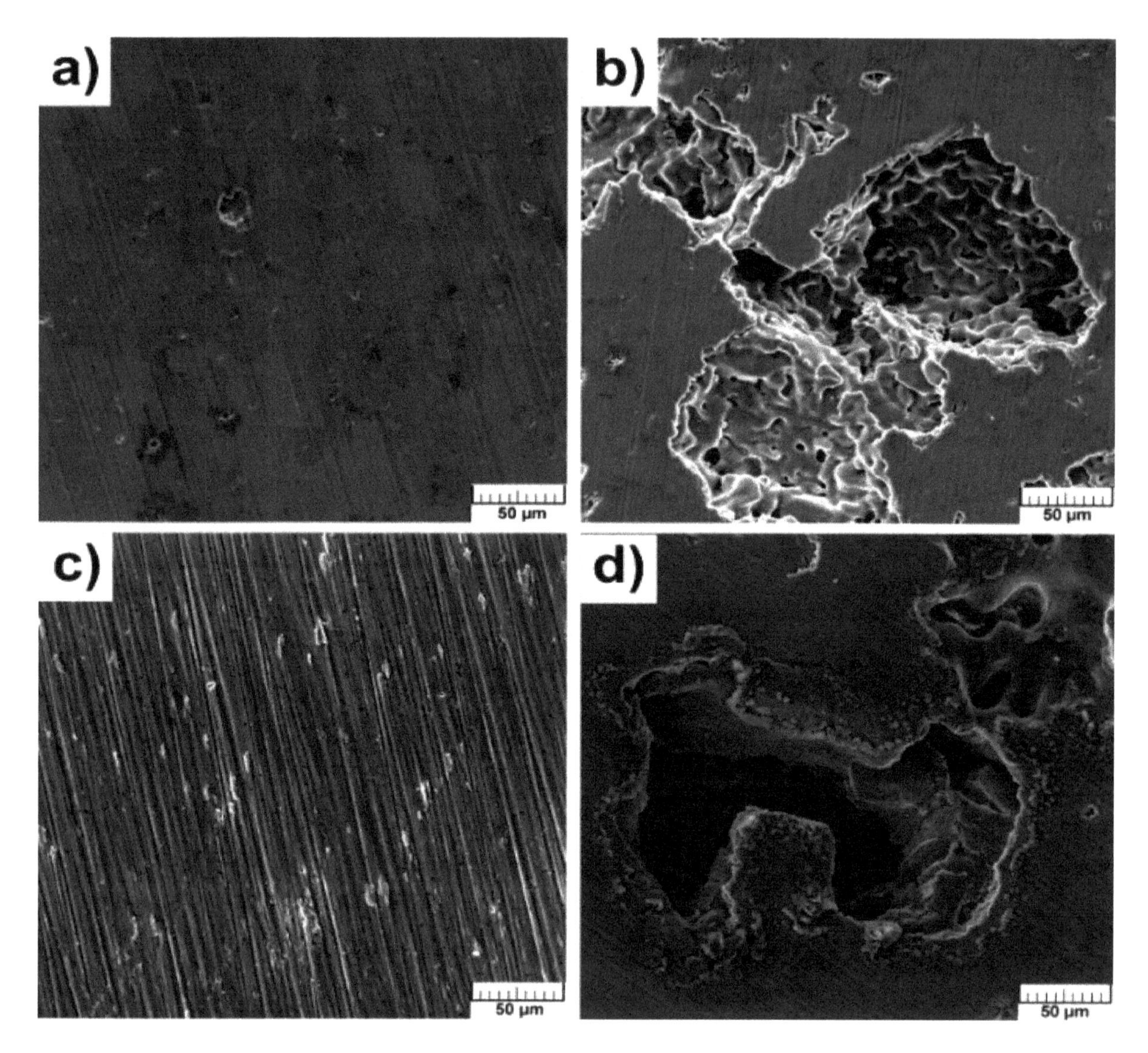

Figure 12. Microstructure of alloys after exposition in SBF: (**a**) TiSi2; (**b**) TiSi2 + 20 wt.% PA; (**c**) Ti; (**d**) Ti + 40 wt.% PA.

When the samples exhibit bioactive behavior, the concentration of ions $(PO_4)^{3-}$ and Ca^{2+} decreases in SBF due to the precipitation of the Ca-P phase. It can be assumed that alloys with Si will not be bioactive according to results of SEM analysis and pH values. Figure 13a–d clearly show that no studied samples are bioactive because the concentration of ions $(PO_4)^{3-}$ and Ca^{2+} increases in time. The observed declines were associated with the adsorption of ions on the surface with their subsequent release into solution. As is known, titanium is a bioinert material [18], and this work showed that the pores did not affect bioactive behavior through the absorption of ions with the subsequent precipitation of required phases. It is still necessary to solve the problem of the bioactivity of studied samples and investigate possible methods to improve it. This obstruction can be successfully resolved by the titania sol-gel coating which was tested on TiSi alloys and presented in [40].

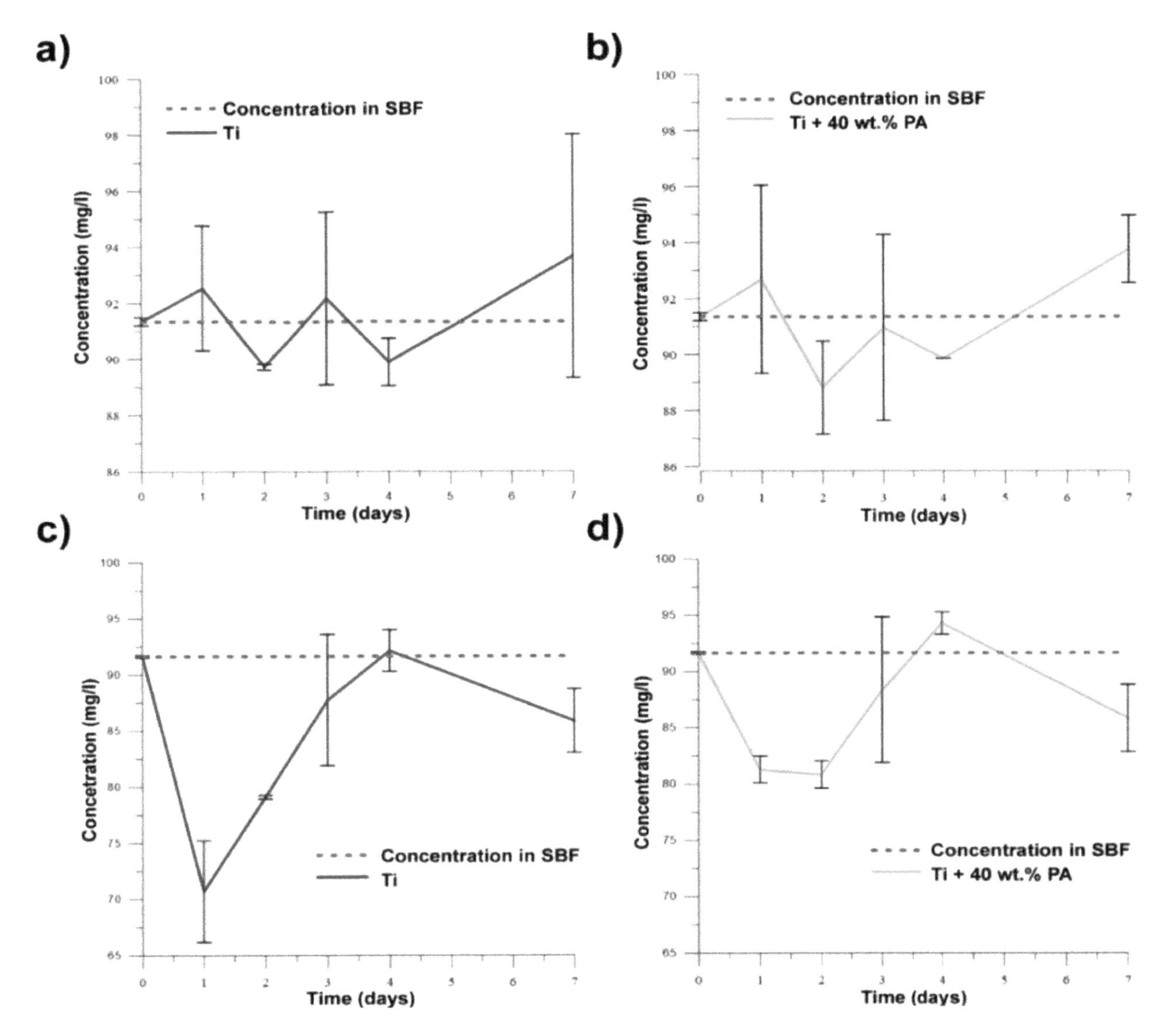

Figure 13. The changes in concentration of $(PO_4)^{3-}$ for (**a**) Ti; (**b**) Ti + 40 wt.% PA and of Ca^{2+} for (**c**) Ti; (**d**) Ti + 40 wt.% PA in SBF.

3.4. In Vitro Cytotoxicity Testing

The in vitro cytotoxicity of samples TiSi2, TiSi2 + 20 wt.% Pa, Ti and Ti + 40 wt.% PA was tested according to the ISO standard. The murine fibroblasts L929 were exposed to extracts of the samples for 24 h. The materials are considered as cytocompatible (not toxic) when the metabolic activity of the cells does not decrease below 70% of the negative control, which is illustrated as the dashed line in Figure 14. As can be seen, none of the extracts caused a decrease in metabolic activity after 24 h incubation with the cells. Thus, according to the elution test, the samples can be considered as cytocompatible. Neither silicon addition nor PA implemented the cytotoxicity and, therefore, the developed materials seem promising for biomedical application.

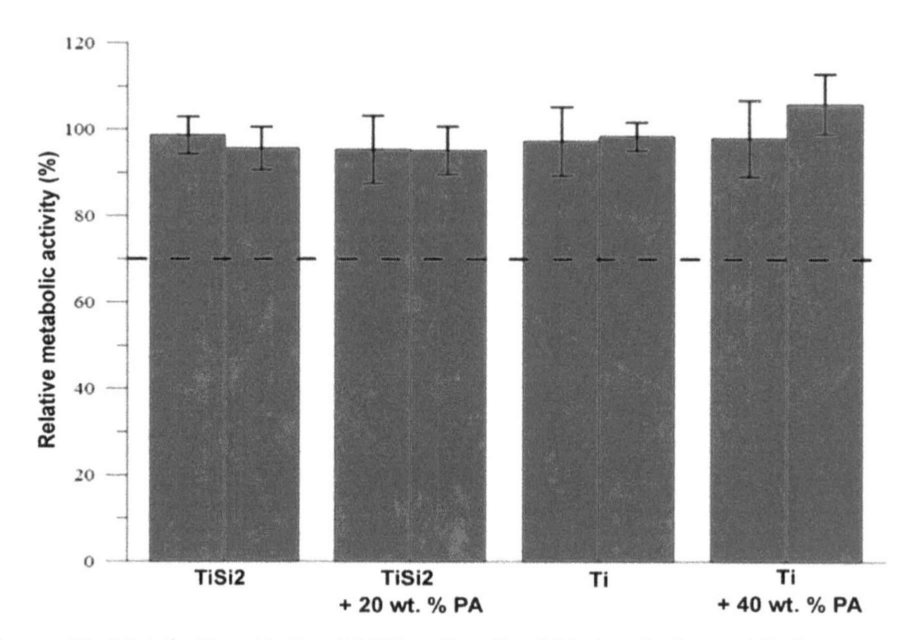

Figure 14. Metabolic activity of L929 cells after 24 h incubation with the extracts.

4. Conclusions

The porous alloys were prepared successfully by powder metallurgy. All Ti-Si samples with pore-forming agent contained pores with the size of 100–500 μm suitable for bones ingrowth but most of them were found in Ti + PA. Titanium with PA also exhibited better mechanical characteristics and plasticity than TiSi2 + 20 wt.% PA alloy. Higher values of yield strength belong to Ti + PA in both used tests (the compression and the three-point flexural tests). However, the most important parameter for augmentation or spinal replacements is Young's modulus. The values of the Young's modulus of TiSi2 + 20 wt.% PA (30 ± 1 GPa) and Ti + 40 wt.% PA (28 ± 1) were close to the Young's modulus of human cortical bone. In the case of these alloys, the volume porosity was similar—47 ± 1% (TiSi2 + 20 wt.% PA) and 49 ± 1% (Ti + 40 wt.% PA). The corrosion resistance was affected by the presence of silicides in microstructure causing the corrosion along the grain boundaries. For this reason, titanium with PA addition samples possessed better corrosion resistance. Moreover, the pores filled by electrolyte created a more aggressive environment due to the worse exchange of electrolyte. Bioactivity and cytotoxicity were not observed. The Ti + PA and Ti + 30 wt.% PA were determined as the most suitable for porous implants.

Author Contributions: Conceptualization, A.Š., J.K. and P.N.; Methodology, A.Š. and P.N.; Formal Analysis, A.Š. and J.K., Investigation, A.Š., J.K., P.S., J.M.; J.P., E.J., E.G. and D.R.; Data Curation, A.Š., J.K., P.S. and J.P.; Writing—Original Draft Preparation, A.Š.; Writing—Review and Editing, A.Š. and P.N. All authors have read and agreed to the published version of the manuscript.

References

1. Dabrowski, B.; Swieszkowski, W.; Godlinski, D.; Kurzydlowski, K.J. Highly porous titanium scaffolds for orthopaedic applications. *J. Biomed. Mater. Res. Part B Appl. Biomater.* **2010**, *95*, 53–61. [CrossRef] [PubMed]
2. Li, J.P.; Li, S.H.; Van Blitterswijk, C.A.; de Groot, K. A novel porous Ti6Al4V: Characterization and cell attachment. *J. Biomed. Mater. Res. Part A* **2005**, *73*, 223–233. [CrossRef] [PubMed]
3. Ryan, G.; Pandit, A.; Apatsidis, D.P. Fabrication methods of porous metals for use in orthopaedic applications. *Biomaterials* **2006**, *27*, 2651–2670. [CrossRef] [PubMed]
4. Geetha, M.; Singh, A.K.; Asokamani, R.; Gogia, A.K. Ti based biomaterials, the ultimate choice for orthopaedic implants—A review. *Prog. Mater. Sci.* **2009**, *54*, 397–425. [CrossRef]
5. Lee, H.; Jang, T.-S.; Song, J.; Kim, H.-E.; Jung, H.-D. Multi-scale porous Ti6Al4V scaffolds with enhanced strength and biocompatibility formed via dynamic freeze-casting coupled with micro-arc oxidation. *Mater. Lett.* **2016**, *185*, 21–24. [CrossRef]

6. Prymak, O.; Bogdanski, D.; Köller, M.; Esenwein, S.A.; Muhr, G.; Beckmann, F.; Donath, T.; Assad, M.; Epple, M. Morphological characterization and in vitro biocompatibility of a porous nickel–titanium alloy. *Biomaterials* **2005**, *26*, 5801–5807. [CrossRef] [PubMed]
7. Školáková, A.; Novák, P.; Salvetr, P.; Moravec, H.; Šefl, V.; Deduytsche, D.; Detavernier, C. Investigation of the Effect of Magnesium on the Microstructure and Mechanical Properties of NiTi Shape Memory Alloy Prepared by Self-Propagating High-Temperature Synthesis. *Metall. Mater. Trans. A* **2017**, *48*, 3559–3569. [CrossRef]
8. Salvetr, P.; Školáková, A.; Novák, P.; Vavřík, J. Effect of Si Addition on Martensitic Transformation and Microstructure of NiTiSi Shape Memory Alloys. *Metall. Mater. Trans. A* **2020**, *51*, 4434–4438. [CrossRef]
9. Salvetr, P.; Dlouhý, J.; Školáková, A.; Průša, F.; Novák, P.; Karlík, M.; Haušild, P. Influence of Heat Treatment on Microstructure and Properties of NiTi46 Alloy Consolidated by Spark Plasma Sintering. *Materials* **2019**, *12*, 4075. [CrossRef]
10. Bednarczyk, W.; Kawałko, J.; Wątroba, M.; Gao, N.; Starink, M.J.; Bała, P.; Langdon, T.G. Microstructure and mechanical properties of a Zn-0.5Cu alloy processed by high-pressure torsion. *Mater. Sci. Eng. A* **2020**, *776*, 139047. [CrossRef]
11. Ren, F.; Zhu, W.; Chu, K. Fabrication and evaluation of bulk nanostructured cobalt intended for dental and orthopedic implants. *J. Mech. Behav. Biomed. Mater.* **2017**, *68*, 115–123. [CrossRef] [PubMed]
12. Yazdimamaghani, M.; Razavi, M.; Vashaee, D.; Moharamzadeh, K.; Boccaccini, A.R.; Tayebi, L. Porous magnesium-based scaffolds for tissue engineering. *Mater. Sci. Eng. C* **2017**, *71*, 1253–1266. [CrossRef] [PubMed]
13. Bobyn, J.D.; Stackpool, G.J.; Hacking, S.A.; Tanzer, M.; Krygier, J.J. Characteristics of bone ingrowth and interface mechanics of a new porous tantalum biomaterial. *J. Bone Jt. Surgery. Br. Vol.* **1999**, *81*, 907–914. [CrossRef]
14. Levine, B.R.; Sporer, S.; Poggie, R.A.; Della Valle, C.J.; Jacobs, J.J. Experimental and clinical performance of porous tantalum in orthopedic surgery. *Biomaterials* **2006**, *27*, 4671–4681. [CrossRef] [PubMed]
15. Čapek, J.; Machová, M.; Fousová, M.; Kubásek, J.; Vojtěch, D.; Fojt, J.; Jablonská, E.; Lipov, J.; Ruml, T. Highly porous, low elastic modulus 316L stainless steel scaffold prepared by selective laser melting. *Mater. Sci. Eng. C* **2016**, *69*, 631–639. [CrossRef] [PubMed]
16. Abdel-Hady Gepreel, M.; Niinomi, M. Biocompatibility of Ti-alloys for long-term implantation. *J. Mech. Behav. Biomed. Mater.* **2013**, *20*, 407–415. [CrossRef]
17. Khodaei, M.; Valanezhad, A.; Watanabe, I.; Yousefi, R. Surface and mechanical properties of modified porous titanium scaffold. *Surf. Coat. Technol.* **2017**, *315*, 61–66. [CrossRef]
18. Gu, Y.W.; Yong, M.S.; Tay, B.Y.; Lim, C.S. Synthesis and bioactivity of porous Ti alloy prepared by foaming with TiH2. *Mater. Sci. Eng. C* **2009**, *29*, 1515–1520. [CrossRef]
19. Rao, S.; Ushida, T.; Tateishi, T.; Okazaki, Y.; Asao, S. Effect of Ti, Al, and V ions on the relative growth rate of fibroblasts (L929) and osteoblasts (MC3T3-E1) cells. *Bio Med. Mater. Eng.* **1996**, *6*, 79–86. [CrossRef]
20. Assad, M.; Chernyshov, A.V.; Jarzem, P.; Leroux, M.A.; Coillard, C.; Charette, S.; Rivard, C.H. Porous titanium-nickel for intervertebral fusion in a sheep model: Part 2. Surface analysis and nickel release assessment. *J. Biomed. Mater. Res. Part B Appl. Biomater.* **2003**, *64*, 121–129. [CrossRef]
21. Hsu, H.-C.; Wu, S.-C.; Hsu, S.-K.; Li, Y.-C.; Ho, W.-F. Structure and mechanical properties of as-cast Ti-Si alloys. *Intermetallics* **2014**, *47*, 11–16. [CrossRef]
22. Knaislová, A.; Novák, P. Preparation of Porous Biomaterial Based on Ti-Si Alloys. *Manuf. Technol.* **2018**, *18*, 411–417. [CrossRef]
23. Arifvianto, B.; Leeflang, M.A.; Zhou, J. The compression behaviors of titanium/carbamide powder mixtures in the preparation of biomedical titanium scaffolds with the space holder method. *Powder Technol.* **2015**, *284*, 112–121. [CrossRef]
24. Daudt, N.d.F.; Bram, M.; Barbosa, A.P.C.; Laptev, A.M.; Alves, C. Manufacturing of highly porous titanium by metal injection molding in combination with plasma treatment. *J. Mater. Process. Technol.* **2017**, *239*, 202–209. [CrossRef]
25. Müller, L.; Müller, F.A. Preparation of SBF with different HCO3-content and its influence on the composition of biomimetic apatites. *Acta Biomater.* **2006**, *2*, 181–189. [CrossRef]
26. ISO. 10993–5: 2009 Biological evaluation of medical devices—Part 5: Tests for in vitro cytotoxicity. In *International Organization for Standardization*; ISO: Geneva, Switzerland, 2009.

27. Thümmler, F.; Oberacker, R. *An introduction to Powder Metallurgy*; Maney Publishing for IOM3, the Institute of Materials, Minerals and Mining: London, UK, 1993.
28. Pabst, W.; Gregorová, E.; Uhlířová, T. Microstructure characterization via stereological relations—A shortcut for beginners. *Mater. Charact.* **2015**, *105*, 1–12. [CrossRef]
29. Massalski, T.B. Binary alloy phase diagrams. *ASM Int.* **1992**, *3*, 2874.
30. Novák, P.; Kubásek, J.; Šerák, J.; Vojtěch, D.; Michalcová, A. Mechanism and kinetics of the intermediary phase formation in Ti–Al and Ti–Al–Si systems during reactive sintering. *Int. J. Mater. Res.* **2009**, *100*, 353–355. [CrossRef]
31. Trambukis, J.; Munir, Z.A. Effect of Particle Dispersion on the Mechanism of Combustion Synthesis of Titanium Silicide. *J. Am. Ceram. Soc.* **1990**, *73*, 1240–1245. [CrossRef]
32. Riley, D.P. Synthesis and characterization of SHS bonded Ti5Si3 on Ti substrates. *Intermetallics* **2006**, *14*, 770–775. [CrossRef]
33. Riley, D.P.; Oliver, C.P.; Kisi, E.H. In-situ neutron diffraction of titanium silicide, Ti5Si3, during self-propagating high-temperature synthesis (SHS). *Intermetallics* **2006**, *14*, 33–38. [CrossRef]
34. Tang, H.P.; Wang, J.; Qian, M. 28-Porous titanium structures and applications. In *Titanium Powder Metallurgy*; Qian, M., Froes, F.H., Eds.; Butterworth-Heinemann: Boston, MA, USA, 2015; pp. 533–554. [CrossRef]
35. Fatemi, A. Mechanical Properties and Testing of Metallic Materials. In *Ullmann's Encyclopedia of Industrial Chemistry*; The Charleston Company: Denver, CO, USA, 2000. [CrossRef]
36. Yaszemski, M.J.; Payne, R.G.; Hayes, W.C.; Langer, R.; Mikos, A.G. Evolution of bone transplantation: Molecular, cellular and tissue strategies to engineer human bone. *Biomaterials* **1996**, *17*, 175–185. [CrossRef]
37. Oh, I.-H.; Nomura, N.; Masahashi, N.; Hanada, S. Mechanical properties of porous titanium compacts prepared by powder sintering. *Scr. Mater.* **2003**, *49*, 1197–1202. [CrossRef]
38. Ding, W.; Wang, Z.; Chen, G.; Cai, W.; Zhang, C.; Tao, Q.; Qu, X.; Qin, M. Oxidation behavior of low-cost CP-Ti powders for additive manufacturing via fluidization. *Corros. Sci.* **2021**, *178*, 109080. [CrossRef]
39. Tao, Q.; Wang, Z.; Chen, G.; Cai, W.; Cao, P.; Zhang, C.; Ding, W.; Lu, X.; Luo, T.; Qu, X.; et al. Selective laser melting of CP-Ti to overcome the low cost and high performance trade-off. *Addit. Manuf.* **2020**, *34*, 101198. [CrossRef]
40. Horkavcová, D.; Novák, P.; Fialová, I.; Černý, M.; Jablonská, E.; Lipov, J.; Ruml, T.; Helebrant, A. Titania sol-gel coatings containing silver on newly developed TiSi alloys and their antibacterial effect. *Mater. Sci. Eng. C* **2017**, *76*, 25–30. [CrossRef] [PubMed]

2

Effects of Strain Rate and Measuring Temperature on the Elastocaloric Cooling in a Columnar-Grained $Cu_{71}Al_{17.5}Mn_{11.5}$ Shape Memory Alloy

Hui Wang, Haiyou Huang * and Jianxin Xie

Key Laboratory for Advanced Materials Processing of the Ministry of Education, Institute for Advanced Materials and Technology, University of Science and Technology Beijing, Beijing 100083, China; wangh9130@163.com (H.W.); jxxie@ustb.edu.cn (J.X.)

* Correspondence: huanghy@mater.ustb.edu.cn

Abstract: Solid-state refrigeration technology based on elastocaloric effects (eCEs) is attracting more and more attention from scientists and engineers. The response speed of the elastocaloric materials, which relates to the sensitivity to the strain rate and measuring temperature, is a significant parameter to evaluate the development of the elastocaloric material in device applications. Because the Cu-Al-Mn shape memory alloy (SMA) possesses a good eCE and a wide temperature window, it has been reported to be the most promising elastocaloric cooling material. In the present paper, the temperature changes (ΔT) induced by reversible martensitic transformation in a columnar-grained $Cu_{71}Al_{17.5}Mn_{11.5}$ SMA fabricated by directional solidification were directly measured over the strain rate range of 0.005–0.19 s^{-1} and the measuring temperature range of 291–420 K. The maximum adiabatic ΔT of 16.5 K and a lower strain-rate sensitivity compared to TiNi-based SMAs were observed. With increasing strain rate, the ΔT value and the corresponding coefficient of performance (COP) of the alloy first increased, then achieved saturation when the strain rate reached 0.05 s^{-1}. When the measuring temperature rose, the ΔT value increased linearly while the COP decreased linearly. The results of our work provide theoretical reference for the design of elastocaloric cooling devices made of this alloy.

Keywords: shape memory alloy; columnar grain; Cu-Al-Mn; elastocaloric effect; strain rate; measuring temperature

---

1. Introduction

The elastocaloric effect (eCE) refers to the thermal response of a given material to external uniaxial stress, which is commonly quantified by the isothermal entropy change (ΔS) and adiabatic temperature change (ΔT). Elastocaloric refrigeration based on eCEs, which is a new type of solid-state refrigeration, has drawn significant attention in recent years owing to its higher coefficient of performance (COP), its lower device costs, and its eco-friendliness as a promising alternative to conventional vapor compression, as well as to its downscaling ability for microcooling [1–5]; it is regarded as the most applicable solid-state refrigeration technology [6]. Shape memory alloys (SMAs) can undergo a reversible martensitic transformation (MT) by applying external stress, which is accompanied by a large latent heat absorption and release. These features make them important elastocaloric materials and thus draw momentous attention.

The refrigeration capability (RC) of a given elastocaloric material can be evaluated by directly measuring the adiabatic temperature change of the phase transformation or deformation. Some related research indicated that a high ΔT value of more than 10 K has been directly measured in Ti-Ni- [7,8], Ni-Fe- [9] and Cu-based SMAs [10], which has laid a good material foundation for the development and application of elastocaloric refrigeration technology. At present, an increasing number of promising elastocaloric refrigeration devices are being developed [11]. In addition to evaluating the RC, the responding speed of SMAs is also a significant factor that influences the system cooling efficiency of the solid refrigeration system. Some researchers have carried out a series of research on the influence of strain rates on the eCE in Ti-Ni-based SMAs, such as $Ti_{49.1}Ni_{50.5}Fe_{0.4}$ foils and $Ti_{50.4}Ni_{49.6}$ films by Ossmer et al. [12,13], $Ti_{51.1}Ni_{48.9}$ wires by Tušek et al. [8], $Ti_{55.4}Ni_{44.6}$ wires by Tobush et al. [14], a $Ti_{47.25}Ni_{45}Cu_5V_{2.75}$ block by Schmidt et al. [15], and so on. Compared with the Ti-Ni-based SMAs, Cu-based SMAs also have a large ΔT value, which has been proved by Xu et al. They found a ΔT value of 12–13 K in columnar-grained Cu-Al-Mn SMAs [10], covering a wide temperature range of more than 100 K, which, combined with a low applying stress [10,16] and low material costs, make the columnar-grained Cu-Al-Mn SMAs a promising material for solid-state refrigeration. In this paper, over the strain rate from 0.005 to 0.19 s^{-1} and the test temperature range from 291 to 393 K, the temperature changes induced by reversible MT of columnar-grained $Cu_{71}Al_{17.5}Mn_{11.5}$ SMAs have been systematically measured, the effects of the strain rate ($\dot{\varepsilon}$) and measuring temperature (T_A) on the eCEs have been obtained, and the influence mechanism is discussed. The results of this study can provide a theoretical reference for the design and application of elastocaloric cooling devices that are made of Cu-based SMAs.

2. Materials and Methods

A $Cu_{71}Al_{17.5}Mn_{11.5}$ ingot with a columnar-grained microstructure was prepared by directional solidification [17]. At first, the ingot was annealed at 1073 K for 5 min followed by quenching into ice water to obtain a single β_1 phase. Then, the ingot was aged at 473 K for 15 min to stabilize the MT temperatures. Dog bone-shaped tensile samples with a gauge size of 25 mm $\times$ 6 mm $\times$ 2 mm were cut out from the ingot, with their longitudinal direction along the solidification direction (SD). The surface of the tensile specimen was polished with 500–3000 # sandpaper and then subjected to mechanical polishing and electrolytic polishing. The texture characterization was conducted by electron back-scattered diffraction (EBSD) with a minimum misorientation resolution of 2°. The electrolytic polishing solution was as follows: 250 mL of H_3PO_4, 250 mL of alcohol, 50 mL of glycerol, 5 g of urea, and 500 mL of H_2O, with a voltage of 10 V and duration from 80 to 120 s. The transformation temperatures and latent heat were determined by NETZSCH 404F3 using differential scanning calorimetry (DSC, Mettler-Toledo, Zurich, Switzerland) under a nitrogen inert gas flow of 10 $mL{\cdot}s^{-1}$ with a heating/cooling rate of 10 K/min. Tensile tests were conducted on a Mechanical Testing System (MTS) testing machine (Wister Industrial Equipment cooperation Limited, Shenzhen, China) equipped with a thermostatic chamber. During the tensile test, all samples were loaded at a constant strain rate of 0.05 s^{-1} to the maximum strain of 10% and were then unloaded at different strain rates to zero stress. In this work, the measuring temperature range was from 291 to 420 K, and the unloading strain rate range was from 0.005 to 0.19 s^{-1}. The temperature of the sample in the tensile cycle was monitored with a *K*-type thermocouple welded on the center of the sample surface. The authors used a set of self-built equipment for data acquisition and a MATLAB program (v7.0, MathWorks, Natick, MA, USA, 2012) to display the measured temperature.

3. Results and Discussion

The $Cu_{71}Al_{17.5}Mn_{11.5}$ SMA sample was composed of a single austenite phase β_1 with $L2_1$ crystallographic structure at room temperature. When a stress applied in the sample was higher than the critical stress of MT, a stress-induced MT of $\beta_1 \rightarrow \beta_1'$ occurred and the β_1' martensite had an 18R ordered structure. The phase structure and transformation behavior have been determined by X-ray diffraction (XRD, SmartLab, Rigaku Corporation, Matsubara-cho, Akishima-shi, Tokyo, Japan) and transmission electron microscopy (TEM, F20, FEI, Hillsboro, OR, USA) operated at 200 kV at room temperature in previous work by authors and other researchers [18–22].

The DSC curve during the heating/cooling process of the as-quenched sample is shown in Figure 1. An exothermic peak corresponding to the MT during cooling and an endothermic peak corresponding to the reverse austenitic transformation during heating can be clearly observed. The MT and reverse transformation temperatures can be obtained from the curve: the MT starting temperature was M_s = 247 K, the MT finishing temperature was M_f = 235 K, the reverse transformation starting temperature was A_S = 253 K and the reverse transformation finishing temperature was A_f = 265 K; the thermal hysteresis was 18 K.

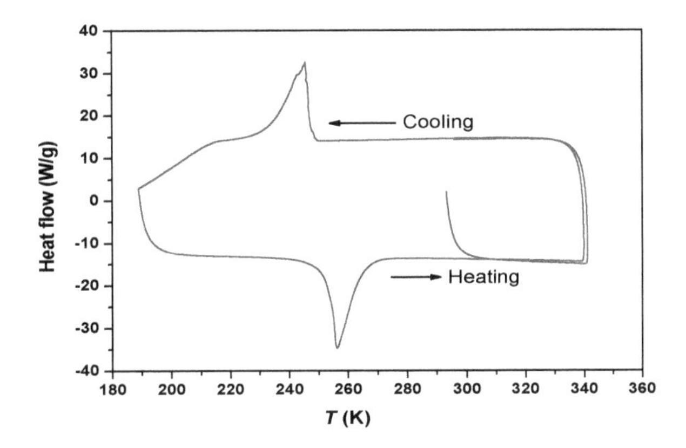

Figure 1. Differential scanning calorimetry (DSC) curve of the columnar-grained $Cu_{71}Al_{17.5}Mn_{11.5}$ shape memory alloy (SMA; heating/cooling rate of 10 K/min).

The EBSD orientation map and the inverse pole figures (Figure 2) illustrate that the columnar-grained Cu-Al-Mn sample has a strong <001> oriented texture along the SD (solidification direction).

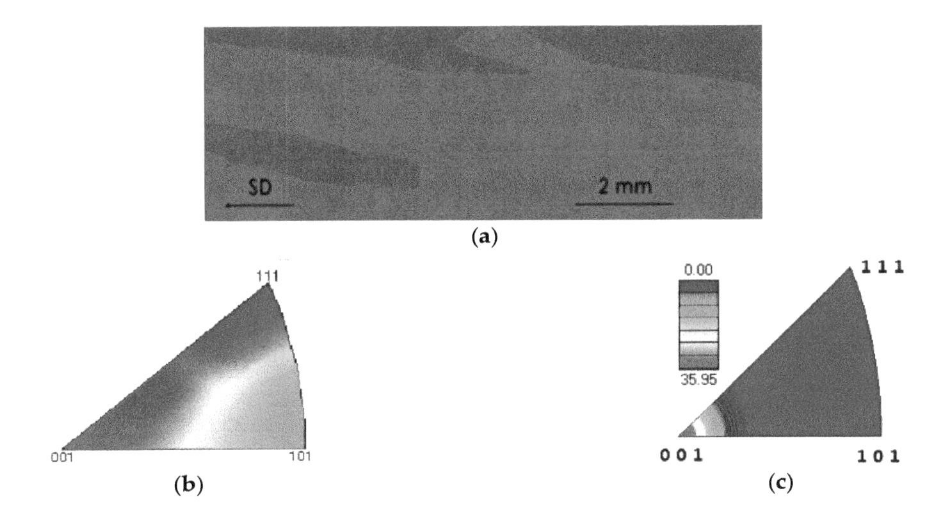

Figure 2. (**a**) Electron back-scattered diffraction (EBSD) quasi-colored orientation map along solidification direction (SD); (**b**) the reference stereographic triangle; (**c**) inverse pole figure along the SD, which illustrate that the columnar-grained $Cu_{71}Al_{17.5}Mn_{11.5}$ SMA sample has a strong <001>-oriented texture along the SD.

The tensile stress-strain curves for the columnar-grained $Cu_{71}Al_{17.5}Mn_{11.5}$ SMA are shown in Figure 3; these were tested at the same loading strain rate $\dot{\varepsilon}_l$ of 0.05 s^{-1} while gradually increasing the unloading strain rates $\dot{\varepsilon}_u$ from 0.005 to 0.19 s^{-1}. Specifically, nine strain rates of 0.005, 0.01, 0.05, 0.07, 0.096, 0.1, 0.13, 0.15, and 0.19 s^{-1} were chosen. It should be noted that in order to avoid the experimental deviation from different samples, the test was carried out using the same sample at each measuring temperature. For all the test conditions in Figure 3, when the measuring temperature was $\leq$393 K, an almost 100% strain recovery could be observed after the samples underwent a loading strain of 10%, indicating an excellent superelasticity of the columnar-grained $Cu_{71}Al_{17.5}Mn_{11.5}$ SMA. With increasing tensile cycles, both the critical stresses of MT and reverse transformation were decreased gradually. Because the loading strain rate is constant, the stress of MT decreasing in loading processes can be attributed to the influence of cycle numbers, which may be related to the fatigue effect. However, the stress of reverse transformation in unloading processes may be affected by loading processes, strain rates or cycle numbers.

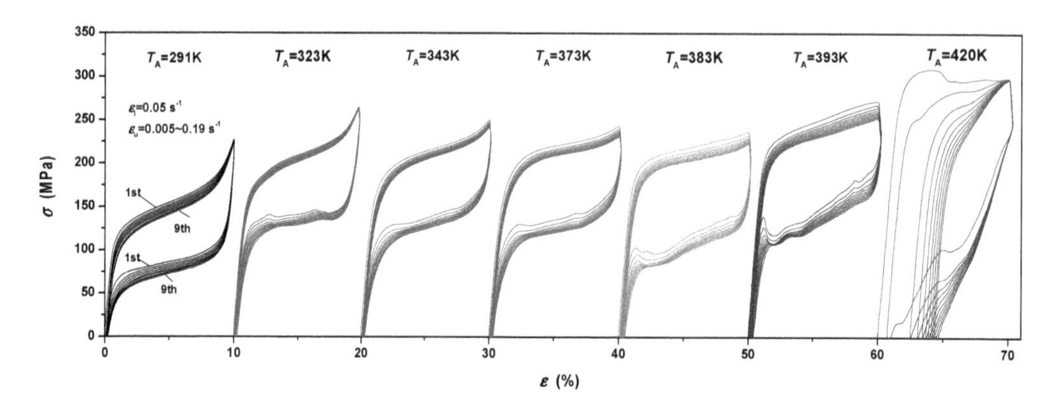

Figure 3. Stress-strain curves for the columnar-grained $Cu_{71}Al_{17.5}Mn_{11.5}$ shape memory alloys (SMAs) with different unloading strain rates ($\dot{\varepsilon}_u$ = 0.005–0.19 s^{-1}) at different measuring temperatures.

In this work, the influence of loading processes can be excluded as a result of the same loading strain and loading strain rate. In order to clarify the influence of cycle numbers or strain rate on the unloading curves, a comparative tensile test with 10 cycles was carried on at a constant unloading rate of 0.05 s^{-1}, and the stress-strain curves are shown in Figure 4a. Comparing the stress-strain curve tested at 291 K in Figure 3 and the curve in Figure 4a, the reduction of the transformation stress has closed relations with the increase in cycle numbers. In other words, the strain rate has little effect on the transformation stress of columnar-grained $Cu_{71}Al_{17.5}Mn_{11.5}$ SMA, which is different from the phenomenon that was observed in Ti-Ni alloys [14,15]. For Ti-Ni alloys, with increasing strain rates, the stress-strain loop enlarges clearly; that is, the loading stress increases and unloading stress decreases. The strain rate sensitivity of the transformation stress in SMAs is related to the capability of stress relaxation, which is caused by martensite nucleation and growth during the MT process. Under constant strain conditions, the MT of Ti-Ni SMAs requires 3 min to induce a stress relaxation of 50 MPa [15]. In other words, when the strain rate exceeds 0.02 s^{-1} of Ti-Ni SMAs, the stress will clearly increase as a result of stress relaxation hysteresis. Figures 3 and 4a indicate that the phenomenon of transformation stress increase is not observed in columnar-grained $Cu_{71}Al_{17.5}Mn_{11.5}$ SMAs until the strain rate reaches 0.19 s^{-1}, which means that the stress relaxation capacity of columnar-grained $Cu_{71}Al_{17.5}Mn_{11.5}$ SMAs is more than 10 times that of Ti-Ni SMAs. Therefore, the strain rate sensitivity of columnar-grained $Cu_{71}Al_{17.5}Mn_{11.5}$ SMAs is dramatically lower than that of Ti-Ni SMAs.

In order to study the effect of measuring temperature on tensile stress-strain curves of the columnar-grained $Cu_{71}Al_{17.5}Mn_{11.5}$ SMA, a series of tensile cycle tests were carried out between the temperature range from 291 to 420 K, and the results are subsequently drawn in Figure 3. A comparison of these stress-strain curves indicates that as the measuring temperature rises, the MT stress increases gradually. When the measuring temperature reaches 420 K, the recoverable strain of the sample decreases rapidly with the increase in the number of tensile cycles. The microstructure observation found that a large number of dislocations were observed in the sample, indicating that the tensile stress was high enough to start the dislocation slip. This phenomenon implies the deformation mechanism begins to change from superelastic deformation induced by phase transformation to permanent plastic deformation caused by a dislocation slip when the sample deforms at 420 K, which can be defined as the critical temperature of stress-induced MT in columnar-grained $Cu_{71}Al_{17.5}Mn_{11.5}$ alloys. Therefore, the upper limit of the temperature window of the eCE of columnar-grained $Cu_{71}Al_{17.5}Mn_{11.5}$ alloys is less than 420 K. In other words, the temperature window of the columnar-grained $Cu_{71}Al_{17.5}Mn_{11.5}$ alloy is 265–393 K. The width of the temperature window is 128 K.

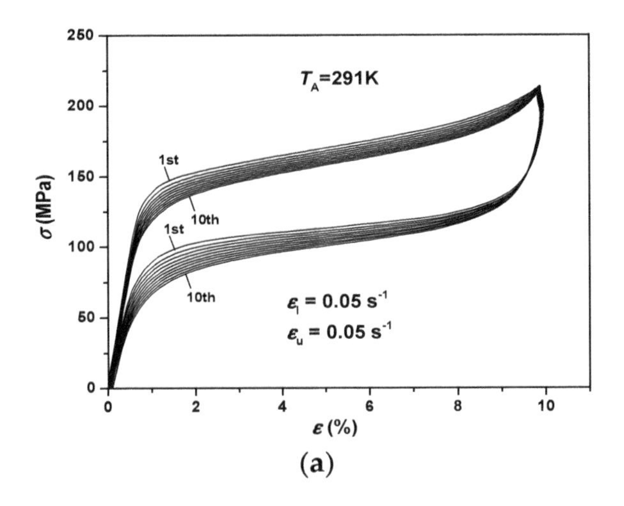

(a)

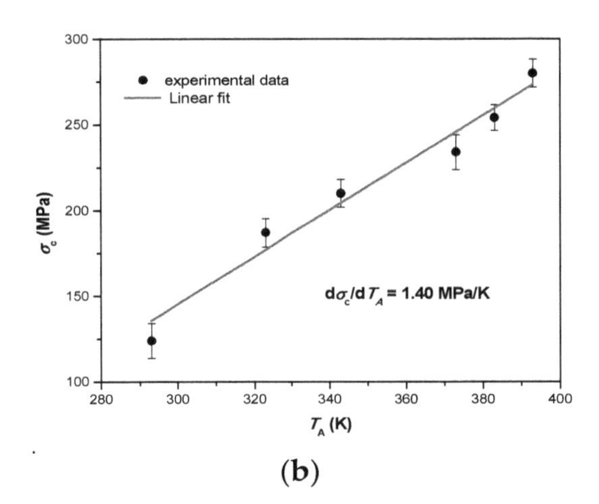

(b)

Figure 4. (**a**) Stress-strain curves with constant loading rate ($\varepsilon_l = 0.05\ s^{-1}$) at 291 K; (**b**) T_A dependence of σ_c for the columnar-grained $Cu_{71}Al_{17.5}Mn_{11.5}$ shape memory alloy (SMA).

On the basis of the first cycle of the measured stress–strain curve in the temperature range of 291–393 K, the MT critical stress (σ_c) as a function of T_A is plotted in Figure 4b. Figure 4b indicates that σ_c linearly increases with the increase of measuring temperature in the range of 291–393 K, and $d\sigma_c/dT_A = 1.40$ MPa/K for the columnar-grained $Cu_{71}Al_{17.5}Mn_{11.5}$ alloy, which can be determined by linear fitting.

Figure 5 indicates the temperature change of the columnar-grained $Cu_{71}Al_{17.5}Mn_{11.5}$ SMA samples in loading-unloading cycles at different measuring temperatures. It can be seen from Figure 5a–d that at a constant strain rate (0.05 s^{-1}), the ΔT values measured from both loading and unloading processes for different cycles are almost unchanged in the temperature range of 291–393 K, which indicates that no evident influence of cycle numbers on ΔT was observed after 10 tensile cycles at a constant strain rate; this implies a good stability of the eCE in the columnar-grained $Cu_{71}Al_{17.5}Mn_{11.5}$ SMA. When the strain rate is less than 0.05 s^{-1}, $|\Delta T|$ increases with increasing $\dot{\varepsilon}_u$. When the strain rate reaches 0.05 s^{-1} or above, $|\Delta T|$ achieves saturation. For instance, at $T_A = 291$ K, when $\dot{\varepsilon}_u$ increases from 0.005 to 0.05 s^{-1}, ΔT changes from 7.53 to 11.21 K. After $\dot{\varepsilon}_u$ reaches 0.05 s^{-1}, the ΔT values remain constant within the range of 11.21–11.51 K. For different T_A values, the variation trends of the T vs. t curves from the samples with increasing $\dot{\varepsilon}_u$ are similar.

In addition, the phenomenon of temperature irreversibility can also be found from Figure 5; that is, the absolute values of the temperature change between loading and unloading are not equal. For example, when the strain rate is 0.05 s^{-1} at 291 K, the loading temperature rises are 14.8–15.3 K (Figure 5a) and 13.7–13.9 K (Figure 5b), while the unloading temperature drops are 13.2–13.5 K (Figure 3a) and 11.2 K (Figure 5b). The absolute value of the unloading temperature drop is 1.6–2.7 K, which is lower than the loading temperature rise. Irreversibility is caused by the existence of the hysteresis area during the loading-unloading process. The formation of the hysteresis area is directly related to the frictional origin during the transformation. These frictions include the interfacial friction between martensite and austenite, the interaction between the phase interface and grain boundary, or other defects [12,14]. When the measuring temperature reached 420 K, as a result of the occurrence of irreversible deformation caused by a dislocation slip, the return strain gradually reduced with the increasing stretching cycles, resulting in a gradually reduced temperature change, as shown in Figure 5e.

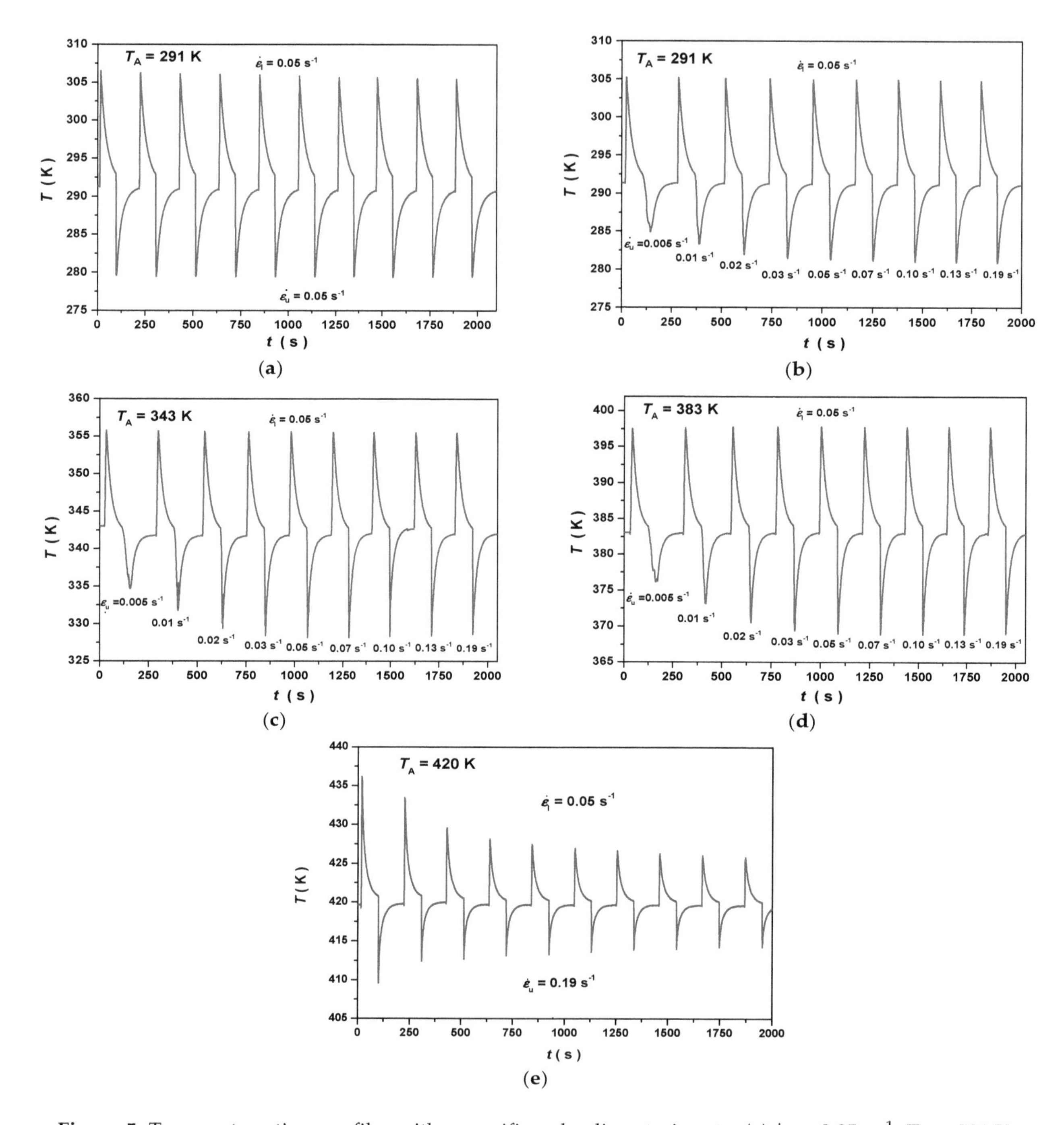

Figure 5. Temperature-time profiles with a specific unloading strain rate: (**a**) $\dot{\varepsilon}_u = 0.05\ s^{-1}$, $T_A = 291$ K; (**e**) $\dot{\varepsilon}_u = 0.19\ s^{-1}$, $T_A = 420$ K and with various unloading strain rates ($\dot{\varepsilon}_u = 0.005$–$0.19\ s^{-1}$) at $T_A = 291$ K (**b**); $T_A = 343$ K (**c**); and $T_A = 383$ K (**d**) for the columnar-grained $Cu_{71}Al_{17.5}Mn_{11.5}$ shape memory alloy (SMA) samples ($\dot{\varepsilon}_l = 0.05\ s^{-1}$ for all samples).

On the basis of the T-t profiles (four typical profiles are shown in Figure 5), a high $|\Delta T|$ value of 11.3–16.5 K can be obtained in the columnar-grained $Cu_{71}Al_{17.5}Mn_{11.5}$ SMA during unloading processes, covering a wide temperature range of more than 100 K. The entropy change of the phase transformation associated with the elastocaloric cooling can be estimated as

$$\Delta S \approx -\frac{\Delta T}{T_A} C_p \tag{1}$$

where C_p is the heat capacity, measured to be 455 J/kg·K for the $Cu_{71}Al_{17.5}Mn_{11}$ SMA [10]. Strictly speaking, when calculating the adiabatic temperature change or entropy change, because of the coexistence of two phases during transformation, we should consider that $C_p = xC_p{}^A + (1 - x)C_p{}^M$, where x is the fraction of the authentic phase, and $C_p{}^A$ and $C_p{}^M$ are the heat capacity of the authentic phase and martensitic phase, respectively. However, $C_p{}^A$ and $C_p{}^M$ are approximately equal for SMAs [23]. Therefore, in the actual measurement and calculation, we supposed $C_p \approx C_p{}^A \approx C_p{}^M$. In this paper, we experimentally measured the $C_p{}^A$ value and used it in the calculation. According to the ΔT values measured above, the maximum ΔS can be estimated to be 19.5 J/(kg·K). Additionally, ΔS can also be calculated from the Clausius-Clapeyron equation:

$$\Delta S = -\frac{1}{\rho}\frac{d\sigma_c}{dT_A}\varepsilon \quad (2)$$

where ρ is the density and ε is the transformation strain. For the columnar-grained $Cu_{71}Al_{17.5}Mn_{11.5}$ SMA, $d\sigma_c/dT_A$ and ε can be determined to be 1.40 MPa/K from Figure 4b and 8.3% from Figure 3, and $\rho = 7.40 \times 10^3$ kg/m^3 [24]. The calculated ΔS value from the Clausius-Clapeyron equation is about 15.7 J/(kg·K), which is smaller than the calculation results based on ΔT. In addition, the theoretical maximum of the isothermal entropy change also can be calculated to be 25.0 J/kg·K by the latent heat of phase transformation determined by DSC measurement [10]. Therefore, the entropy change estimated on the basis of the experimental data in this paper approached ~78% of its theoretical value. It is generally believed that the latent heat of phase transformation determined by the DSC method corresponds to a completed entropy change from 100% phase transformation without any loss, thus often called the theoretical entropy change. The entropy change value estimated on the basis of the experimentally measured ΔT value is less than the theoretical entropy change, because of the existence of internal frictions induced by the phase interface frictions and the interactions between phase migration and defects. In addition, in the actual tensile deformation processes, incomplete MT may be another probable cause of the smaller entropy change value estimated by stress-strain curves. According to the above discussion, the energy loss induced by internal frictions is about 5.5 J/(kg·K) for the columnar-grained $Cu_{71}Al_{17.5}Mn_{11.5}$ SMA.

The $|\Delta T|$ value and the COP for unloading processes in the columnar-grained $Cu_{71}Al_{17.5}Mn_{11.5}$ SMA as a function of $\dot{\varepsilon}$ and T_A are summarized in Figure 6. The COP of the material, which describes the cooling efficiency, is defined by the ratio of cooling power (ΔQ) to input work (ΔW) [25]:

$$\text{COP} = \Delta Q/\Delta W \quad (3)$$

where ΔQ can be estimated from the latent heat, which is $\Delta T_{ad} \times C_p$ (ΔT_{ad} is the adiabatic temperature change), and ΔW can be obtained by integrating the area enclosed by the stress hysteresis loop (Figure 3). Figure 6a indicates that all $|\Delta T|$ vs. $\dot{\varepsilon}$ curves show a similar variation trend. At first, the $|\Delta T|$ value increases with increasing $\dot{\varepsilon}$. When the strain rate reaches 0.05 s^{-1}, the $|\Delta T|$ value remains almost unchanged, implying it achieves saturation. The critical strain rate $\dot{\varepsilon}_c$ corresponding to the onset of $|\Delta T|$ saturation is less than that of Ti-Ni alloy, which was reported as 0.2 s^{-1} [13]. In other words, it is easier to reach near-adiabatic conditions using the columnar-grained $Cu_{71}Al_{17.5}Mn_{11.5}$ SMA. In the process of MT (reverse MT), the homogeneity of martensite (austenite) nucleation and extension is a noteworthy factor to influence $\dot{\varepsilon}_c$. The more homogeneously the MT occurs in the sample, the lower the $\dot{\varepsilon}_c$ value [12]. The columnar-grained Cu-Al-Mn SMA has a homogeneous contribution of stress/strain in the whole sample as a result of a high deformation and transformation compatibility among grains [26]. Therefore, the martensite (austenite) can nucleate and grow homogeneously in the columnar-grained samples. Furthermore, the Cu-based SMAs have a higher thermal conductivity compared to the Ti-Ni alloy, which also helps to obtain a uniform temperature distribution within a very short period of time. The above two reasons show that the columnar-grained $Cu_{71}Al_{17.5}Mn_{11.5}$ SMA has a low $\dot{\varepsilon}_c$, which can reduce the design difficulty of refrigeration devices.

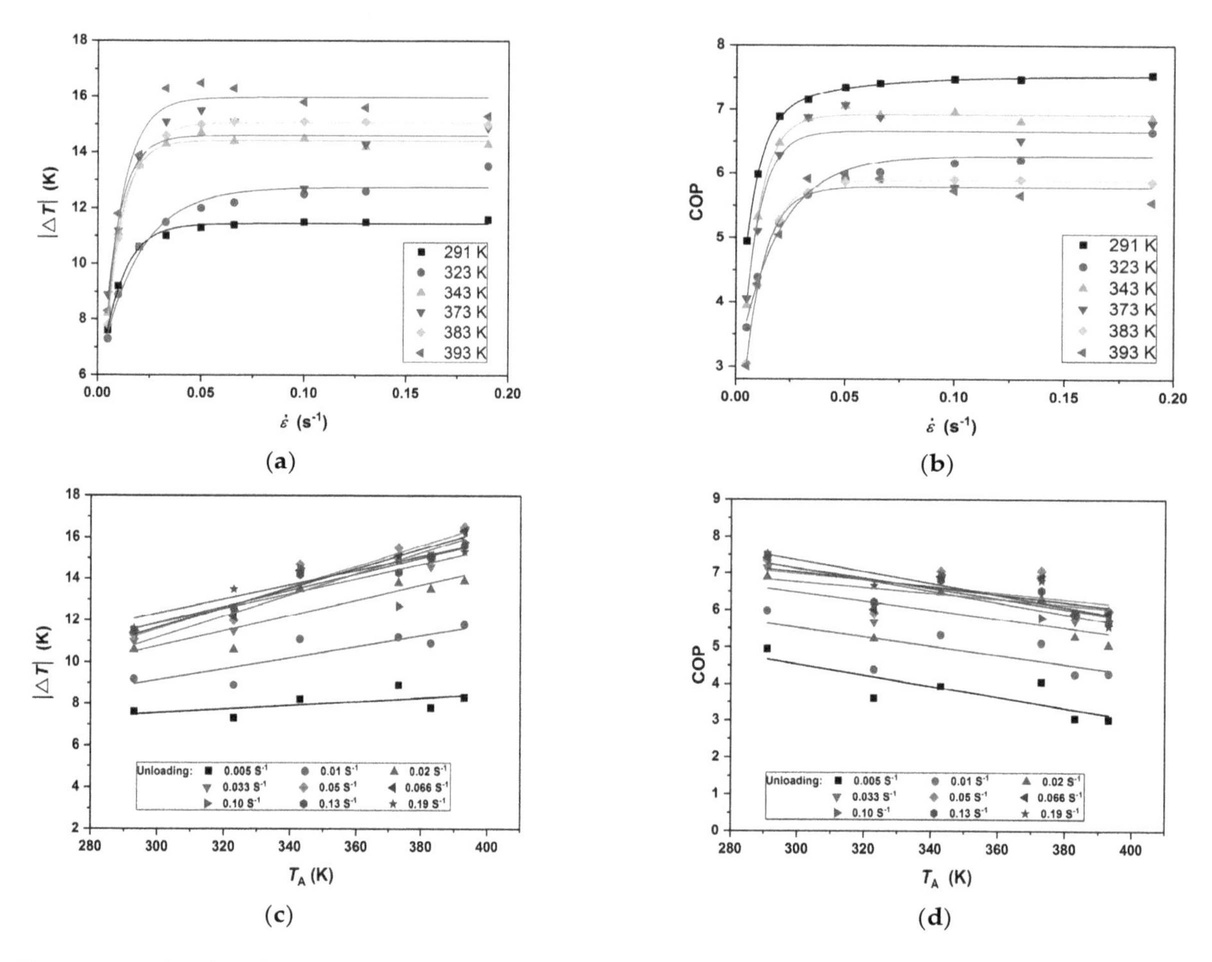

Figure 6. The $|\Delta T|$ and coefficient of performance (COP) values for unloading process in the columnar-grained $Cu_{71}Al_{17.5}Mn_{11.5}$ shape memory alloy (SMA) as a function of $\dot{\varepsilon}$ and T_A: (**a**) $|\Delta T|$ vs. $\dot{\varepsilon}$; (**b**) COP vs. $\dot{\varepsilon}$; (**c**) $|\Delta T|$ vs. T_A; (**d**) COP vs. T_A.

Figure 6b indicates that the variation trend of COP with increasing strain rate is consistent with $|\Delta T|$, while the variations of $|\Delta T|$ and COP with T_A are the opposite, as shown in Figure 6c,d. For refrigeration device or system design, a higher $|\Delta T|$ and COP are expected. For the columnar-grained $Cu_{71}Al_{17.5}Mn_{11.5}$ SMA, in order to achieve a stable and high refrigeration capability and COP, the applied strain rates should be more than 0.05 s^{-1}.

4. Conclusions

In summary, the effects of strain rates and the measuring temperature on the elastocaloric cooling in a columnar-grained $Cu_{71}Al_{17.5}Mn_{11.5}$ SMA were experimentally investigated over the strain rate range of 0.005–0.19 s^{-1} and the measuring temperature range of 291–393 K. With the increasing stain rate, the ΔT and COP values of the alloy increase firstly and then achieve saturation when the strain rate reaches 0.05 s^{-1}, which indicates a lower strain rate sensitivity of refrigeration capability compared to Ti-Ni-based SMAs (about 0.2 s^{-1}). The relatively low strain rate sensitivity of the columnar-grained $Cu_{71}Al_{17.5}Mn_{11.5}$ SMA is attributed to two reasons. The first is a high deformation and phase transformation compatibility among grains in the columnar-grained microstructure, which causes a homogenous stress/strain distribution in the entire samples. The other is a high thermal conductivity of Cu-based SMAs, which also helps to obtain a uniform temperature distribution in a very short period of time. In addition, a maximum adiabatic ΔT value of 16.5 K with corresponding

ΔS of 19.5 J/(kg·K) and a wide operational temperature window (ω_T > 100 K) were directly measured in the experiment. The above results demonstrate that the columnar-grained $Cu_{71}Al_{17.5}Mn_{11.5}$ SMA is a promising candidate of elastocaloric materials with the advantages of a high refrigeration capability in a wide strain-rate range and operational temperature window, which are beneficial for the design and application to refrigeration devices.

Acknowledgments: This work was supported by the National Key Research and Development Program of China (Grant No. 2016YFB0700500) and the National Natural Science Foundation of China (Grant No. 51574027). Hui Wang thanks the State Key Laboratory of Technologies in Space Cryogenics Propellants, Technical Institute of Physics and Chemistry, Chinese Academy of Sciences for providing experimental equipment.

Author Contributions: Hui Wang and Haiyou Huang conceived and designed the experiments; Hui Wang performed the experiments; Hui Wang and Haiyou Huang analyzed the data and wrote the paper; Haiyou Huang and Jianxin Xie supported the writing of the paper. All authors participated in the discussions of the results.

References

1. Moya, X.; Kar-Narayan, S.; Mathur, N.D. Caloric materials near ferroic phase transitions. *Nat. Mater.* **2014**, *13*, 439–450. [CrossRef] [PubMed]
2. Gschneidnerjr, K.A.; Pecharsky, V.K.; Tsokol, A.O. Recent developments in magnetocaloric materials. *Rep. Prog. Phys.* **2005**, *68*, 1479–1539. [CrossRef]
3. Mischenko, A.S.; Zhang, Q.; Scott, J.F.; Whatmore, R.W.; Mathur, N.D. Giant electrocaloric effect in thin-film $PbZr_{0.95}Ti_{0.05}O_3$. *Science* **2006**, *311*, 1270–1271. [CrossRef] [PubMed]
4. Mañosa, L.; González-Alonso, D.; Planes, A.; Bonnot, E.; Barrio, M.; Tamarit, J.L.; Aksoy, S.; Acet, M. Giant solid-state barocaloric effect in the Ni-Mn-In magnetic shape-memory alloy. *Nat. Mater.* **2010**, *9*, 478–481. [CrossRef] [PubMed]
5. Bonnot, E.; Romero, R.; Mañosa, L.; Vives, E.; Planes, A. Elastocaloric effect associated with the martensitic transition in shape-memory alloys. *Phys. Rev. Lett.* **2008**, *100*, 125901. [CrossRef] [PubMed]
6. Goetzler, W.; Zogg, R.; Young, J.; Johnson, C. *Energy Savings Potential and RD&D Opportunities for Non-Vapor-Compression HVAC Technologies*; U.S. Department of Energy (Navigant Consulting Inc.): Washington, DC, USA, 2014.
7. Cui, J.; Wu, Y.; Muehlbauer, J.; Hwang, Y.; Radermacher, R.; Fackler, S.; Wuttig, M.; Takeuchi, I. Demonstration of high efficiency elastocaloric cooling with large ΔT using NiTi wires. *Appl. Phys. Lett.* **2012**, *101*, 1175–1178. [CrossRef]
8. Tušek, J.; Engelbrecht, K.; Mikkelsen, L.; Pryds, N. Elastocaloric effect of Ni-Ti wire for application in a cooling device. *J. Alloys Compd.* **2015**, *117*, 10–19. [CrossRef]
9. Pataky, G.J.; Ertekin, E.; Sehitoglu, H. Elastocaloric cooling potential of NiTi, Ni_2FeGa, and CoNiAl. *Acta Mater.* **2015**, *96*, 420–427. [CrossRef]
10. Xu, S.; Huang, H.Y.; Xie, J.; Takekawa, S.; Xu, X.; Omori, T.; Kainuma, R. Giant elastocaloric effect covering wide temperature range in columnar-grained $Cu_{71.5}Al_{17.5}Mn_{11}$ shape memory alloy. *APL Mater.* **2016**, *4*, 106106. [CrossRef]
11. Qian, S.; Geng, Y.; Wang, Y.; Radermacher, R. A review of elastocaloric cooling: Materials, cycles and system integrations. *Int. J. Refrig.* **2016**, *64*, 1–19. [CrossRef]
12. Ossmer, H.; Miyazaki, S.; Kohl, M. The elastocaloric effect in TiNi-based foils. *Mater. Today Proc.* **2015**, *2*, S971–S974. [CrossRef]
13. Ossmer, H.; Lambrecht, F.; Gültig, M.; Chluba, C.; Quandt, E.; Kohl, M. Evolution of temperature profiles in Ti-Ni films for elastocaloric cooling. *Acta Mater.* **2014**, *81*, 9–20. [CrossRef]
14. Tobushi, H.; Shimeno, Y.; Hachisuka, T.; Tanaka, K. Influence of strain rate on superelastic properties of TiNi shape memory alloy. *Mech. Mater.* **1998**, *30*, 141–150. [CrossRef]
15. Schmidt, M.; Schütze, A.; Seelecke, S. Elastocaloric cooling processes: The influence of material strain and strain rate on efficiency and temperature span. *Appl. Phys. Lett.* **2016**, *4*, 10–19. [CrossRef]
16. Manosa, L.; Jarque-Farnos, S.; Vives, E.; Planes, A. Large temperature span and giant refrigerant capacity in elastocaloric Cu-Zn-Al shape memory alloys. *Appl. Phys. Lett.* **2013**, *103*, 211904. [CrossRef]
17. Liu, J.L.; Huang, H.Y.; Xie, J.X. Superelastic anisotropy characteristics of columnar-grained Cu-Al-Mn shape memory alloys and its potential applications. *Mater. Des.* **2015**, *85*, 211–220. [CrossRef]
18. Xu, S.; Huang, H.; Xie, J.; Kimura, Y.; Xu, X.; Omori, T.; Kainuma, R. Dynamic recovery and superelasticity of columnar-grained Cu-Al-Mn shape memory alloy. *Metals* **2017**, *7*, 141. [CrossRef]

19. Liu, J.L.; Chen, Z.H.; Huang, H.Y.; Xie, J.X. Microstructure and superelasticity control by rolling and heat treatment in columnar-grained Cu-Al-Mn shape memory alloy. *Mater. Sci. Eng. A* **2017**, *696*, 315–322. [CrossRef]
20. Dutkiewicz, J.; Kato, H.; Miura, S.; Messerschmidtet, U.; Bartsch, M. Structure changes during pseudoelastic deformation of CuAlMn single crystals. *Acta Mater.* **1996**, *44*, 4597–4609. [CrossRef]
21. Kato, H.; Dutkiewicz, J.; Miura, S. Superelasticity and shape memory effects in Cu-23 at. % Al-7 at. % Mn alloy single crystals. *Acta Metall. Mater.* **1994**, *42*, 1359–1365. [CrossRef]
22. Ahlers, M. Martensite and equilibrium phases in Cu-Zn and Cu-Zn-Al alloys. *Prog. Mater. Sci.* **1986**, *30*, 135–186. [CrossRef]
23. Mañosa, L.; Planes, A. Materials with giant mechanocaloric effects: Cooling by strength. *Adv. Mater.* **2017**, *29*, 1603607. [CrossRef] [PubMed]
24. Sutou, Y.; Koeda, N.; Omori, T.; Kainuma, R.; Ishida, K. Effect of aging on bainitic and thermally induced martensitic transformations in ductile Cu-Al-Mn-based shape memory alloys. *Acta Mater.* **2009**, *57*, 5748–5758. [CrossRef]
25. Ossmer, H.; Chluba, C.; Krevet, B.; Quandt, E.; Rohde, M.; Kohl, M. Elastocaloric cooling using shape memory alloy films. *J. Phys. Conf. Ser.* **2013**, *476*, 012138. [CrossRef]
26. Liu, J.L.; Huang, H.Y.; Xie, J.X. The roles of grain orientation and grain boundary characteristics in the enhanced superelasticity of $Cu_{71.8}Al_{17.8}Mn_{10.4}$ shape memory alloys. *Mater. Des.* **2014**, *64*, 427–433. [CrossRef]

3

Effect of Natural Aging on the Stress Corrosion Cracking Behavior of A201-T7 Aluminum Alloy

Mien-Chung Chen [1], Ming-Che Wen [2], Yang-Chun Chiu [1], Tse-An Pan [1], Yu-Chih Tzeng [3] and Sheng-Long Lee [1,*]

1 Institute of Material Science and Engineering, National Central University, Taoyuan 320, Taiwan; mianzhongchen@gmail.com (M.-C.C.); albert77918@yahoo.com.tw (Y.-C.C.); peterpan.ck@gmail.com (T.-A.P.)
2 Department of Mechanical Engineering, National Central University, Taoyuan 320, Taiwan; j0e9rr3y0@gmail.com
3 Department of Power Vehicle and Systems Engineering, Chung-Cheng Institute of Technology, National Defense University, Taoyuan 334, Taiwan; a0932467761@gmail.com
* Correspondence: shenglon@cc.ncu.edu.tw

Abstract: The effect of natural aging on the stress corrosion cracking (SCC) of A201-T7 alloy was investigated by the slow strain rate testing (SSRT), transmission electron microscopy (TEM), scanning electron microscopy (SEM), differential scanning calorimetry (DSC), conductivity, and polarization testing. The results indicated that natural aging could significantly improve the resistance of the alloys to SCC. The ductility loss rate of the unaged alloy was 28%, while the rates for the 24 h and 96 h aged alloys were both 5%. The conductivity of the as-quenched alloy was 30.54 (%IACS), and the conductivity of the 24 h and 96 h aged alloys were decreased to 28.85 and 28.65. After T7 tempering, the conductivity of the unaged, 24 h, and 96 h aged alloys were increased to 32.54 (%IACS), 32.52 and 32.45. Besides, the enthalpy change of the 24 h and 96 h aged alloys increased by 36% and 37% compared to the unaged alloy. The clustering of the solute atoms would evidently be enhanced with the increasing time of natural aging. Natural aging after quenching is essential to improve the alloy's resistance to SCC. It might be due to the prevention of the formation of the precipitation free zone (PFZ) after T7 tempering.

Keywords: Al-Cu-Mg-Ag alloy; natural aging; stress corrosion cracking; SSRT; PFZ

---

1. Introduction

A201 (Al-4.5Cu-0.3Mg-0.7Ag) is a heat treatable aluminum alloy, which has the highest strength among the casting aluminum alloys, so it has been used in the aerospace and military industries for many years [1] The primary strengthening phases of A201 are θ' and Ω, both having a similar composition to that of $CuAl_2$ [2,3]. The crystal structure of θ' is tetragonal and with a = b = 0.414 nm and c = 0.580 nm, forming large rectangular or octagonal plates parallel to the $\{100\}_\alpha$ plane of the matrix α phase [4]. The Ω phase has a face-centered orthorhombic structure, with a = 0.496 nm, b = 0.859 nm and c = 0.848 nm, which forms hexagonal plate-like precipitates on the $\{111\}_\alpha$ plane of the matrix α phase [5–8].

To enhance the mechanical properties, especially the tensile strength, a T6 temper treatment (solution heat treated then artificially aged) is usually applied in heat treatable alloys [1]. However, for high strength Al-Cu-Mg (2XX series) or Al-Zn-Mg-Cu (7XXX series) alloys, T6 temper treatment is not recommended because it will increase the susceptibility of the alloy to stress corrosion cracking (SCC) [9–14]. SSC can occur when alloys are simultaneously subjected to stress and corrosive environments. Burleigh [15] specified three SCC mechanisms for aluminum alloys, including anodic dissolution, hydrogen-induced cracking, and the brittle passive film's rupturing. Speidel [16] have indicated that anodic dissolution is the primary mechanism of SCC in Al-Cu alloys. Misra [17] showed that Al-Cu alloys formed a precipitate free zone (PFZ) along the grain boundary following artificial aging, and this zone acts as an anode relative to the base of alloy. Eventually, under a corrosive environment, the grain boundary corrodes quickly and resulting in grain boundary cracking of the alloy.

T7 tempering (solution heat treatment then overaging) is recommended to lower the susceptibility of high strength Al-Cu-Mg (2XX series) or Al-Zn-Mg-Cu (7XXX series) alloys to SCC [1]. In the AA7075 (Al-Zn-Mg-Cu) alloy, for example, the primary strengthening phase is η ($MgZn_2$), which has the lowest potential compared to the α matrix and PFZ [18]. The T7 temper coarsens the precipitation, resulting in a discontinuous structure along the grain boundary, thereby decreasing the alloy's susceptibility to SCC [19]. However, for a high strength Al-Cu-Mg alloy, PFZ has the lowest potential than the $CuAl_2$ and α matrix. The inhibition of the formation of PFZ is the primary way to improve the resistance of the Al-Cu-Mg alloy to SCC [20].

Although the effect of aging on the SCC behavior of high strength aluminum alloys had been investigated for decades [13–21], and these works mainly focused on the effect of single artificial aging on the SCC, such as T6 (peak aging) or T7 (over aging) treatment. However, the lack of research on multiple heat treatments (combined natural aging with artificial aging) is the primary purpose of this work. Hence, the influence of natural aging on the SCC behavior, microstructure, and mechanical properties of A201-T7 alloy were investigated in this work to find the feasibility of multiple heat treatments. The results can provide a reference for the development of high-performance alloys with lower SCC suspicious while the mechanical properties could be maintained.

2. Materials and Methods

2.1. Melting and Casting

The specimens for testing were prepared as follows. High purity aluminum ingots (99.9 wt.%) were first melted in an electric resistance furnace using a graphite crucible. Suitable amounts of pure Cu, Mg, Ag, Al-75Mn, and Al-60Ti master alloys were then added. Pure argon was used for degassing the melt for 40 min. After being held for 10 min at 700 °C, it was poured into a 125 mm × 100 mm × 25 mm steel mold preheated to 300 °C. Table 1 shows the experimental alloy's chemical composition as determined by inductively coupled plasma optical emission spectrometry.

Table 1. Chemical composition of the experimental alloy (wt.%).

Alloy	Cu	Mg	Ag	Ti	Fe	Si	Al
A201	4.5 (0.1) *	0.3 (0.05)	0.7 (0.05)	0.3 (0.05)	0.05 (0.01)	0.03 (0.01)	Balance

* Standard deviations are listed in parentheses.

2.2. Heat Treatment of As-Cast Alloys

Table 2 shows the heat treatment cycles used to produce the as-cast alloy. First, they were solution treated at 510 °C for 2 h and then 530 °C for 20 h. Subsequently, water quenched (WQ) and immediately subjected to natural aging for different lengths of time (unaged, 24 h, 96 h). Finally, artificial aging was performed with T7 tempering at 190 °C for 5 h.

Table 2. Heat treatment cycles for preparation of as-cast alloys.

Alloy Notation	Solution Treatment	Natural Aging	Artificial Aging
NA0d	510 °C/2 h + 530 °C/20 h + WQ	none	190 °C/5 h
NA1d	510 °C/2 h + 530 °C/20 h + WQ	24 h	190 °C/5 h
NA4d	510 °C/2 h + 530 °C/20 h + WQ	96 h	190 °C/5 h

2.3. Slow Strain Rate Test

Slow strain rate testing (SSRT) was carried out in a 3.5% NaCl solution at pH = 7 using a tensile testing rate of 1.25×10^{-7} s^{-1}. The test specimens prepared according to ASTM B557M [22] had dimensions of 4mm in diameter and 20 mm gauge length. The occurrence of suspected stress corrosion cracking was determined from the ductility loss rate [$(E_{scc} - E_{air})/E_{air}$)] and ultimate tensile strength loss rate [$(UTS_{scc} - UTS_{air})/UTS_{air}$)]. This is the ratio of elongation and tensile strength in 3.5% NaCl compared to that in air.

2.4. Materials Characterization

A Sigmascope SMP10 electrical conductivity meter was used to measure the alloys' electrical conductivity under different heat treatment conditions. The unit of electrical conductivity were percentages of the international annealed copper standard (%IACS). Sample disks 3 mm in diameter and 1mm thick were prepared for the differential scanning calorimetry (DSC) experiments. A SEIKO-DSC6200 differential scanning calorimeter (Chiba, Japan) at a heating rate of 10 °C/min was used to capture the DSC traces. The T7 state specimens were prepared for transmission electron microscopy (FEI-TEM, Tecnai, G2-F20, Hillsboro, OR, USA) by twin-jet electro-polishing in a solution of 30% HNO_3 methanol at −30 °C and 12 V. SSRT fractography was observed by scanning electron microscopy (FEI-SEM, JEOL, JSM-7800F Prime, Akishima, Japan). The Rockwell hardness B scale was used to measure the hardness of the experimental alloys.

2.5. Polarization Testing

Anodic potentiodynamic polarization experiments were performed in a 3.5% NaCl solution to determine the breakdown potentials of A201-T7 alloys prepared with various natural aging times. Silver-Silver chloride was used as the reference electrode and platinum as the auxiliary electrode. Each 15 mm diameter sample (working electrode) was exposed to the solution for 48 h before the start of the measurement and then potentiodynamically polarized from −630 mV below the open circuit potential (OCP) to a potential above the breakdown potential at 0.5 mV·s^{-1}. The corrosion rates and polarization resistance were determined by the polarization resistance method and Stern-Geary equation. Besides, the polarization test was repeated three times in order to check the reproducibility.

3. Results and Discussion

3.1. Electrical Conductivity

As shown in Table 3, the alloys' electrical conductivity decreased with an increase in the natural aging time. It was because the clustering of solute atoms led to a lattice distortion of the α matrix [23]. The longer the natural aging time was, the worse the lattice distortion became. As a result, the electrical conductivity of the alloys became lower. The changes in conductivity [$(CAN-C_0)/C_0 \times 100\%$] of the NA1d and NA4d alloys of 5.5% and 6.2%, indicated the flattening of clustering with an extension of the aging time from 24 h to 96 h.

Table 3. Electrical conductivity of alloys under different heat treatment conditions.

Alloy Notation	IACS (%)			Percentage Change (%)	
	As-Quenched (C_0)	Natural Aging (C_{NA})	T7 (C_{T7})	$\frac{C_{NA}-C_0}{C_0} \times 100\%$	$\frac{C_{T7}-C_{NA}}{C_{NA}} \times 100\%$
NA0d	30.54 (0.12) *	30.54 (0.12)	32.54 (0.24)	-	6.5
NA1d	30.54 (0.12)	28.85 (0.11)	32.52 (0.11)	−5.5	12.7
NA4d	30.54 (0.12)	28.65 (0.09)	32.45 (0.09)	−6.2	13.4

* Standard deviations are listed in parentheses.

3.2. DSC Analysis

An enlarged view of the DSC traces for the experimental alloys in the range of 150 °C to 350 °C is presented in Figure 1. The exothermic peak at 250 °C represents the precipitation of θ′ and Ω strengthening phase [24]. The enthalpy changes of the NA0d, NA1d, and NA4d alloys were 1.35, 1.83, and 1.85 W/g, respectively, showing that natural aging could significantly increase the exothermic heating. However, there was no obvious increase in exothermic heat while the increased period of natural aging time from 24 h to 96 h.

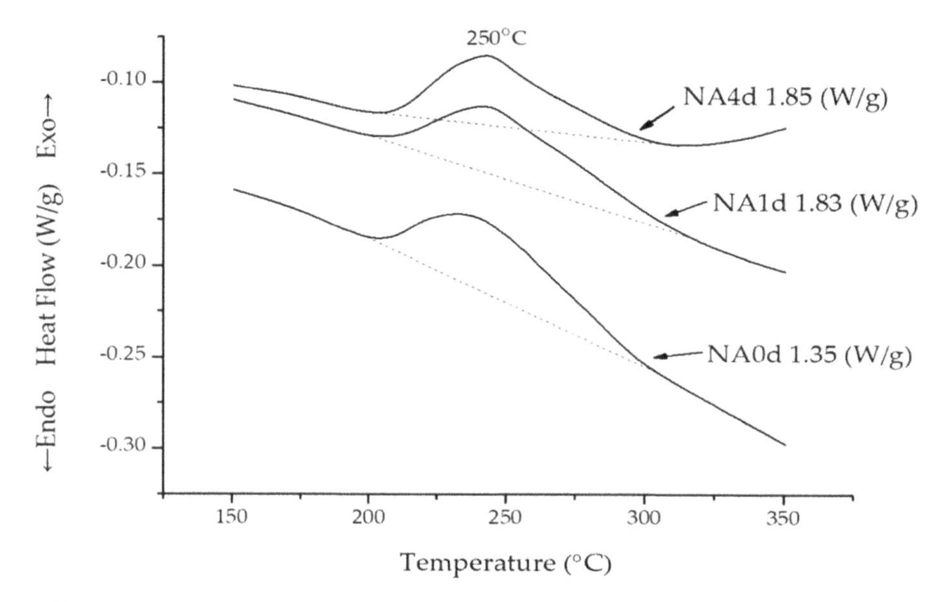

Figure 1. Differential scanning calorimetry (DSC) profile of different naturally aged A201 alloys.

3.3. Mechanical Properties

The mechanical properties of the alloys produced under different heat treatment conditions are shown in Table 4. The hardness of the alloy after solution treatment then water quenching was 33.5HRB, which increased gradually following natural aging. The percentage changes [$(H_{NA}-H_0)/H_0 \times 100\%$] of the alloys, NA1d and NA4d, were 40% and 54%, indicating that there was no obvious promotion of the clustering of solute atoms with an increase in the natural aging time from 24 h to 96 h, which was consistent with the electrical conductivity and DSC analysis results.

Table 4. Mechanical properties of alloys under different heat treatment conditions.

Alloy Notation	Hardness		Tensile Test				Percentage Change (%)	
	As-Quenched	Natural Aging	T7				$\frac{H_{NA}-H_0}{H_0} \times 100\%$	$\frac{H_{T7}-H_{NA}}{H_{NA}} \times 100\%$
	H_0 (HRB)	H_{NA} (HRB)	H_{T7} (HRB)	YS (MPa)	UTS (MPa)	EL (%)		
NA0d	33.5 (2.2) *	33.5 (2.2)	71.1 (2.1)	320 (2.6)	398 (2.7)	3.5 (0.2)	-	112
NA1d	33.5 (2.2)	46.8 (1.5)	71.6 (1.6)	325 (2.0)	396 (2.8)	3.6 (0.1)	40	53
NA4d	33.5 (2.2)	54.5 (1.7)	70.6 (2.5)	318 (3.2)	397 (3.0)	3.5 (0.2)	54	30

* Standard deviations are listed in parentheses.

After T7 tempering, with the precipitation of the strengthening phases θ′ and Ω, there was an increase in the hardness of all the alloys, NA0d, NA1d, and NA4d, to 71HRB. Evidenced that the clustering of the solute atoms remained at the same level regardless of whether natural aging was adopted or not, and that lattice distortion would be eliminated after T7 tempering. The percentage changes [$(H_{T7}-H_{NA})/H_{NA} \times 100\%$] of the alloys, NA0d, NA1d, and NA4d, were 112%, 53%, and 30%. Obviously, the hardness did not increase as much with the extension of the natural aging time from 24 h to 96 h. In addition, there was no difference in the yielding stress (YS), ultimate tensile stress (UTS), or elongation (EL) whether natural aging was adopted or not. The YS, UTS, and EL of the three alloys were approximately 320 MPa, 397 MPa, and 3.5%, respectively, after T7 tempering.

3.4. Polarization Test

The polarization curves of A201-T7 alloys after different natural aging (unaged, 24 h, 96 h) are shown in Figure 2. No passivation areas could be found after the samples were immersed in a 3.5% NaCl solution for 48 h before the test. The electrochemical analysis results were shown in Table 5. The corrosion potential, rates, and polarization resistance were represented as E_{corr} (V), I_{corr} (A/cm^2), and R_p (Ω/cm^2), separately. The unaged alloy exhibited a lower corrosion potential and lower corrosion rates than the aged alloys. The corrosion potential of the NA0d alloy was −0.71 V, lower than for NA1d (−0.60 V), and NA4d (−0.59 V). The corrosion rates of the alloys, NA0d, NA1d, and NA4d, were 3.91×10^{-5}, 5.94×10^{-5}, and 6.77×10^{-5} (A/cm^2), respectively, as determined by the polarization resistance method. Correspondingly, the polarization resistance of the naturally aged alloys was lower than that of the unaged alloy due to the clearly seen of active region in Figure 2. It is worth noting that the SCC resistance may be decreased with the long-term experiment. The polarization resistance of the unaged allot was 828 (Ω/cm^2), while the 24 h and 96 h aged alloy were 633 and 604 (Ω/cm^2). However, there was no obvious decrease in the polarization resistance of the A201-T7 alloys with an increase in the natural aging time from 24 h to 96 h.

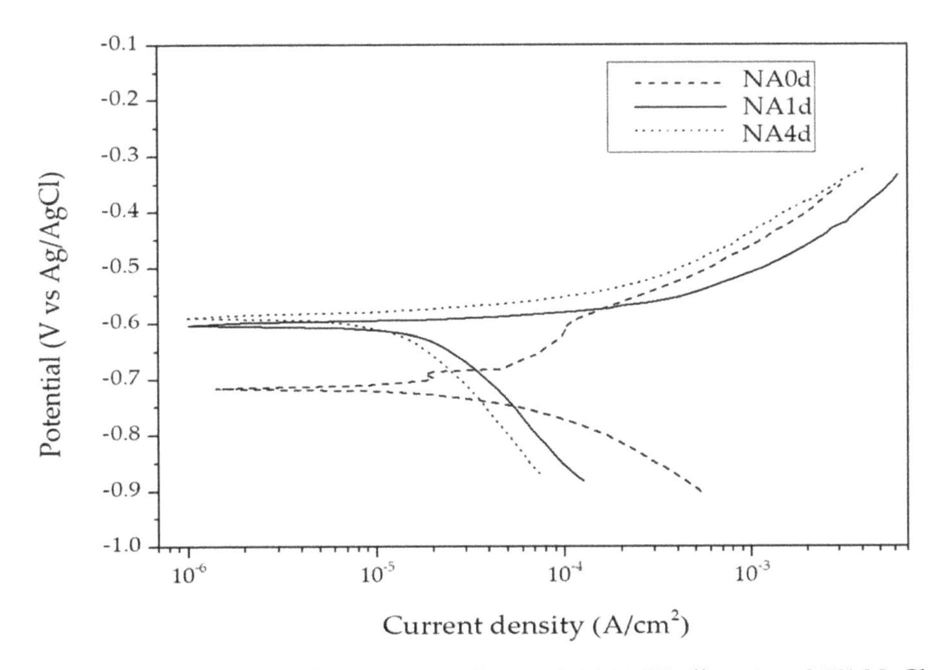

Figure 2. Polarization curves of different naturally aged A201-T7 alloys in a 3.5% NaCl solution.

Table 5. Parameters of polarization test of different naturally aged A201-T7 alloys in a 3.5% NaCl solution.

Alloy Notation	E_{corr} (V)	I_{corr} (A/cm^2)	R_p (Ω/cm^2)
NA0d	−0.71 (0.05) *	3.91×10^{-5} (1.2×10^{-5})	828 (61)
NA1d	−0.60 (0.04)	5.94×10^{-5} (3.3×10^{-6})	633 (35)
NA4d	−0.59 (0.05)	6.77×10^{-5} (2.8×10^{-6})	604 (28)

* Standard deviations are listed in parentheses.

3.5. Slow Strain Rate Testing

The results of slow strain rate testing of the A201-T7 alloys are presented in Table 6. Natural aging did not affect their elongation in air, approximately 3.5% for all three alloys. However, when it came to the saltwater environment, the elongation of the NA0d alloy decreased to 2.6%. In comparison, that of the NA1d and NA4d alloys remained about 3.5%, indicating that natural aging could significantly improve the alloy's resistance to SCC. The losses of ductility of NA1d and NA4d were 5.4% and 5.7%, respectively, while the loss in the unaged alloy could be as much as to 27.8%. Moreover, the loss of strength also showed the same tendency. The loss of strength of the unaged alloy was 15%, while the 24 h and 96 h aged alloys showed losses of 2.5% and 3.0%, respectively.

Table 6. Slow strain rate testing results of A201-T7 alloys.

Alloy Notation	EL in Air E_{air} (%)	EL in Salt Water E_{scc} (%)	UTS in Air UTS_{air} (MPa)	UTS in Salt Water UTS_{scc} (MPa)	$\frac{E_{scc}-E_{air}}{E_{air}} \times 100\%$	$\frac{UTS_{scc}-UTS_{air}}{UTS_{air}} \times 100\%$
NA0d	3.6 (0.1) *	2.6 (0.3)	399 (2.9)	339 (2.1)	−27.8	−15.0
NA1d	3.7 (0.2)	3.5 (0.2)	395 (2.6)	385 (2.8)	−5.4	−2.5
NA4d	3.5 (0.1)	3.3 (0.1)	397 (2.8)	385 (2.3)	−5.7	−3.0

* Standard deviations are listed in parentheses.

The SSRT results indicated that natural aging before T7 tempering was essential, for it had the great benefit of increased resistance of the A201-T7 alloy to SCC. It also showed that aging for 24 h was sufficient. Extending the aging time further had no additional benefit. We would also like to remind the reader that the slow strain rate testing (SSRT) might not be suitable to determine the SCC behavior of alloys in the latest research due to the sub-critical cracking and to invalidate the SSRT results [25,26].

An examination of Figure 3 shows the fracture surface of the A201-T7 alloys after slow strain rate testing. During SSRT in air, the fracture surfaces of the alloys, NA0d, NA1d, and NA4d, were similar with many dimples of different shapes and sizes observed, implying that the fracture mechanism was ductile fracturing. As a result, natural aging did not affect the elongation in the air. However, the fracture surfaces of the alloys were quite different when SSRT was conducted in the 3.5% NaCl solution. For the unaged alloy, nearly no dimples were observed, and the fracture mechanism was brittle fracturing. For the 24 h and 96 h aged alloys, cleavages and dimples were observed in the sample, and the fracture mechanism was a combination of ductile and brittle fracturing. The results were consistent with the mechanical properties presented in Table 4.

3.6. Transmission Electron Microscopy (TEM) Characterization

The Schematic diagram of the calculated diffraction pattern of Ω and θ' strengthening phases along $[011]_\alpha$ zone axis were shown in Figure 4 [27]. Strengthening phases Ω and θ' have a similar composition to $CuAl_2$, but with a different crystal structure. As shown in Figure 5d, the selected area electron diffraction patterns (SAED) along $[011]_\alpha$ zone axis evidenced Ω and θ' precipitation strengthening phases exist in A201 alloys after T7 tempered. The bright streaks and $(10\bar{1})$ diffraction spot of Ω phase and (100) diffraction spot of θ' phase were found at $(200)_\alpha$ position. The TEM bright field images of the grain boundaries of A201-T7 alloys shown in Figure 5 reveal the presence of precipitation at the grain boundaries. The needle-like strengthening phases Ω and θ' inside the grains can also be seen. The primary precipitation phase of A201-T7 alloys was Ω phases, and minor θ' phases with different orientation can also be observed. It seems that natural aging did not affect the size of precipitation phases, Ω and θ', after T7 heat treatment, just as shown in Figure 5d–f. The maximum length of Ω and θ' phase of three different natural aged alloys (unaged, 24 h, 96 h) were around 100 nm and 70 nm, separately. The NA0d alloy had significant prefer-precipitation during T7 tempering [28]. The oversaturation of solute atoms near the grain boundaries would diffuse to the grain boundaries, forming coarse and discontinuous precipitation phases along with a 200 nm wide precipitation free zone (PFZ) as shown in Figure 5a. However, for NA1d and NA4d, no obvious PFZ could be found at the grain boundaries, as shown in Figure 5b,c. This might be due to the clustering of solute atoms (Cu, Mg), which occurred during natural aging, which would directly transfer to the precipitation phases (Ω and θ') during T7 tempering, thus inhibiting the formation of PFZ.

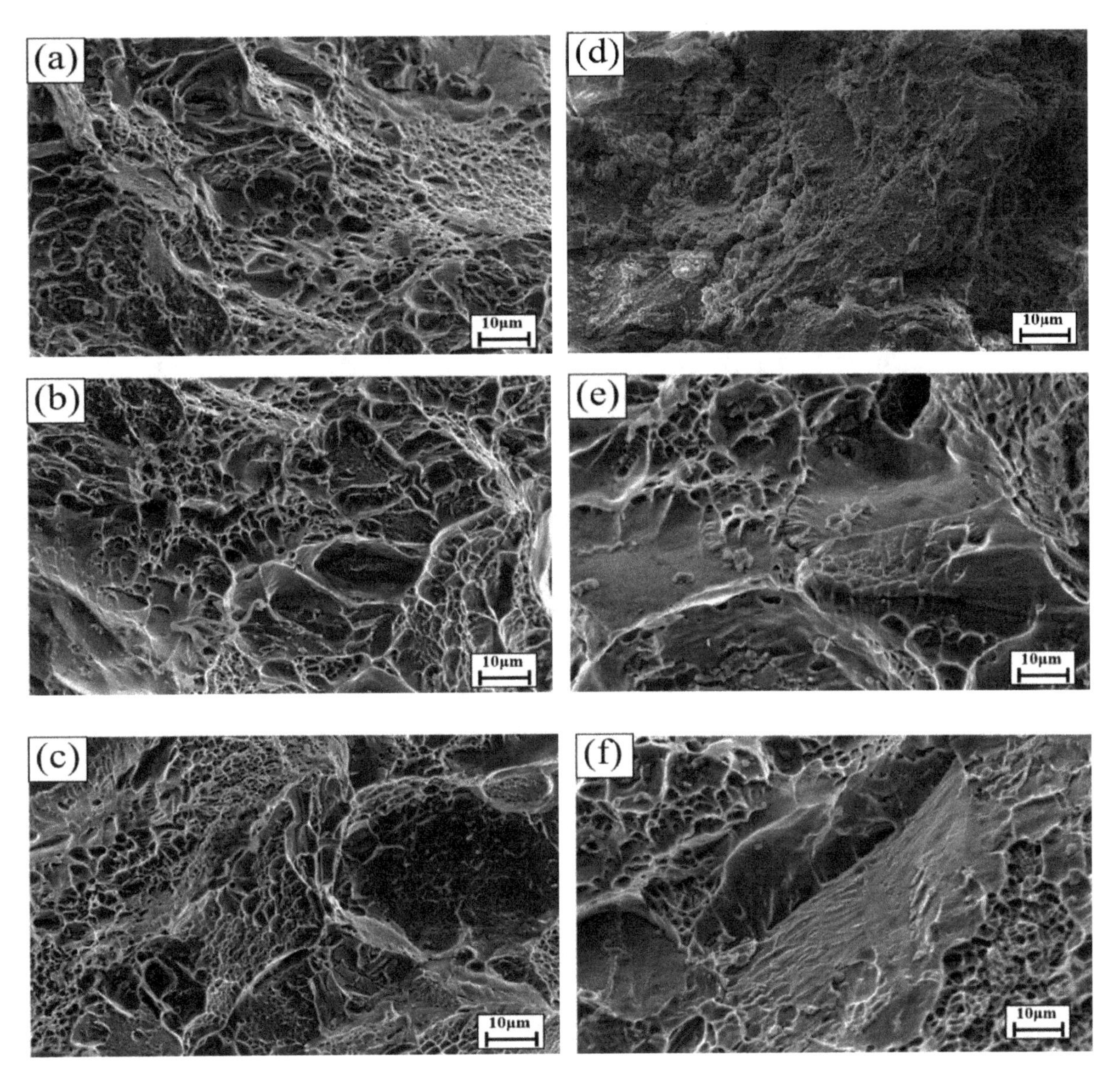

Figure 3. Fracture surface of A201-T7 alloys after slow strain rate testing: (**a**) NA0d sample in air; (**b**) NA1d sample in air; (**c**) NA4d sample in air; (**d**) NA0d sample in salt water; (**e**) NA1d sample in salt water; (**f**) NA4d sample in salt water.

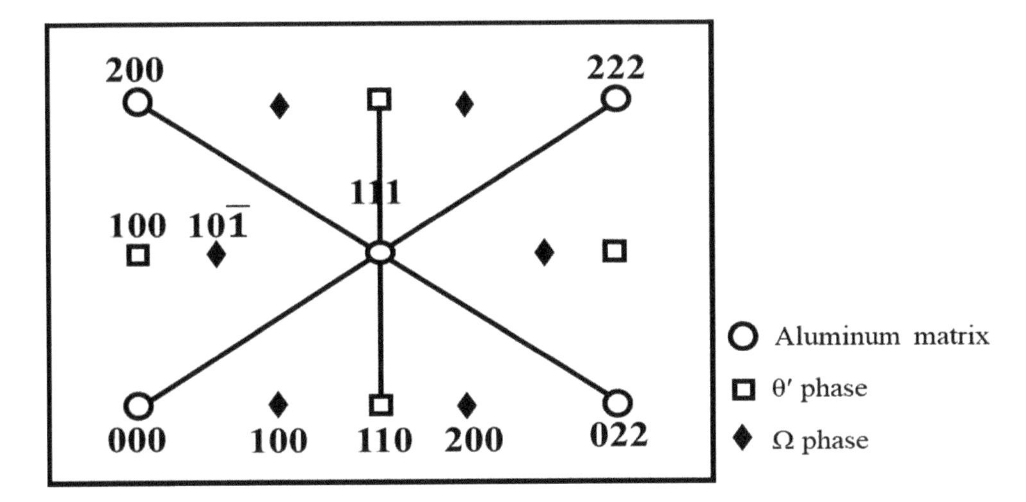

Figure 4. Schematic diagram of the diffraction pattern of θ′ and Ω strengthening phases along $[011]_\alpha$ zone in A201-T7 alloy.

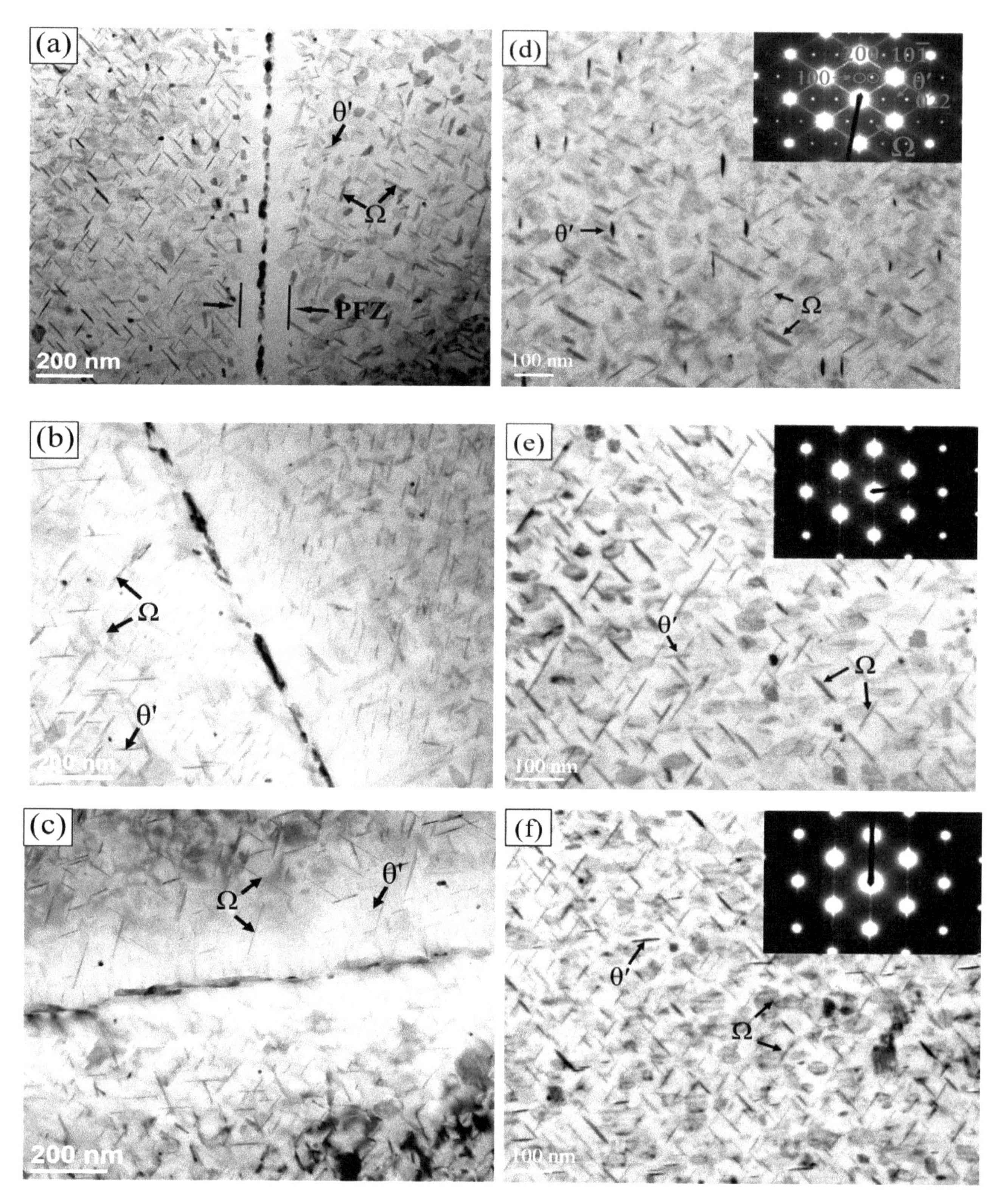

Figure 5. TEM bright field images and the selected area electron diffraction pattern (SAED) of: (**a**,**d**) Alloy NA0d; (**b**,**e**) Alloy NA1d; (**c**,**f**) Alloy NA4d after T7 tempering.

It is worth noting that the PFZ has the lowest potential compared to $CuAl_2$ phases (θ′ and Ω) and α matrix in the 2XX series and 2XXX series Al-Cu-Mg alloys [29]. To lower the galvanic corrosion effect, the inhibition of the formation of PFZ can help prevent SCC [30]. The discussion above is supported by the polarization test and slow strain rate test results, indicating that naturally aged alloys have better SCC resistance than unaged alloys.

4. Conclusions

The effects of natural aging on stress corrosion cracking in A201-T7 alloy were investigated in this study and the following conclusions can be drawn:

(1) For the as-quenched alloy, the conductivity decrease and the hardness increase during natural aging. However, the conductivity and mechanical properties (hardness, strength, and elongation) were unaffected by natural aging after T7 tempering.
(2) Natural aging improved the resistance of A201-T7 alloys to SCC. 24 h aging was sufficient. Extending the aging time provided no additional benefit.
(3) In the unaged alloys, PFZ existed and brittle fractures could be found on the SCC fracture surface; for the aged alloys, no PFZ existed, but a combination of fracture types with cleavages and dimples could be observed on the fracture surface.

Author Contributions: Data curation, Writing—original draft preparation, M.-C.C.; Conceptualization; Methodology, M.-C.W.; Performed experiment, Formal analysis, T.-A.P.; Investigation; Validation, Y.-C.C.; Project administration, Resources, Y.-C.T.; Supervision, Funding acquisition, Review and editing the draft, S.-L.L. All authors have read and agreed to the published version of the manuscript.

Acknowledgments: The author would like to thank Tsai-Fu Chung from National Taiwan University for the TEM observation.

References

1. Davis, J.R. *ASM Specialty Handbook: Aluminum and Aluminum Alloys*; ASM International: Materials Park, OH, USA, 1994.
2. Chester, R.J.; Polmear, I.J. TEM Investigation of Precipitates in Al-Cu-Mg-Ag and Al-Cu-Mg Alloys. *Micron* **1980**, *11*, 311–312. [CrossRef]
3. Kim, K.D.; Zhou, B.C.; Wolverton, C. Interfacial Stability of θ′/Al in Al-Cu Alloys. *Scr. Mater.* **2019**, *159*, 99–103. [CrossRef]
4. Phillips, V.A. High Resolution Electron Microscope Observations on Precipitations in Al-3.0% Cu Alloy. *Acta Mater.* **1975**, *23*, 751–767. [CrossRef]
5. Purnendu, K.M. Influence of Micro-alloying with Silver on Microstructure and Mechanical Properties of Al-Cu Alloy. *Mater. Sci. Eng. A* **2018**, *722*, 99–111.
6. Knowles, K.M.; Stobbs, W.M. The Structure of {111} Age-Hardening Precipitates in Al-Cu-Mg-Ag Alloys. *Acta Cryst.* **1988**, *44*, 207–227. [CrossRef]
7. Ivan, Z.; Rustam, K. Aging Behavior of an Al-Cu-Mg Alloy. *J. Alloys Compd.* **2018**, *759*, 108–119.
8. Saeed, K.M.; Mohsen, K.; Roland, L. Mechanical Behavior and Texture Development of Over-aged and Solution Treated Al-Cu-Mg Alloy during Multi-directional Forging. *Mater. Charact.* **2018**, *135*, 221–227.
9. Muddle, B.C.; Polmear, I.J. The Precipctate Ω Phase in Al-Cu-Mg-Ag Alloys. *Acta Metal.* **1989**, *37*, 777–789. [CrossRef]
10. Garg, A.; Chang, Y.C.; Howe, J.M. Precipitation of the Ω Phase in an Al-4.0Cu-0.5Mg Alloy. *Scr. Mater.* **1990**, *24*, 677–680. [CrossRef]
11. Hu, Y.C.; Liu, Z.Y.; Zhao, Q.; Bai, S.; Liu, F. P-Texture Effect on the Fatigue Crack Propagation Resistance in an Al-Cu-Mg Alloy Bearing a Small Amount of Silver. *Materials* **2018**, *11*, 2481. [CrossRef]
12. ASTM B917. *Standard Practice for Heat Treatment of Aluminum-Alloy Castings form All Processes*; ASTM International: West Conshohocken, PA, USA, 2020.
13. Rajan, K.; Wallace, W.; Beddoes, J.C. Microstructure Study of a High Strength Stress-Corrosion Resistant 7075 Aluminum Alloy. *J. Mater. Sci.* **1982**, *17*, 2817–2848. [CrossRef]
14. Islam, M.U.; Wallace, W. Retrogression and Reaging Response of 7475 Aluminum Alloy. *Met. Technol.* **1983**, *10*, 386–392. [CrossRef]
15. Burleigh, T.D. The Postulated Mechanisms for Stress Corrosion Cracking of Aluminum Alloys. *Corrosion* **1991**, *47*, 89–98. [CrossRef]
16. Speidel, M.O.; Hyatt, M.V. *Advances in Corrosion Science and Technology*; Springer: Boston, MA, USA, 1972.
17. Misra, M.S.; Oswalt, K.J. Corrosion Behavior of Al-Cu-Mg-Ag (201) Alloy. *Met. Eng. Q.* **1976**, *16*, 39–44.

18. Alexander, I.I.; Zhang, B.; Wang, J.Q.; Han, E.H.; Ke, W.; Peter, C.O. SVET and SIET Study of Galvanic Corrosion of Al/$MgZn_2$ in Aqueous Solutions at Different pH. *J. Electrochem. Soc.* **2018**, *165*, 180–194.
19. Shi, Y.J.; Pan, Q.L. Influence of Alloyed Sc and Zr, and Heat Treatment on Microstructures and Stress Corrosion Cracking of Al-Zn-Mg-Cu Alloys. *Mater. Sci. Eng. A* **2015**, *621*, 173–181. [CrossRef]
20. Lee, H.J.; Kim, Y.J.; Jeong, Y.; Kim, S.S. Effects of Testing Variables on Stress Corrosion Cracking Susceptibility of Al 2024-T351. *Corros. Sci.* **2012**, *55*, 10–19. [CrossRef]
21. Cabrini, M.; Bocchi, S.; D'Urso, G.; Giardini, C.; Lorenzi, S.; Testa, C.; Pastore, T. Stress Corrosion Cracking of Friction Stir-Welded AA-2024 T3 Alloy. *Materials* **2020**, *13*, 2610. [CrossRef]
22. ASTM B557. *Standard Test Methods for Tension Testing Wrought and Cast Aluminum- and Magnesium-Alloy Products*; ASTM International: West Conshohocken, PA, USA, 2020.
23. Ivanov, R.; Deschamps, A.; Geuser, F.D. Clustering Kinetics during Natural Ageing of Al-Cu Based Alloys with (Mg, Li) Additions. *Acta Mater.* **2018**, *157*, 186–195. [CrossRef]
24. Chang, C.H.; Lee, S.L.; Lin, J.C.; Yeh, M.S.; Jeng, R.R. Effect of Ag Content and Heat Treatment on The Stress Corrosion Cracking of Al-4.6Cu-0.3Mg alloy. *Mater. Chem. Phys.* **2005**, *91*, 454–462. [CrossRef]
25. Martínez-Pañeda, E.; Harris, Z.D.; Fuentes-Alonso, S.; Scully, J.R.; Burns, J.T. On the suitability of slow strain rate tensile testing for assessing hydrogen embrittlement susceptibility. *Corros. Sci.* **2019**, *163*, 108291. [CrossRef]
26. Nikolaos, D.A.; Christina, C.; Panagiotis, S.; Stavros, K.K. Synergy of Corrosion-induced Micro-cracking and Hydrogen Embrittlement on The Structural Integrity of Aluminum Alloy (Al-Cu-Mg) 2024. *Corros. Sci.* **2017**, *121*, 32–42.
27. Beffort, O.; Solenthaler, C.; Uggowitzer, P.J.; Speidel, M.O. High toughness and high strength spray-deposited AlCuMgAg-base alloys for use at moderately elevated temperatures. *Mater. Sci. Eng. A* **1995**, *191*, 121–134. [CrossRef]
28. Wang, H.S.; Jiang, B.; Zhang, J.Y.; Wang, N.H.; Yi, D.Q.; Wang, B.; Liu, H.Q. The Precipitation Behavior and Mechanical Properties of Cast Al-4.5Cu-3.5Zn-0.5Mg Alloy. *J. Alloys Compd.* **2018**, *768*, 707–713. [CrossRef]
29. Qi, H.; Liu, X.Y.; Liang, S.X.; Zhang, X.L.; Cui, H.X.; Zheng, L.Y.; Gao, F.; Chen, Q.H. Mechanical Properties and Corrosion Resistance of Al-Cu-Mg-Ag Heat-resistant Alloy Modified by Interrupted Aging. *J. Alloys Compd.* **2016**, *657*, 318–324. [CrossRef]
30. Liu, X.Y.; Li, M.J.; Gao, F.; Liang, S.H.; Zhang, X.L.; Cui, H.X. Effects of Aging Treatment on The Intergranular Corrosion Behavior of Al-Cu-Mg-Ag alloy. *J. Alloys Compd.* **2015**, *639*, 263–267. [CrossRef]

4

Evolution of Microstructure and Mechanical Properties of a CoCrFeMnNi High-Entropy Alloy during High-Pressure Torsion at Room and Cryogenic Temperatures

Sergey Zherebtsov [1], Nikita Stepanov [1], Yulia Ivanisenko [2], Dmitry Shaysultanov [1], Nikita Yurchenko [1], Margarita Klimova [1,*] and Gennady Salishchev [1]

[1] Laboratory of Bulk Nanostructured Materials, Belgorod State University, 308015 Belgorod, Russia; zherebtsov@bsu.edu.ru (S.Z.); stepanov@bsu.edu.ru (N.S.); shaysultanov@bsu.edu.ru (D.S.); yurchenko_nikita@bsu.edu.ru (N.Y.); salishchev@bsu.edu.ru (G.S.)

[2] Karlsruhe Institute of Technology, Institute of Nanotechnology, 76021 Karlsruhe, Germany; julia.ivanisenko@kit.edu

* Correspondence: klimova_mv@bsu.edu.ru

Abstract: High-pressure torsion (HPT) is applied to a face-centered cubic CoCrFeMnNi high-entropy alloy at 293 and 77 K. Processing by HPT at 293 K produced a nanostructure consisted of (sub)grains of ~50 nm after a rotation for 180°. The microstructure evolution is associated with intensive deformation-induced twinning, and substructure development resulted in a gradual microstructure refinement. Deformation at 77 K produces non-uniform structure composed of twinned and fragmented areas with higher dislocation density then after deformation at room temperature. The yield strength of the alloy increases with the angle of rotation at HPT at room temperature at the cost of reduced ductility. Cryogenic deformation results in higher strength in comparison with the room temperature HPT. The contribution of Hall–Petch hardening and substructure hardening in the strength of the alloy in different conditions is discussed.

Keywords: high-entropy alloys; high-pressure torsion; microstructure evolution; twinning; mechanical properties

1. Introduction

The concept of high-entropy alloys (HEAs) as a mixture of more than five metallic elements in equimolar proportions was proposed first by Yeh et al. in 2004 [1]. Due to increased configurational entropy, the HEAs were expected to exist in the form of a single solid solution phase while formation of intermetallic compounds and secondary phases would be suppressed. In reality, the microstructure of alloys of multiple principle elements is usually more complicated and can consist of a mixture of different phases including intermetallic ones [2]. Among many proposed HEA compositions, CoCrFeMnNi alloy indeed has a single phase disordered fcc microstructure [3–5]. Due to such a 'model' structure, this alloy is currently one of the most studied HEAs so far.

It was found that the CoCrFeMnNi alloy demonstrated very high elongation (7080%), but rather low yield strength of 200 MPa at room temperature [6,7]. Strengthening of HEA can be achieved via microstructure refinement. A reduction of a grain size from 150 to 5 μm in the CoCrFeMnNi alloy increased the yield strength by a factor of two while maintained good ductility [6]. Severe plastic deformation of the CoCrFeNiMn alloy via high-pressure torsion (HPT) expectably refined the microstructure to a grain size d~40–50 nm and increased the microhardness of the alloy by a factor of ~3 [8–10]. The ultimate tensile strength of the alloy after HPT was found to be ~2000 MPa [8].

The formation of the nanostructure was attributed to fragmentation of nanotwins [8] and to accelerated atomic diffusivity under the shear strain during HPT [9]. However, the microstructure and mechanical properties evolution of the alloy was insufficiently studied. For instance, the evolution of tensile properties during HPT had not been studied yet. Besides, considerably higher intensity of deformation twinning in the alloy at cryogenic temperature was observed during tension [6,11] or rolling [12]. Increased susceptibility to twinning can significantly accelerate the microstructure refinement, as it was shown earlier [12–14]. However, no information on attempts of HPT of the CoCrFeMnNi HEA under cryogenic conditions was found in the literature.

Therefore, the aim of the present work was to investigate in details the microstructure and mechanical properties evolution of the high-entropy CoCrFeMnNi alloy during high-pressure torsion at room temperature and to evaluate the effect of temperature decrease to 77 K on structure and properties of the alloy.

2. Materials and Methods

The equiatomic alloy with the composition of CoCrFeMnNi was produced by arc melting of the components in high-purity argon inside a water-cooled copper cavity. The purities of the alloying elements were above 99.9 %. To ensure chemical homogeneity, the ingots were flipped over and re-melted at least five times. The produced ingots of the CoCrFeMnNi alloy had dimensions of about $6 \times 15 \times 60$ mm^3. Homogenization annealing was carried out at 1000 °C for 24 h [15,16]. Prior to homogenization, the samples were sealed in vacuumed (10^{-2} Torr) quartz tubes filled with titanium chips to prevent oxidation. After annealing, the tubes were removed from the furnace and the samples were cooled inside the vacuumed tubes down to room temperature. After the homogenization procedure, the alloy was cold rolled with ~80% height reduction and then annealed at 850 °C for 1 h to produce a recrystallized structure. The alloy after such a treatment is hereafter referred to as the initial condition in the current study.

For HPT processing discs measuring 15 mm diameter × 0.8 mm thick were cut from the recrystallized specimens, then ground and mechanically polished. The samples were subjected to HPT at 293 K (to 90, 180, or 720° of the anvils turn) or 77 K (to 180°) under a pressure of 4.3 GPa in a Bridgman anvil-type unit with a rate of 1 rpm using a custom-built computer-controlled HPT device (W. Klement GmbH, Lang, Austria). The corresponding shear strain level γ was calculated as [17]

$$\gamma = \frac{2\pi N r}{h} \tag{1}$$

where N is the number of revolutions, r is the radius, and h is the thickness of the specimen.

The microstructure of the alloys was studied using transmission (TEM) and scanning (SEM) electron microscopy and electron back-scattered diffraction (EBSD) analysis. For SEM observations the specimens were mechanically polished in water with different SiC papers and a colloidal silica suspension; the final size of the Al_2O_3 abrasive was 0.04 μm. The samples were examined after polishing without etching. The SEM back-scattered electron (BSE) images of microstructures in the initial condition were obtained using a FEI Quanta 3D microscope. EBSD was conducted in a FEI Nova NanoSEM 450 field emission gun SEM equipped with a Hikari EBSD detector and a TSL OIM™ system version 6.0. EBSD examinations with the step size of 50 nm were carried out in an axial section in three different characteristic areas: (i) in the central part of the disc, i.e., within ±0.5 mm of the center; (ii) at half of a radius, i.e., at ~3.5 mm from the center; (iii) at the edge, i.e., at ~6–7 mm from the center. The points with confidence index (CI) < 0.1 were excluded from the analysis; these points are shown with black color to avoid any artificial interference in the microstructure. In the presented inverse pole figure (IPF) maps, the high angle (>15°) and low angle (2–15°) boundaries are shown respectively as black and white lines.

TEM examination was performed in the mid-thickness shear plane of deformed samples in the vicinity (~1 mm away) of the edge. The samples for TEM analysis were prepared by conventional twin-jet electro-polishing of mechanically pre-thinned to 100 μm foils, in a mixture of 90% CH_3COOH and 10% $HClO_4$ at the 27 V potential at room temperature. TEM investigations were performed using JEOL JEM-2100 apparatus at accelerating voltage of 200 kV.

The dislocation density was determined using X-ray diffraction (XRD) profiles analysis obtained by a RIGAKU diffractometer with Cu Kα radiation at 45 kV and 35 mA as it was done earlier for the similar alloy [18]. The value of the dislocation density, ρ, was calculated using the equation [19]

$$\rho = \frac{3\sqrt{2\pi}\,\langle\varepsilon_{50}^2\rangle}{Db} \quad (2)$$

where $\langle\varepsilon_{50}^2\rangle$ is microstrains, D is the crystallite size, and b is the Burgers vector. The microstrains, $\langle\varepsilon_{50}^2\rangle$, and the crystallite size, D, values were estimated on the basis of the Williamson–Hall plot [20], using the equation

$$\frac{\beta_s cos\Theta}{\lambda} = \frac{2\langle\varepsilon_{50}^2\rangle sin\Theta}{\lambda} + \frac{K}{D} \quad (3)$$

where β_s is the corrected full width at half maximum (FWHM) of the selected Kα_1 reflection of the studied material, Θ is the Bragg angle of the selected reflection, λ is the Kα_1 wavelength, and K is Scherrer constant. In the present study, the FWHM values and the positions of fcc (111) and (222) reflections were determined and used for further calculations. The instrumental broadening was determined from the FWHM values of the annealed silicon powder.

Mechanical properties of the alloy after HPT were determined through tensile tests of flat dog-bone specimens with gauge measuring 5 mm length × 1 mm width × 0.6 mm thickness. The samples for the tensile tests were cut from the HPT-processed discs using an electric discharge machine so that the gage was located at a distance of ~3.5 mm from the sample center. The specimens were then pulled at a constant crosshead speed of 3 mm/min (strain rate of 0.01 s^{-1}) in a custom built computer controlled tensile stage for miniature samples to fracture. The stress–strain curves were obtained using a high-precision laser extensometer.

3. Results

In the initial condition the CoCrFeNiMn alloy had an fcc single-phase microstructure (Figure 1a). An SEM-BSE image shows a homogeneous microstructure with a grain size of ~15 μm (Figure 1b). Numerous annealing twins were found inside grains. Some pores visible as black dots can also be found in the microstructure.

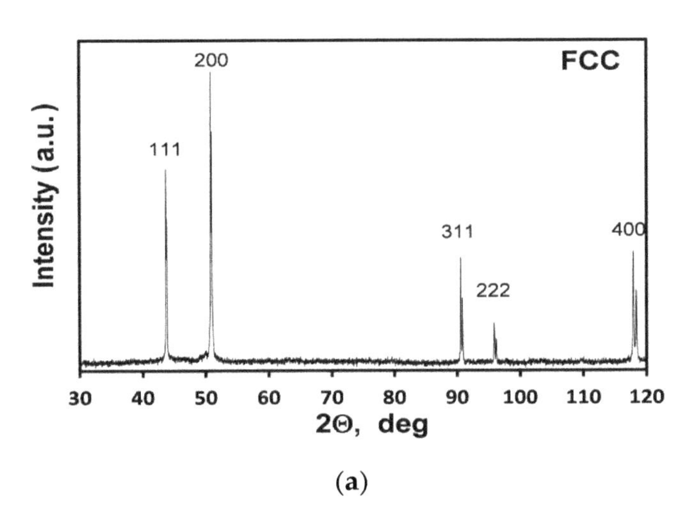

(a)

(b)

Figure 1. Microstructure of the initial condition of the CoCrFeNiMn alloy: (**a**) XRD pattern; (**b**) SEM-BSE image.

The influence of HPT on the microstructure depended essentially on the distance from the specimen center (Figure 2). In comparison with the initial condition, a rotation of 90° at room temperature resulted in rather small changes in the microstructure (Figure 2a) associated with the formation of low-angle boundaries in some grains and appearance of individual deformation twins. It should be noted that boundaries of both deformation and annealing twins are crystallographycally identical. However, deformation twins are usually much thinner than annealing twins and intersect grains. The thickness of deformation twins in a single phase fcc microstructure is usually several tens of nanometers [12,21], whereas the thickness of annealing twins can reach several micrometers [22]. Some of the deformation twins are indicated with arrows in Figure 2b,c. The analysis of the EBSD data showed that observed (both annealing and deformation) twins belong to (111) <112> system with twin/matrix misorientation of 60° around <111>. An increase in the number of deformation twins and the development of substructure (sometimes appearing as black, poorly distinguished areas mainly in the vicinity of twin or grain boundaries) with the increase of imposed strain can be observed (Figure 2b,c).

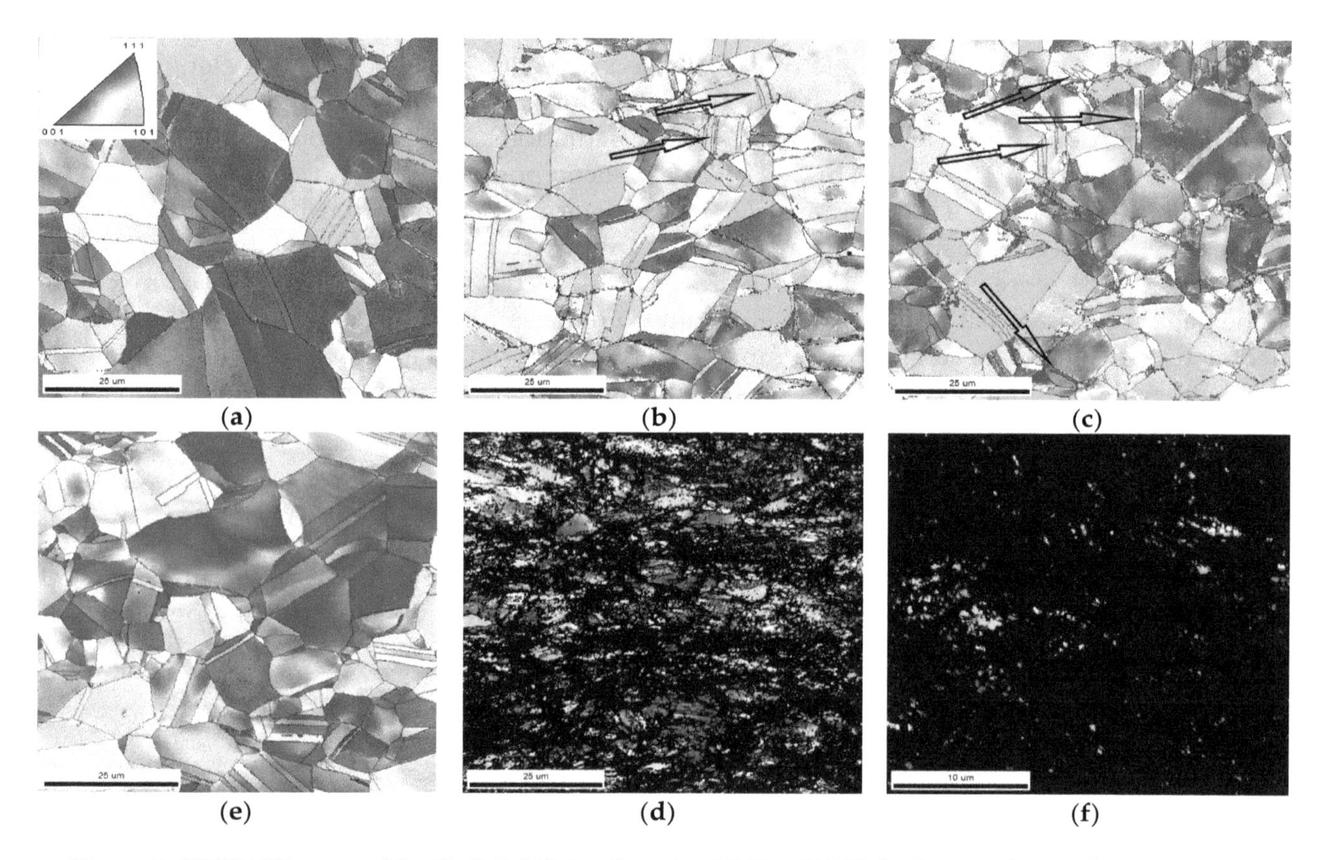

Figure 2. EBSD IPF maps of the CoCrFeNiMn alloy after HPT at 293 K: (**a**,**d**) central part of the specimen; (**b**,**e**) half-radius; (**c**,**f**) edge of the specimens; (**a**–**c**) 90° and (**d**–**f**) 180° rotation. The corresponding true shear strain is (**a**) 1.1; (**b**) 8.4; (**c**) 13.5; (**d**) 2.2; (**e**) 16.8; (**f**) 26.9. The color code is inserted in Figure 2a. The deformation twins are indicated with arrows in Figure 2b,c.

The central part of the specimens after rotation for 180° (Figure 2d) had a microstructure very similar to that observed after 90° rotation; i.e., only little changes can be observed in comparison with the initial microstructure. This is possibly due to relatively low true strain in the central part of both specimens ($\varepsilon \leq 2.2$). However, much more pronounced changes were observed at larger strain (Figure 2e,f). The microstructure at the half-radius consisted of relatively large areas of 5–7 μm with developed substructure and small fragments of ~0.5 μm (Figure 2e). Due to a high level of internal stresses which prevents obtaining identifiable Kikuchi patterns, a considerable part of the EBSD map was presented by black dot areas with a low CI. The comparison with the IPF map of the edge part of the specimen after the rotation for 90° (Figure 2c) suggests that the true shear strain of ~15 is required to produce a significant fraction of such highly-deformed areas.

At the edge of the specimen, the average size of the visible fragments decreased to ~0.5 μm (Figure 2f). However, the black dot areas with low CI occupied the majority of the scanned area thereby suggesting a very high level of internal stresses and considerable microstructure refinement. This finding is in good agreement with an increase in the true shear strain to 26.9. Detailed TEM analysis of the microstructure of those severely deformed areas is given below. Note that EBSD analysis of the sample after the rotation for 720° has produced IPF maps almost completely occupied with black dots both at the center, half-radius, and edge of the specimen that most likely associated with a considerable increase in homogeneity of the microstructure refinement. The corresponding images are not shown as they do not present any meaningful information.

The effect of a decrease in deformation temperature to 77 K was studied after the rotation for 180° (Figure 3). Microstructure changes in comparison with the specimen deformed at room temperature (Figure 2d-f) depended considerably on the examined part of the disk and the imposed strain. Microstructure of the central part of the specimen deformed at 77 K (Figure 3a) was evidentially more refined than that at room temperature (Figure 2d) and consisted of rather large areas of ~8 μm which contained individual twins and relatively poor developed substructure. Low quality of the EBSD map in between these areas indicated quite a high level of internal stresses, most likely due to strain localization. Small fragments < 1 μm in diameter were observed in severely deformed areas. Higher strain level obtained at the half of radius (ε = 16.8, Figure 3b) resulted in a microstructure which is quite similar to that at room temperature (Figure 2e). At the edge of specimens the microstructure after HPT at 77 K (Figure 3c) had an obvious metallographic texture with fragments elongated along the shear direction. The size of the majority of the fragments was quite small (0.2–0.4 μm). However, big fragments > 2 μm with poorly developed substructure were also observed in the microstructure. In addition, the fraction of visible fragments was greater than that after room temperature HPT (Figure 2f) that can suggest lower level of internal stresses in the former case.

(a) (b) (c)

Figure 3. Microstructure of the CoCrFeNiMn alloy after HPT at temperature of 77 K to 180°: (**a**) central part; (**b**) half-radius; (**c**) edge. The corresponding true shear strain is (**a**) 2.2; (**b**) 16.8; (**c**) 26.9.

TEM micrographs of the CoCrFeNiMn alloy microstructure after HPT at room temperature is shown in Figure 4. After HPT rotation for 90° at 293 K (Figure 4a,b) the microstructure at the edge of specimens consisted of crossing deformation twins of 50–100 nm width (corresponding diffraction pattern inserted in Figure 4a) and high dislocation density. Twins either spaced 0.5–1 μm apart (Figure 4a) or cluster together with the formation of twin bundles (Figure 4b). After the rotation for 180° at room temperature the microstructure at the edge of specimens was considerably refined (Figure 4c,d). Although the microstructure can be mainly described as a cellular one with a very high dislocation density (Figure 4c), very small grains of ~50 nm can also be recognized at larger magnification. Twin boundaries were not detected in the microstructure. However, a chain of small elongated (sub)grains in some places (indicated with dotted lines in Figure 4d) suggested that these grains can originated from twins. Rotation for 720° resulted in the formation of a homogeneous ultra-fine microstructure consisted of irregular dislocation pile-ups of different shapes and sizes and very small grains with a size of ~20–60 nm (Figure 4e,f). The diffraction pattern consisted of rings with many diffracted beams indicating the presence of many small (sub)grains with mainly high misorientation angle boundaries within the selected field of view (insert on Figure 4e).

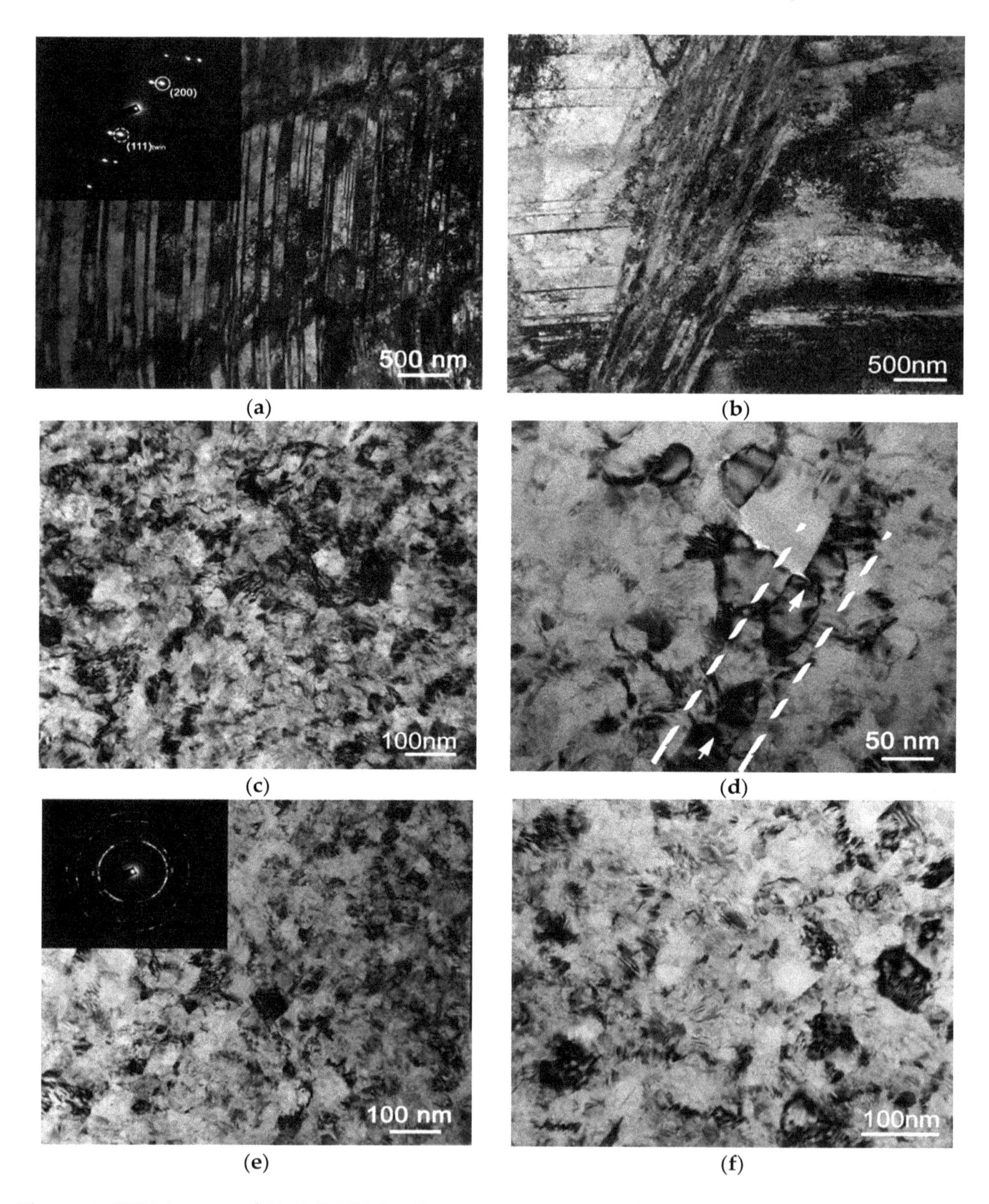

Figure 4. TEM images of CoCrFeNiMn alloy microstructure at the edge of specimens after HPT at temperature of 293 K to: (**a**,**b**) 90° (true strain of ≈12); (**c**,**d**) 180° (true strain ≈25); (**e**,**f**) 720° (true strain ≈50). Elongated (sub)grains and transverse subboundaries are indicated with dash lines and arrows, respectively in Figure 4d.

In contrast to the room-temperature deformation (Figure 4c,d), the microstructure observed after HPT at 77 K was found to be highly inhomogeneous (Figure 5). It consisted of areas with twin bundles and structure with high dislocation density and subgrain orientation (Figure 5a). The selected area diffraction pattern (insert in Figure 5a) was obtained from a crystallite with zone axis [$0\bar{6}2$]; within these crystallite areas, with both twin and subgrain structure can be observed. The reflexes characterizing the twin orientation are indexed on the diffraction pattern. The matrix reflexes show azimuthal spread, thereby suggesting the presence of small misorientations inside the crystallite. An example of such microstructure is shown in the Figure 5b. The microstructure revealed sites of different shapes and sizes with low-angle misorientation between them; meanwhile, no grains were observed in the microstructure (Figure 5b).

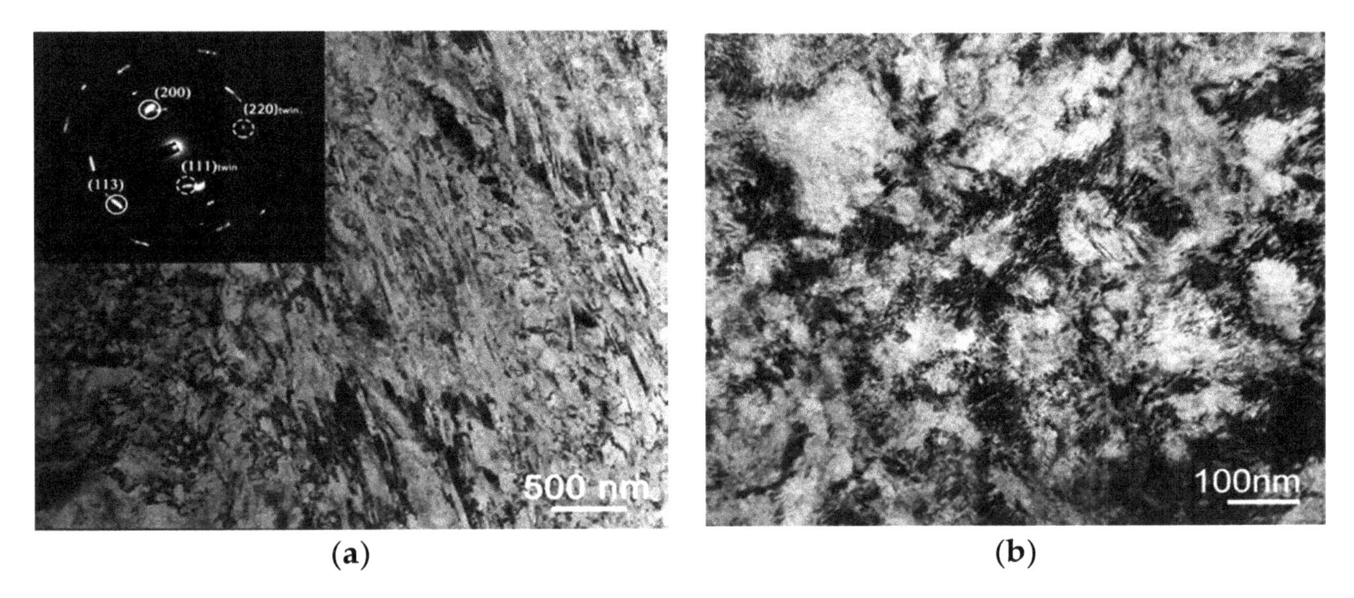

Figure 5. TEM image of the CoCrFeNiMn alloy microstructure at the edge of specimens after HPT at temperature of 77 K to 180° (true strain ~25). (**a**) low magnification; (**b**) high magnification.

According to the XRD analysis, the dislocation density after cryogenic deformation was 2.9×10^{15} m^{-2}. In comparison, after similar processing at room temperature the dislocation density was 9.14×10^{14} m^{-2}, i.e., three-fold increase in dislocation density due to decrease of HPT temperature occurred. In turn, the imposed strain at room temperature had weakly affected the dislocation density. For example, the dislocation density values after rotations for 90° and 720° were 1.48×10^{15} m^{-2} and 8.86×10^{14} m^{-2}, respectively, i.e., they were rather similar to each other and to dislocation density after rotation for 180° at room temperature.

Tensile stress–strain curves of the alloy are shown in Figure 6. The resulting mechanical properties including yield strength (YS), ultimate tensile strength (UTS), uniform elongation (UE), and total elongation (TE), are summarized in Table 1. Already after a rotation for 90° the alloy had a high ultimate tensile strength approaching 1 GPa but rather limited ductility of ~12%. Increasing HPT rotation from 90° to 180° resulted in a moderate rise in strength and a pronounced decrease in ductility to ~5%. At further increase in strain to two full rotations (720°) the strength of the alloy increased considerably (approximately by a factor of 2) while the ductility decreased moderately. A decrease in temperature of HPT from 293 to 77 K considerably increased the strength of the alloy but not affected ductility much. All flow curves exhibited a peak flow stress at the initial stages of deformation followed by a moderate decrease in flow stress. The uniform elongation of the alloy was rather small in all cases which is quite typical of severely deformed materials [23]. However, after strain localization (at the maximum flow stress), when the plastic deformation had become concentrated in the neck, the alloy had showed quite pronounced additional elongation.

Table 1. Tensile mechanical properties [1] of the CoCrFeNiMn alloy after HPT at 293 and 77 K.

HPT Condition		YS, MPa	UTS, MPa	UE, %	TE, %
Temperature, K	Rotation, °				
293	90	860	981	1.4	11.9
	180	1134	1197	0.5	6.4
	720	1834	2069	1.4	7.4
77	180	1442	1596	1.3	10.4

[1] YS—yield stress, UTS—ultimate tensile stress, UE—uniform elongation, TE—total elongation.

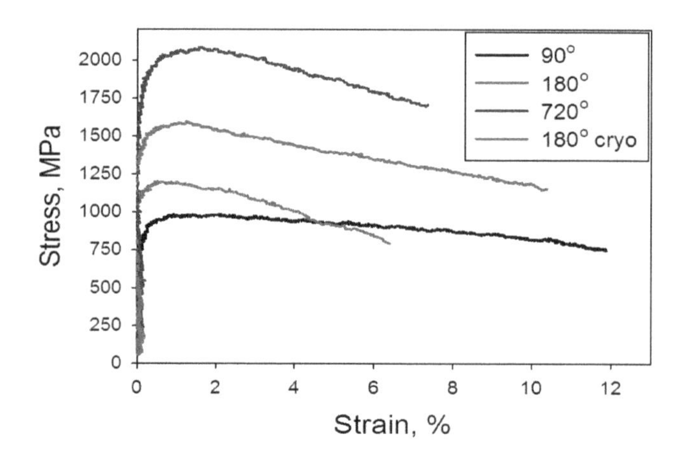

Figure 6. Engineering stress–strain curves obtained during tensile tests of the CoCrFeNiMn alloy after HPT at 293 and 77 K.

4. Discussion

The results of the present work clearly demonstrate that the microstructure development in the CoCrFeNiMn alloy during HPT strongly depended on both the imposed strain (i.e., the angle of rotation and the distance from the center of the discs) and deformation temperature. At room temperature, the formation of ultra-fine grained (UFG) structure on the periphery of the HPT discs occurred quite fast: 50 nm-sized (sub)grains were observed after a 180° turn (Figure 4c,d). Fast kinetics of microstructure evolution in the alloy during HPT has been already reported [9]. However, it should be noted that the microstructure developed in the beginning of HPT-straining was highly heterogeneous (Figure 2d–f). In the center of the discs, initial grains with weakly developed deformation twins/substructure had survived even after rotation for 180° (Figure 2d). This is an obvious result of a strong gradient of the imposed strain along the radius of the discs and is typical of the HPT process [17]. Meanwhile, according to the present results, homogenous UFG microstructure can be produced after the rotation for 720°.

Formation of UFG microstructure after HPT at room temperature is well documented for various metallic materials [17] including high-entropy alloys [8–10,24–26]. The microstructure evolution during HPT is usually associated with the formation of low-angle grain boundaries at low strains and transformation of some of these low-angle subgrain boundaries into high-angle grain boundaries at higher strains giving rise to a considerable microstructure refinement. However, in the materials that deform by both twinning and slip the kinetics of microstructure refinement can be enhanced due to intensive twin formation [12–14]. Thin twin laths which are separated from the matrix by high-angle boundaries readily transform into a chain of grains during deformation. This process is particularly relevant during initial stages of deformation when the number of twins and the density of high-angle twin boundaries are the highest [13]. It is already established that mechanical twinning can occur in the CoCrFeNiMn alloy during plastic deformation at room temperature, however, the extent and the role of twinning in microstructure development remains quite debatable [11,12,27,28].

The results of the present work indicate that twinning during HPT develops at the edge of specimens till rather small strains (rotation for ~90°, Figure 4a,b). The occurrence of twinning during HPT observed in the present study is in agreement with the results of previous investigations [8,9]. At later stages of strain (rotation for 180°), an equiaxed structure composed of a mixture of dislocation cells and grains was observed (Figure 4c). In general the process of the equiaxed fine-grained structure formation can be associated with: (i) deformation-induced twinning; (ii) fragmentation of twin laths by secondary twinning and/or transverse subboundaries (indicated with arrows in Figure 4d); (iii) formation of low-angle boundaries in those parts of microstructure which were not involved into twinning; and (iv) further increase of their misorientation to the high-angle range due to interaction with lattice dislocation like it was observed in different metallic materials—including titanium, copper based alloys, austenitic steel and some others [17,29–32]. Finally, these processes are likely to be responsible for the microstructure refinement at large strain (Figure 4e,f). This behavior is similar to that observed during room-temperature rolling of the same alloy [12]. Therefore, fast kinetics of the microstructure refinement in the CoCrFeNiMn alloy during HPT can be attributed to extensive deformation twinning which readily occurs at the early stages of deformation.

The effect of twinning on microstructure refinement should be sensitive to deformation temperature. Since the critical resolved shear stress (CRSS) for twinning in various metals and alloys depends weakly on temperature compared to that for slip [33], a decrease in deformation temperature from 293 to 77 K was believed to have resulted in more intensive twinning and increased kinetics of microstructure refinement [12,13]. However, a decrease in deformation temperature in case of HPT of the CoCrFeNiMn alloy leaded to the formation of a highly inhomogeneous microstructure composed of a mixture of twinned and fragmented areas (Figure 5). This is possibly a result of considerable increase in strength (approximately by a factor of 2 [6]) due to the lower temperature which increased strain localization and could provoke slippage of anvils at HPT and, thus, reduction of the actually imposed strain. However, the dislocation density estimated using XRD (Table 2) after cryo-deformation was considerably higher than that after room temperature HPT. This finding does not agree with the results of the dislocation density measurements during rolling at 77 and 293 K [12]. During rolling, active twinning at 77 K promoted microstructure refinement. Possibly, dislocation mechanisms play more important role in cryo-HPT of the CoCrFeNiMn alloy than during HPT at room temperature, and therefore, the formation of homogeneous UFG structure occurs slower under cryogenic conditions.

The microstructure analysis is in reasonable agreement with the results of mechanical properties. The observed microstructure refinement during HPT at room temperature was accompanied by a considerable rise in strength and some decrease in ductility (Figure 6). It should be noted, however, that the difference in the microstructure in terms of (sub)grain size and dislocation density (Table 2) was negligible between the specimens processed for 180° or 720° turns. Meanwhile, the alloy after rotation for 720° was found to be approximately 30% stronger than that after strain to 180°. Also, strength of the cryo-deformed alloy was substantially higher than that of the specimen after room temperature HPT despite highly inhomogeneous, coarser microstructure.

The contributions of the most relevant hardening mechanisms in strength of the deformed alloy can be expressed as

$$YS = \sigma_0 + \sigma_\rho + \sigma_{H-P} \tag{4}$$

where σ_0 denotes the friction stress, σ_ρ is the substructure hardening and σ_{H-P} is the Hall–Petch hardening. The substructure hardening σ_ρ can be expressed as

$$\sigma_\rho = M\alpha Gb\sqrt{\rho} \tag{5}$$

where M is the average Taylor factor, α is a constant, G is the shear modulus, b is the Burgers vector, and ρ is the dislocation density. The Hall–Petch contribution to the strength is typically of the form

$$\sigma_{H-P} = K_y\, d^{\frac{-1}{2}} \tag{6}$$

in which K_y is the Hall–Petch coefficient and d is the grain size.

The dislocation density was found to be quite similar for the specimens processed for 180° or 720° at room temperature (Table 2). This result is in agreement with the equality of measured size of cells or subgrains in both structures since it is well established that $d_{sub} \sim \sqrt{\rho}$ [34], where d_{sub}—is the average size of subgrains. Therefore, it can be assumed that the only relevant factor which changes during HPT of the alloy from 180° to 720° strain is the grain size. Here the "grain" term is used to denote the crystallites bordered by high ($\geq 15°$) angle boundaries. The evolution of grain size during deformation was evaluated using equation (1); the input parameters for calculation were: $K_y = 0.494$ MPa $m^{1/2}$ and $\sigma_0 = 125$ MPa (both parameters were taken from [6]), $M = 3$, $\alpha = 0.2$, $G = 81$ GPa [35], $b = 2.54 \times 10^{-10}$ m [28]. Estimated dislocation densities and yield stress are tabulated in Table 2.

Table 2. Input parameters and results of calculation for the evaluation of contributions of the Hall–Petch and substructure hardening mechanisms.

Strain, ° (Temperature, K)	Dislocation Density, m^{-2}	Calculated Grain Size, nm	σ_{H-P}, MPa	σ_ρ, MPa	Predicted YS, MPa
90 (293)	1.48×10^{15}	2500	311	475	911
180 (293)	9.14×10^{14}	580	646	373	1144
720 (293)	8.86×10^{14}	140	1317	367	1809
180 (77)	2.9×10^{15}	625	625	664	1414

The best fit of experimental results was obtained when the grain sizes take on values shown in Table 2. Although microstructural investigation did not obviously support these findings, the obtained data seems quite reasonable. At the initial stages of deformation the fraction of the high-angle boundaries is rather low and consists of initial grain boundaries and boundaries of deformation twins. Therefore, the effective grain size should be approximately equal to the average space between twin boundaries. This similarity can indeed be observed in the microstructure.

During further deformation, new grains developed in place of subgrains within initial grains. This is a result of gradual transformation of geometrically necessary boundaries, which separate microvolumes with different combinations of slip systems [36], from low-angle subboundaries into high-angle grain boundaries. The fraction of high-angle grain boundaries gradually increased, thereby decreasing the effective grain size. At the final stages of HPT all sub-boundaries are expected to be transformed into boundaries; this situation was nearly reached after the rotation for 720° when the calculated grain size was only three times larger than the measured size of subgrains. Taking into account some decrease in strain along the radius (the yield stress was measured exactly at the middle of radius, while TEM investigations were performed closer to the edge of the discs) than the point where the microstructure was analyzed) this approximation can be considered quite reasonable.

The results of the calculation show that the Hall–Petch hardening contribution increased with strain while the substructure hardening maintained at approximately the same level (Table 2, Figure 7). Thus, the Hall–Petch hardening contribution becomes much more important than the substructure hardening in samples rotated for angles larger than 180°. This result is in good agreement with data reported earlier for the nanocrystalline HEA [37].

It is worth noting that, according to the calculations, the effective grain size in the CoCrFeNiMn alloy after HPT at 293 and 77 K is approximately equal (Table 2). Higher strength of the sample deformed in cryogenic condition can be ascribed to the contribution of substructure (dislocation) hardening. After the rotation for 180° at cryogenic temperature, the contribution of the substructure hardening is comparable to that of the Hall–Petch hardening (Figure 7).

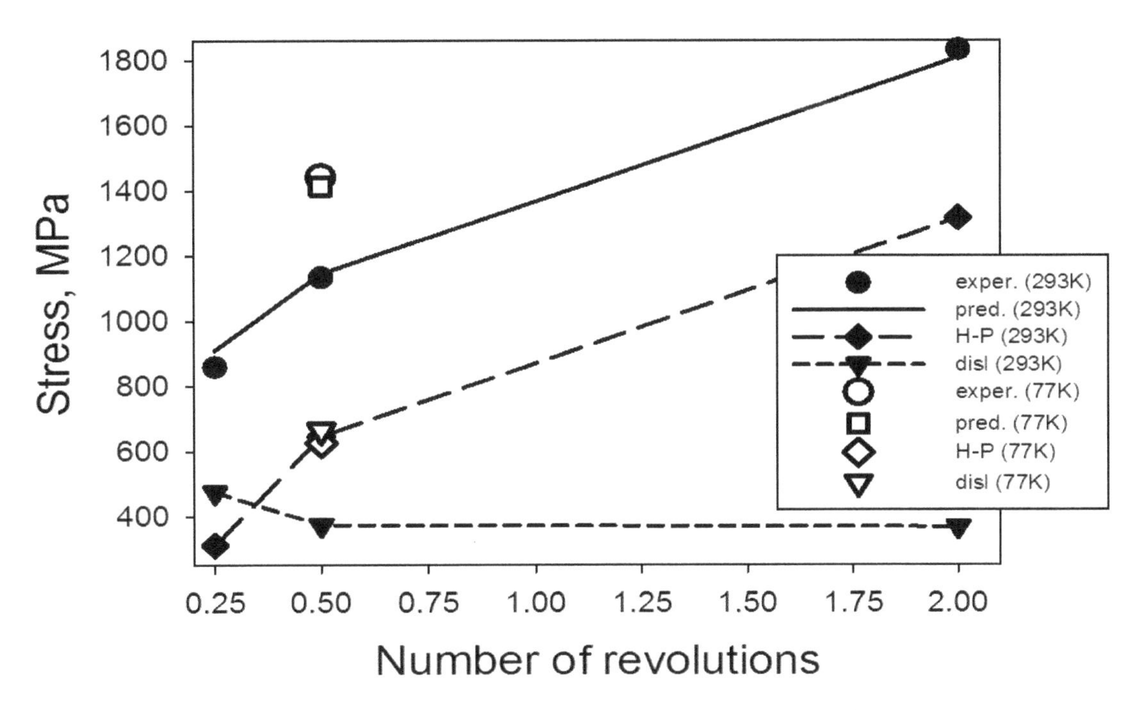

Figure 7. The contributions of different hardening mechanisms in strength of CoCrFeNiMn alloy during HPT at 293 K and 77 K. Exper.—experimental yield stress; pred.—predicted yield stress, obtained from. Equation (4); H-P—Hall-Petch hardening; disl—substructure hardening. Temperatures of HPT are indicated in brackets.

5. Conclusions

Microstructure evolution and mechanical properties of CoCrFeMnNi high-entropy alloy during high-pressure torsion (HPT) at 293 and 77 K was studied. Following conclusions were made:

(1) HPT processing at room temperature results in the formation of in inhomogeneous severely deformed microstructure with (sub)grains of ~50 nm already after the rotation for 180°. The microstructure evolution was associated with intensive deformation-induced twinning, and substructure development resulted in a gradual microstructure refinement.

(2) HPT at 77 K produced more heterogeneous structure in comparison with the room-temperature deformation. The dislocation density was much higher after cryogenic deformation.

(3) Tensile strength of the alloy after HPT at 293 K was found to be strongly dependent on HPT strain. The ultimate tensile strength increased from 981 to 2069 MPa when the rotation angle at HPT increased from 90° to 720°. In all examined conditions, the alloy exhibited limited ductility.

(4) A decrease of HPT temperature from 293 to 77 K resulted in higher tensile strength. A rotation for 180° resulted in the ultimate tensile strength of 1193 MPa and 1596 MPa after processing at room and cryogenic temperature, respectively.

(5) An increase of yield strength of the alloy with an increase of the angle of rotation can mostly be ascribed to a contribution of Hall–Petch strengthening. In turn, higher strength of the alloy after HPT at 77 K was attributed to substructure (dislocation) hardening.

Acknowledgments: This study was supported by Russian Foundation for Basic Research (grant no. 16-38-60061). Authors are thankful to J. Beach for performing tensile tests. The authors are also grateful to the personnel of the Joint Research Center, "Technology and Materials", Belgorod State National Research University, for their assistance with the instrumental analysis.

Author Contributions: Sergey Zherebtsov, Yuliya Ivanisenko, Nikita Stepanov, and Gennady Salishchev conceived and designed the experiments. Dmitry Shaysultanov, Nikita Yurchenko, Margarita Klimova and Yuliya Ivanisenko performed the experiments. Gennady Salishchev, Nikita Stepanov and Sergey Zherebtsov analyzed the data and wrote the paper.

References

1. Yeh, J.-W.; Chen, S.-K.; Lin, S.-J.; Gan, J.-Y.; Chin, T.-S.; Shun, T.-T.; Tsau, C.-H.; Chang, S.-Y. Nanostructured High-Entropy Alloys with Multiple Principal Elements: Novel Alloy Design Concepts and Outcomes. *Adv. Eng. Mater.* **2004**, *6*, 299–303. [CrossRef]
2. Miracle, D.B.; Senkov, O.N. A critical review of high entropy alloys and related concepts. *Acta Mater.* **2017**, *122*, 448–511. [CrossRef]
3. Cantor, B.; Chang, I.T.H.; Knight, P.; Vincent, A.J.B. Microstructural development in equiatomic multicomponent alloys. *Mater. Sci. Eng. A* **2004**, *375*, 213–218. [CrossRef]
4. Otto, F.; Yang, Y.; Bei, H.; George, E.P.P. Relative effects of enthalpy and entropy on the phase stability of equiatomic high-entropy alloys. *Acta Mater.* **2013**, *61*, 2628–2638. [CrossRef]
5. Laurent-Brocq, M.; Akhatova, A.; Perrière, L.; Chebini, S.; Sauvage, X.; Leroy, E.; Champion, Y. Insights into the phase diagram of the CrMnFeCoNi high entropy alloy. *Acta Mater.* **2015**, *88*, 355–365. [CrossRef]
6. Otto, F.; Dlouhý, A.; Somsen, C.; Bei, H.; Eggeler, G.; George, E.P. The influences of temperature and microstructure on the tensile properties of a CoCrFeMnNi high-entropy alloy. *Acta Mater.* **2013**, *61*. [CrossRef]
7. Gali, A.; George, E.P. Tensile properties of high- and medium-entropy alloys. *Intermetallics* **2013**, *39*, 74–78. [CrossRef]
8. Schuh, B.; Mendez-Martin, F.; Völker, B.; George, E.P.P.; Clemens, H.; Pippan, R.; Hohenwarter, A.; Völker, B.; George, E.P.P.; Clemens, H.; et al. Mechanical properties, microstructure and thermal stability of a nanocrystalline CoCrFeMnNi high-entropy alloy after severe plastic deformation. *Acta Mater.* **2015**, *96*, 258–268. [CrossRef]
9. Lee, D.-H.; Choi, I.-C.; Seok, M.-Y.; He, J.; Lu, Z.; Suh, J.-Y.; Kawasaki, M.; Langdon, T.G.; Jang, J. Nanomechanical behavior and structural stability of a nanocrystalline CoCrFeNiMn high-entropy alloy processed by high-pressure torsion. *J. Mater. Res.* **2015**, *30*, 2804–2815. [CrossRef]
10. Heczel, A.; Kawasaki, M.; Lábár, J.L.; Jang, J.; Langdon, T.G.; Gubicza, J. Defect structure and hardness in nanocrystalline CoCrFeMnNi High-Entropy Alloy processed by High-Pressure Torsion. *J. Alloy. Compd.* **2017**, *711*, 143–154. [CrossRef]
11. Laplanche, G.; Kostka, A.; Horst, O.M.M.; Eggeler, G.; George, E.P.P. Microstructure evolution and critical stress for twinning in the CrMnFeCoNi high-entropy alloy. *Acta Mater.* **2016**, *118*, 152–163. [CrossRef]
12. Stepanov, N.; Tikhonovsky, M.; Yurchenko, N.; Zyabkin, D.; Klimova, M.; Zherebtsov, S.; Efimov, A.; Salishchev, G. Effect of cryo-deformation on structure and properties of CoCrFeNiMn high-entropy alloy. *Intermetallics* **2015**, *59*, 8–17. [CrossRef]
13. Zherebtsov, S.V.; Dyakonov, G.S.; Salem, A.A.; Sokolenko, V.I.; Salishchev, G.A.; Semiatin, S.L. Formation of nanostructures in commercial-purity titanium via cryorolling. *Acta Mater.* **2013**, *61*, 1167–1178. [CrossRef]
14. Zherebtsov, S.V.; Dyakonov, G.S.; Salishchev, G.A.; Salem, A.A.; Semiatin, S.L. The Influence of Grain Size on Twinning and Microstructure Refinement during Cold Rolling of Commercial-Purity Titanium. *Metall. Mater. Trans. A* **2016**, *47*, 5101–5113. [CrossRef]
15. Salishchev, G.A.; Tikhonovsky, M.A.; Shaysultanov, D.G.; Stepanov, N.D.; Kuznetsov, A.V.; Kolodiy, I.V.; Tortika, A.S.; Senkov, O.N. Effect of Mn and V on structure and mechanical properties of high-entropy alloys based on CoCrFeNi system. *J. Alloy. Compd.* **2014**, *591*, 11–21. [CrossRef]
16. Stepanov, N.D.; Shaysultanov, D.G.; Salishchev, G.A.; Tikhonovsky, M.A.; Oleynik, E.E.; Tortika, A.S.; Senkov, O.N. Effect of V content on microstructure and mechanical properties of the CoCrFeMnNiVx high entropy alloys. *J. Alloy. Compd.* **2015**, *628*, 170–185. [CrossRef]
17. Zhilyaev, A.P.; Langdon, T.G. Using high-pressure torsion for metal processing: Fundamentals and applications. *Prog. Mater. Sci.* **2008**, *53*, 893–979. [CrossRef]
18. Stepanov, N.D.; Shaysultanov, D.G.; Chernichenko, R.S.; Yurchenko, N.Y.; Zherebtsov, S.V.; Tikhonovsky, M.A.; Salishchev, G.A. Effect of thermomechanical processing on microstructure and mechanical properties of the carbon-containing CoCrFeNiMn high entropy alloy. *J. Alloy. Compd.* **2017**, *693*, 394–405. [CrossRef]
19. Smallman, R.E.; Westmacott, K.H. Stacking faults in face-centred cubic metals and alloys. *Philos. Mag.* **1957**, *2*, 669–683. [CrossRef]
20. Williamson, G.K.; Hall, W.H. X-ray line broadening from filed aluminium and wolfram. *Acta Metall.* **1953**, *1*, 22–31. [CrossRef]

21. Klimova, M.; Zherebtsov, S.; Stepanov, N.; Salishchev, G.; Haase, C.; Molodov, D.A. Microstructure and texture evolution of a high manganese TWIP steel during cryo-rolling. *Mater. Charact.* **2017**, *132*, 20–30. [CrossRef]
22. Odnobokova, M.; Tikhonova, M.; Belyakov, A.; Kaibyshev, R. Development of S3n CSL boundaries in austenitic stainless steels subjected to large strain deformation and annealing. *J. Mater. Sci.* **2017**, *52*, 4210–4223. [CrossRef]
23. Zherebtsov, S.; Kudryavtsev, E.; Kostjuchenko, S.; Malysheva, S.; Salishchev, G. Strength and ductility-related properties of ultrafine grained two-phase titanium alloy produced by warm multiaxial forging. *Mater. Sci. Eng. A* **2012**, *536*, 190–196. [CrossRef]
24. Tang, Q.H.; Huang, Y.Y.; Huang, Y.Y.; Liao, X.Z.; Langdon, T.G.; Dai, P.Q. Hardening of an $Al_{0.3}CoCrFeNi$ high entropy alloy via high-pressure torsion and thermal annealing. *Mater. Lett.* **2015**, *151*, 126–129. [CrossRef]
25. Shahmir, H.; He, J.; Lu, Z.; Kawasaki, M.; Langdon, T.G. Effect of annealing on mechanical properties of a nanocrystalline CoCrFeNiMn high-entropy alloy processed by high-pressure torsion. *Mater. Sci. Eng. A* **2016**, *676*, 294–303. [CrossRef]
26. Moon, J.; Qi, Y.; Tabachnikova, E.; Estrin, Y.; Choi, W.-M.; Joo, S.-H.; Lee, B.-J.; Podolskiy, A.; Tikhonovsky, M.; Kim, H.S. Deformation-induced phase transformation of $Co_{20}Cr_{26}Fe_{20}Mn_{20}Ni_{14}$ high-entropy alloy during high-pressure torsion at 77 K. *Mater. Lett.* **2017**, *202*, 86–88. [CrossRef]
27. Joo, S.-H.; Kato, H.; Jang, M.J.; Moon, J.; Tsai, C.W.; Yeh, J.W.; Kim, H.S. Tensile deformation behavior and deformation twinning of an equimolar CoCrFeMnNi high-entropy alloy. *Mater. Sci. Eng. A* **2017**, *689*, 122–133. [CrossRef]
28. Jang, M.J.; Ahn, D.-H.; Moon, J.; Bae, J.W.; Yim, D.; Yeh, J.-W.; Estrin, Y.; Kim, H.S. Constitutive modeling of deformation behavior of high-entropy alloys with face-centered cubic crystal structure. *Mater. Res. Lett.* **2017**, *5*, 350–356. [CrossRef]
29. Zherebtsov, S.; Lojkowski, W.; Mazur, A.; Salishchev, G. Structure and properties of hydrostatically extruded commercially pure titanium. *Mater. Sci. Eng. A* **2010**, *527*, 5596–5603. [CrossRef]
30. Salishchev, G.; Mironov, S.; Zherebtsov, S.; Belyakov, A. Changes in misorientations of grain boundaries in titanium during deformation. *Mater. Charact.* **2010**, *61*, 732–739. [CrossRef]
31. Sakai, T.; Belyakov, A.; Kaibyshev, R.; Miura, H.; Jonas, J.J. Dynamic and post-dynamic recrystallization under hot, cold and severe plastic deformation conditions. *Prog. Mater. Sci.* **2014**, *60*, 130–207. [CrossRef]
32. Templeman, Y.; Ben Hamu, G.; Meshi, L. Friction stir welded AM50 and AZ31 Mg alloys: Microstructural evolution and improved corrosion resistance. *Mater. Charact.* **2017**, *126*, 86–95. [CrossRef]
33. Meyers, M.A.; Vöhringer, O.; Lubarda, V.A. The onset of twinning in metals: A constitutive description. *Acta Mater.* **2001**, *49*, 4025–4039. [CrossRef]
34. Hull, D.; Bacon, D.J. *Introduction to Dislocations*; Butterworth-Heinemann: Oxford, UK, 2011; ISBN 9780080966724.
35. Laplanche, G.; Gadaud, P.; Horst, O.; Otto, F.; Eggeler, G.; George, E.P.P. Temperature dependencies of the elastic moduli and thermal expansion coefficient of an equiatomic, single-phase CoCrFeMnNi high-entropy alloy. *J. Alloy. Compd.* **2015**, *623*, 348–353. [CrossRef]
36. Hughes, D.; Hansen, N. Microstructure and strength of nickel at large strains. *Acta Mater.* **2000**, *48*, 2985–3004. [CrossRef]
37. Fu, Z.; Chen, W.; Wen, H.; Zhang, D.; Chen, Z.; Zheng, B.; Zhou, Y.; Lavernia, E.J. Microstructure and strengthening mechanisms in an FCC structured single-phase nanocrystalline $Co_{25}Ni_{25}Fe_{25}Al_{7.5}Cu_{17.5}$ high-entropy alloy. *Acta Mater.* **2016**, *107*, 59–71. [CrossRef]

5

Effect of CO_2 Partial Pressure on the Corrosion Inhibition of N80 Carbon Steel by Gum Arabic in a CO_2 Water Saline Environment for Shale Oil and Gas Industry

Gaetano Palumbo [1,*], Kamila Kollbek [2], Roma Wirecka [2,3], Andrzej Bernasik [3] and Marcin Górny [4]

1 Department of Chemistry and Corrosion of Metals, Faculty of Foundry Engineering, AGH University of Science and Technology, 30-059 Krakow, Poland
2 Academic Centre for Materials and Nanotechnology, AGH University of Science and Technology, Mickiewicza St. 30, 30-059 Kraków, Poland; kamila.kollbek@agh.edu.pl (K.K.); roma.wirecka@fis.agh.edu.pl (R.W.)
3 Department of Condensed Matter Physics, Faculty of Physics and Applied Computer Science, AGH University of Science and Technology, Mickiewicza St. 30, 30-059 Krakow, Poland; bernasik@agh.edu.pl
4 Department of Cast Alloys and Composites Engineering, Faculty of Foundry Engineering, AGH University of Science and Technology, 30-059 Krakow, Poland; mgorny@agh.edu.pl
* Correspondence: gpalumbo@agh.edu.pl

Abstract: The effect of CO_2 partial pressure on the corrosion inhibition efficiency of gum arabic (GA) on the N80 carbon steel pipeline in a CO_2-water saline environment was studied by using gravimetric and electrochemical measurements at different CO_2 partial pressures (e.g., P_{CO_2} = 1, 20 and 40 bar) and temperatures (e.g., 25 and 60 °C). The results showed that the inhibitor efficiency increased with an increase in inhibitor concentration and CO_2 partial pressure. The corrosion inhibition efficiency was found to be 84.53% and 75.41% after 24 and 168 h of immersion at P_{CO_2} = 40 bar, respectively. The surface was further evaluated by scanning electron microscopy (SEM), energy dispersive spectroscopy (EDS), grazing incidence X-ray diffraction (GIXRD), and X-ray photoelectron spectroscopy (XPS) measurements. The SEM-EDS and GIXRD measurements reveal that the surface of the metal was found to be strongly affected by the presence of the inhibitor and CO_2 partial pressure. In the presence of GA, the protective layer on the metal surface becomes more compact with increasing the CO_2 partial pressure. The XPS measurements provided direct evidence of the adsorption of GA molecules on the carbon steel surface and corroborated the gravimetric results.

Keywords: high-pressure CO_2 corrosion; corrosion inhibition; gum arabic; carbon steel N80

1. Introduction

Shale oil and gas are "unconventional" resources of natural oil and gas trapped in fine-grained sedimentary rocks called shale. The rapid expansion of shale oil and gas exploration and the development of a new technology (i.e., hydraulic fracturing (HF) techniques), has seen the popularity of these natural resources to grow over the years. However, after years of exploitation, the oil and gas production in the reservoir declines to result in a major economic challenge for the oil companies. The injection of CO_2 at high pressure into the wellbore is an effective method to increase the oil fields lifetime [1–4]. This process is usually referred to as carbon dioxide flooding enhanced oil recovery (CO_2-EOR). However, CO_2 gas dissolves in the fluid to form the weak carbonic acid, which in turn dissociates into bicarbonate and carbonate anions [4,5]. The presence of this weak acid can lead to severe corrosion attacks on the steel structures [4–6].

Another common problem encountered in the extraction of these natural resources is the use of aggressive fluids with high concentrations of chloride ions (e.g., fracturing fluid) [7]. In the HF process, the fluid usually injected into the wellbore is a neutral water-based chloride solution (up to 4% of potassium chloride) with different additives (i.e., inhibitors of scaling, thickening agents, corrosion inhibitors, etc.) [8]. The literature reports that the presence of a high concentration of chloride ions in a CO_2-containing fluid can exponentially accelerate the dissolution of the steel [6,9].

Carbon and low-alloys steel are often used in the construction of the pipeline in the shale oil and gas industry infrastructures, mainly due to its durability, ductility, high strength, and low cost [7,8,10]. However, due to these harsh operating conditions encountered during the exploitation of these natural resources, the steel is prone to corrode. One practical and relatively cheap method for controlling sweet corrosion in the shale oil and gas industry is the use of corrosion inhibitors. Corrosion inhibitors are substances that added to the solution greatly reduce the dissolution of the metal by forming a protective layer on its surface. The literature reports that over the last decades the use of corrosion inhibitors as a means to mitigate CO_2 corrosion that occurs inside the carbon steel pipelines has received a wide interest. Nitrogen-based compounds such as pyridine derivatives [11] imidazolines [12], benzimidazole derivatives [13], and amines [14] were found to be effective corrosion inhibitors against CO_2 corrosion. However, most of these compounds are reported to be toxic and their synthesis can be very expensive [15,16]. These drawbacks and the increase in environmental awareness have led many researchers to focus on the use of more naturally occurring substances as corrosion inhibitors. Plant extracts substances, such as berberine extract [17], *Momordica charantia* [18], *Gingko biloba* [19] were successfully tested as green corrosion inhibitors in CO_2-saturated saline solutions.

The last trend of research has also seen the use of many naturally occurring polymers as green corrosion inhibitors in various corrosive environments [10,20–22]. They are abundant in nature, environmentally sustainable, and have an appreciable solubility. Additionally, polymers, unlike small molecules, with their multiple adsorption sites for bonding on the metal surface, are expected to show a higher corrosion inhibition efficiency, compared to their monomer counterpart.

Umoren et al. [15] studied the corrosion inhibition effect of two naturally occurring polymers such as carboxymethyl cellulose and chitosan for API 5 L X60 steel in a CO_2 saline solution at P_{CO_2} = 1 bar. The results showed that both inhibitors reduced the corrosion rate of the metal due to the formation of a protective layer on its surface. Singh et al. [23] studied the corrosion inhibition effect of a modified natural polysaccharide (e.g., guar gum + methylmethacrylate) in a 3.5 wt% NaCl solution saturated with CO_2 (e.g., P_{CO_2} = 1 bar) at 50 °C. The authors found that this modified polysaccharide acted like a good corrosion inhibitor for P110 steel with maximum inhibition efficiency found to be 90%. However, most of these studies were carried out at atmospheric pressure (e.g., P_{CO_2} = 1 bar). The CO_2-EOR process can significantly increase the dissolution of the tube. As reported by many studies, the severity of the CO_2 corrosion attack increases with an increase in CO_2 partial pressure due to the increase in acidity of the fluid [4–6]. Therefore, understanding how the CO_2 partial pressure can influence the inhibitory action of certain corrosion inhibitors in CO_2 saline environments is important and can help to minimize the material and economic losses.

Mustafa et al. [4] studied the effect of the CO_2 partial pressure (e.g., 10, 40, and 60 bar) on the corrosion inhibition of an imidazoline-based inhibitor for X52 steel exposed to CO_2 water saline solution at 60 °C. The authors reported that the inhibitor efficiency of the tested inhibitor was observed to be strongly affected by the concentration of inhibitor and CO_2 partial pressure. Ansari et al. [16] studied the influence of a modified chitosan corrosion inhibitor on J55 carbon steel in a 3.5 wt% NaCl solution saturated with CO_2 at 60 bar and 65 °C, reporting a corrosion inhibition efficiency of 95%. Yet, all inhibitors tested so far are labeled either as toxic or are expensive to synthesize.

Gum arabic (GA) is a natural polymer obtained from the Acacia trees of the Leguminosae family [22] and it has been reported to successfully inhibit the corrosion of the steel in different environments [7,21,22,24–28]. Furthermore, GA is often used in the fracturing fluid as a thickening agent to increase the viscosity of the fluid [29]. Therefore, due to the encouraging results presented by these studies and the continuous research of affordable and eco-friendly corrosion inhibitors, this work was undertaken to study the efficacy of GA as an eco-friendly corrosion inhibitor to mitigate high-pressure CO_2 corrosion for carbon steel pipeline in a saline solution. This paper also aims to show that GA not only can be used as a thickening agent in the make-up of the fracturing fluid, but it could also be used as an active component in corrosion inhibitor in the shale gas industry. To this end, the study was performed in an autoclave in the presence and absence of different concentrations of GA, different CO_2 partial pressures, and different temperatures using weight loss and electrochemical measurements. SEM-EDS, GIXRD, and XPS measurements were also employed to characterize the corrosion product layer and to support the gravimetric and electrochemical results.

2. Experimental Procedure

2.1. Materials

The study was carried out on carbon steel (N80) with composition of (weight %): C 0.39%, Mn 1.80%, Si 0.26%, Cu 0.26%, V 0.19%, Cr 0.04%, Ni 0.04%, Al 0.03%, Mo 0.003%, Co 0.002%, Sn 0.004%, S 0.001%, P 0.001% and the remainder Fe. Figure 1 shows that the microstructure of the N80 carbon steel pipeline is composed of perlite and ferrite (α-Fe) phases, where the latter phase accounting for circa 41% of the total. The samples used in this study were machined from pipeline carbon steel, ground with silicon carbide abrasive paper up to 1200 grit, then were ultrasonically washed with distilled water, dried with absolute alcohol.

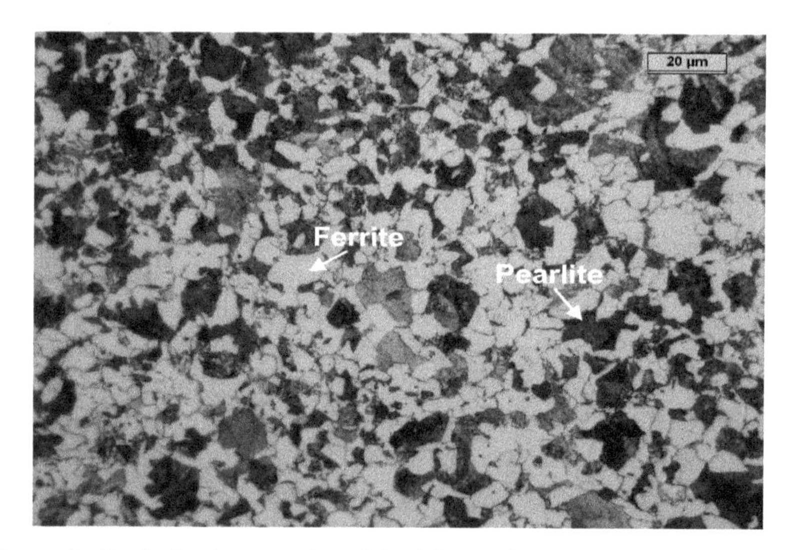

Figure 1. Optical micrographs of the N80 carbon steel microstructures.

All experiments were carried out in 3% of potassium chloride (KCl). Potassium salt was used in this study instead of the more common NaCl, because in the fracturing fluid, the potassium (K^+) ions formed a semi-permeable membrane on the shale rock and therefore, preventing the water from entering the shale.

The tested solution was prepared from reagent grade material potassium chloride (Sigma-Aldrich) and pure deionized water with an electrical resistivity of 0.055 µS/cm at T = 25 °C. The tested inhibitor was purchased from Sigma-Aldrich (Warsaw, Poland). The concentrations of inhibitor solution prepared and used for the study ranged from 0.6 up to 2.0 g L^{-1}.

2.2. Gravimetric Measurements

The gravimetric experiments were carried out in a 1.2 L high-pressure autoclave (PARR instrument) at different CO_2 partial pressures (1, 20, and 40 bar). The coupons were suspended in a 1.0 L solution in the presence and absence of different concentrations of the inhibitor (i.e., from 0.6 up to 2.0 g L^{-1}) at 25 and 60 °C. Before each experiment, the tested solution was deaerated with CO_2 for 2 h under atmospheric pressure and then CO_2 was purged for another 2 h at the tested pressure after the introduction of the samples. After saturation, the pH and conductivity of the tested solution were 4.5 and 60.30 mS cm^{-1} at 1 bar and 25 °C, respectively. To ensure homogeneous mixing, a Teflon-coated blade agitator was used (e.g., 200 rpm). The weight loss was determined by retrieving the coupons after 24 h of immersion by means of an analytical balance with an accuracy of ±0.1 mg. To assess the effect of time, the samples were immersed for 168 h in the presence and absence of 1.0 g L^{-1} of GA at different CO_2 partial pressures (1, 20, and 40 bar). The corrosion products were removed according to the ASTM G1-90 [30], then the specimens were ultrasonically washed with distilled water, dried with absolute alcohol, and reweighed. In each case, the experiment was conducted thrice and the corrosion rate (*CR*) in mm y^{-1} was obtained from the following equation:

$$CR\ (\text{mm y}^{-1}) = \frac{87.6\ \Delta m}{dAt} \tag{1}$$

where, Δm is the weight loss calculated form the difference between the initial (W_i) and the final (W_f) weight (mg). d is the density (7.87 g cm^{-3}), A is the surface of the sample (cm^{-2}) and t is the immersion time (h). The inhibition efficiency (*IE*%) was determined using the following equation [20,22,26]:

$$IE\% = \frac{CR - CR^{\text{inh}}}{CR} \times 100 \tag{2}$$

where CR^{inh} and CR are the corrosion rates of the steel with and without the inhibitor, respectively.

2.3. Electrochemical Experiments

The electrochemical experiments were carried out in a 1.2 L high-pressure autoclave (PARR instrument) at different CO_2 partial pressures (1, 20, and 40 bar) with a conventional three-electrode system. N80 carbon steel specimen was used as a working electrode, a platinum foil as a counter electrode (CE), and a high-pressure 0.1 M KCl Ag/AgCl probe was used as a reference electrode. To ensure homogeneous mixing, a Teflon-coated blade agitator was used (200 rpm). The electrochemical impedance spectroscopy (EIS) and potentiodynamic polarization (PDP) measurements were carried in a Gamry reference 600 potentiostat/galvanostat electrochemical system after the sample was exposed for 24 h in the tested solution, with and without the presence of 1.0 g L^{-1} of GA. The EIS tests were performed over the frequency range of 100 kHz to 10 mHz and amplitude of 10 mV at open circuit potential. The data were then fitted by means of Echem Analyst 5.21 software using the opportune equivalent circuit. The *IE*% was calculated from the polarization resistances (R_p) determined from the fitting process using the following equation [21,26]:

$$IE\% = \frac{R_p^{\text{inh}} - R_p}{R_p^{\text{inh}}} \times 100 \tag{3}$$

where ${R_p}^{\text{inh}}$ and R_p are the values of the polarization resistances in the presence and absence of the inhibitor, respectively. The PDP measurements were carried out at a potential of ±−0.3 V from the OCP and a scan rate of 1 mV s^{-1}. The potentiodynamic parameters were determined by means of Echem Analyst 5.21 software. The values of *IE*% were calculated from the measured i_{corr} values using the relationship [21,26]:

$$IE\% = \frac{i_{\text{corr}} - i_{corr}^{inh}}{i_{\text{corr}}} \times 100 \tag{4}$$

where i_{corr} and $i_{\text{corr}}^{\text{inh}}$ represent the values of the corrosion current densities without and with inhibitor, respectively.

2.4. Surface Analysis

The surface of the samples, prepared as described above, were analyzed in the presence and absence of 1.0 g L^{-1} of GA. After the immersion, the samples were removed and rinsed with deionized water and dried. The surface analysis was carried out by means of different techniques such as a scanning electron microscopy combined with an energy dispersive spectroscopy, grazing incidence X-ray diffraction (GIXRD), and an X-ray photoelectron spectroscopy (XPS). The SEM measurements were carried out by using a JEOL scanning electron microscope. The GIXRD analysis was carried out to further determine the composition of the corrosion products film. Grazing incident X-ray diffraction (GIXRD) with an incident angle of 3° was applied to study samples phase composition. A Panalytical Empyrean X-ray diffractometer in the parallel beam geometry (Goebel mirror in the incident beam optics and parallel plate collimator in the secondary beam optics) with Co lamp (Kα = 1.7902 Å) was used to perform measurements. The samples were scanned with a 0.02° step in the range of 20°– 70° at room temperature. The XPS analysis was carried out in a PHI 5000 VersaProbe II spectrometer with an Al Kα monochromatic X-ray beam. The X-ray source was operated at 25 W and 15 kV beam voltages. Dual-beam charge compensation with 7 eV Ar^+ ions and 1 eV electrons was used to maintain a constant sample surface potential regardless of the sample conductivity. The pass energy of the hemispherical analyzer for the iron (Fe 2p) spectra was fixed at 23.5 eV and for other elements at 46.95 eV. The spectra were charge corrected to the mainline of the carbon C 1 s spectrum set to 284.8 eV.

3. Results and Discussion

3.1. Effect of Pressure and Temperature

3.1.1. Gravimetric Experiments

Figure 2 and Table S1 show the corrosion rate and the variation of the inhibition efficiency obtained at different concentrations of GA and CO_2 partial pressures. It follows from the data that *CR* increases with an increase of CO_2 partial pressure, going from 1.28 to 10.95 mm y^{-1} at P_{CO_2} = 1 bar and P_{CO_2} = 40 bar at 25 °C, respectively.

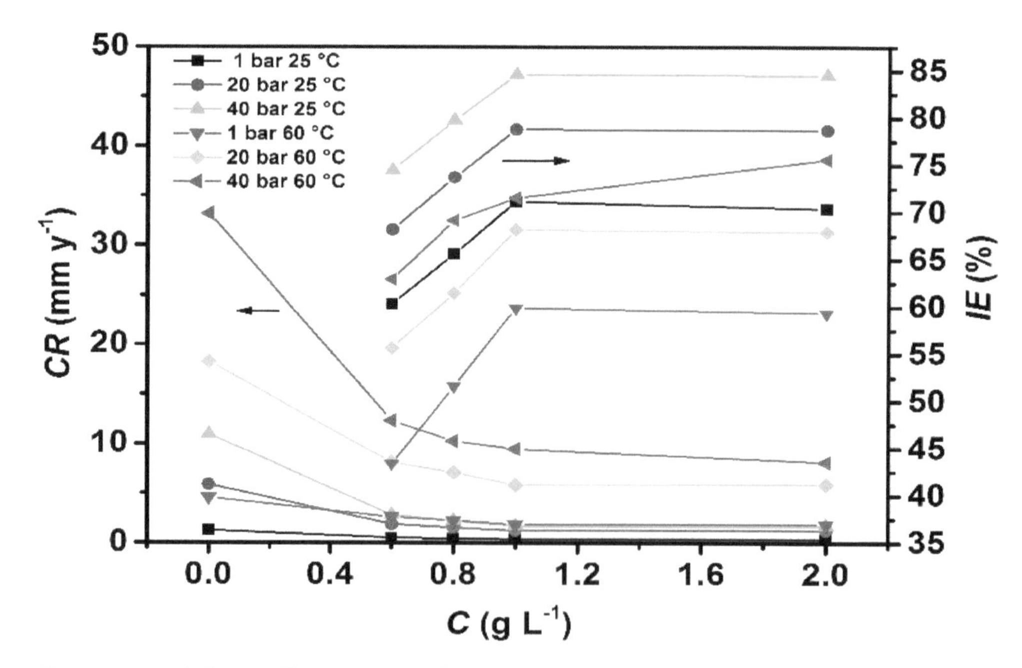

Figure 2. Corrosion inhibitor efficiency at different concentrations of gum arabic (GA) and CO_2 partial pressures after 24 h of immersion.

The solubility of CO_2 in water increases sharply with increasing the pressure of the system [31]. The high corrosion rate observed at higher CO_2 partial pressures can be explained with the increase of the acidity of the solution. In fact, in the presence of CO_2, the weak carbonic acid is formed, which in turn dissociates in HCO_3^- and in CO_3^{2-}, according to the following reactions:

$$CO_2 + H_2O \leftrightarrow H_2CO_3 \quad (5)$$

$$H_2CO_3 \leftrightarrow H^+ + HCO_3^- \quad (6)$$

$$HCO_3^- \leftrightarrow H^+ + CO_3^{2-} \quad (7)$$

The corrosion process in a CO_2 containing solution is controlled by the anodic reaction (Equation (8)) and the three cathodic reactions (Equation (9)–(11)) [4,5]:

$$Fe \rightarrow Fe^{2+} + 2e^- \quad (8)$$

$$2H_2CO_3 + 2e^- \rightarrow H_2 + 2HCO_3^- \quad (9)$$

$$2HCO_3^- + 2e^- \rightarrow H_2 + 2CO_3^{2-} \quad (10)$$

$$2H^+ + 2e^- \rightarrow H_2 \quad (11)$$

The pH of the solution plays an important role in determining the corrosion rate of carbon steel in a CO_2 environment. As the CO_2 partial pressure increases, its solubility also increases, resulting in an increase of the carbonic acid concentration in the solution (Equation (5)). Nesic' predicted that the concentrations of H_2CO_3 in the solution would increase of about 40 times with changing the pressure from P_{CO_2} = 1 bar to P_{CO_2} = 40 bar [31]. Increasing the concentration of carbonic acid leads to an increase in the rate of reduction of carbonic acid and bicarbonate ions (Equations (9) and (10)), and ultimately the anodic dissolution of the steel (Equation (8)) as reported by several studies [4,5,31].

After the addition of the inhibitor, it can be seen that the corrosion rate of the metal is greatly reduced going from 1.28 to 0.37 mm y^{-1}, with a maximum corrosion inhibition efficiency found to be 71.09% at P_{CO_2} = 1 bar, after 24 h of immersion. The data shows that in contrast to the uninhibited solution, an increase in CO_2 partial pressure has a favorable effect on the corrosion rate of the metal in the presence of the inhibitor. It follows from Figure 2 that *IE*, which varies inversely with *CR*, significantly increased after the addition of GA and with CO_2 partial pressure, with a maximum corrosion inhibition efficiency of 78.77% and 84.53% at P_{CO_2} = 20 bar and P_{CO_2} = 40 bar, after 24 h of immersion, respectively [4].

The literature reports that GA [7,21], and in general polysaccharides-like inhibitors [15,20], is mainly adsorbed on the metal surface in acidic condition by weak electrostatic interaction between the protonated inhibitor molecules and the chloride ions adsorbed on the metal surface. In a weak acid solution GA molecules are in equilibrium with their protonated molecules according to the following reaction (see also Section 3.5.1) [7]:

$$GA + xH^+ \leftrightarrow [GAH_x]^{x+}_{(sol)} \quad (12)$$

where $[GAH_x]^{x+}_{(sol)}$ is the protonated inhibitor in the solution. As mentioned before, an increase in CO_2 partial pressure leads to an increase in the acidity of the solution [32]. The higher value of *IE* observed at higher CO_2 partial pressures can be ascribed to the higher concentration of H^+ ions present in the solution, which in turn leads to an increase in the number of protonated inhibitor molecules that can be adsorbed on the metal surface. Moreover, Figure 2 also reveals that *IE* varies with the concentration of the inhibitor until the system reached a state (e.g., 1.0 g L^{-1} of GA), in which it can be said that the inhibitor molecules are in equilibrium with their protonated counterpart. For further increase in GA concentration, *IE* remains almost stable. The results clearly demonstrate that GA has greatly reduced the *CR* of the metal in the tested environment, and the high corrosion inhibition activity of GA was influenced by both its concentration and CO_2 partial pressure. The lower values of *CR* observed in the presence of the inhibitor can be ascribed to its adsorption on the metal surface, covering the metal surface and thereby, blocking the active corrosion sites on its surface [4,7,28]. The gravimetric results are also supported by the SEM analysis presented from Figures 7–9, where it can be seen that the surface coverage increases and the protective layer becomes more compact in the presence of GA and with increasing CO_2 partial pressure.

As the temperature rises, *IE* slightly decreased. This decrease may be due to the combination of two different reasons. For instance, the solubility of CO_2 decreases with increasing the temperature of the solution [31], which can lead to a less acid environment. The pH of the solution increases slightly and therefore shifting the equilibrium reaction Equation (12) to the left. At higher pH, the concentration of H^+ ions in the solution is smaller, which would result in the formation of fewer protonated inhibitor molecules available for the absorption process. Another possible reason may be due to the fact that these types of inhibitors get absorbed via electrostatic interactions (e.g., van der Waal forces) onto the surface of the metal, and it is known that this types of interaction generally grow weaker with an increase in temperature due to larger thermal motion [3,20]. Consequently, an increase in temperature will increase the metal surface kinetic energy, which has a detrimental effect on the adsorption process and encourages desorption processes [15,20].

Table S2 lists the inhibition efficiency of various corrosion inhibitors used to mitigate sweet corrosion obtained at different immersion times and temperatures. It is worth mentioning that most of these inhibitors are labeled either as toxic or are expensive to synthesize. Umoren et al. [15] reported the corrosion inhibition efficiency of a commercial inhibitor to be 87 and 88% at 25 and 60 °C, respectively after 24 h of immersion. The table shows that GA, compared to other studied corrosion inhibitors, and the commercial corrosion inhibitor, can be considered a good environmentally friendly corrosion inhibitor for carbon steel in a CO_2-saturated saline solution. Moreover, since GA is already used as a thickening agent in the make-up of the fracturing fluid, can also work as an active component in corrosion inhibitor in the shale gas industry.

3.1.2. Electrochemical Experiments

The electrochemical experiments such as electrochemical impedance spectroscopy (EIS) and potentiodynamic polarization (PDP) were also employed as a means to support the gravimetric findings. These experiments were carried out at 1.0 g L^{-1} of GA at different CO_2 partial pressures after 24 of immersion. 1.0 g L^{-1} is the concentration in which the tested inhibitor exhibited a maximum in the concentration-efficiency curve.

The EIS measurements were used to evaluate the resistance of the protective layer from the electrochemical angle and are presented in Figure 3. It can be seen from the Bode (Figure 3b) and phase angle plots (Figure 3c) that the system is characterized by two-time constants at low (LF) and high frequencies (HF). The presence of these two-time constants suggests that the electrochemical reaction process of the N80 carbon steel in a CO_2 saturated saline solution is affected by two state variables i.e., the corrosion products layer and/or the protective adsorptive layer, and electric double-layer, as also reported by Dong et al. [33]. For this reason, the EIS plots presented in Figure 3 were fitted with the help of the equivalent circuit (EC) presented in Figure 3d, consisting of the following elements: R_s is the electrolyte resistance. CPE_l and R_l are the constant phase element and the resistance of the layer formed on the metals surface, respectively. CPE_{dl} and R_{ct} are the constant phase element representing the double-charge layer capacitance and the charge transfer resistance, respectively. The EIS parameters are listed in Table 1 and from the small values of χ^2 (i.e., the goodness of fit) it can be said that the EC used to fit the system under investigation was the most appropriate one.

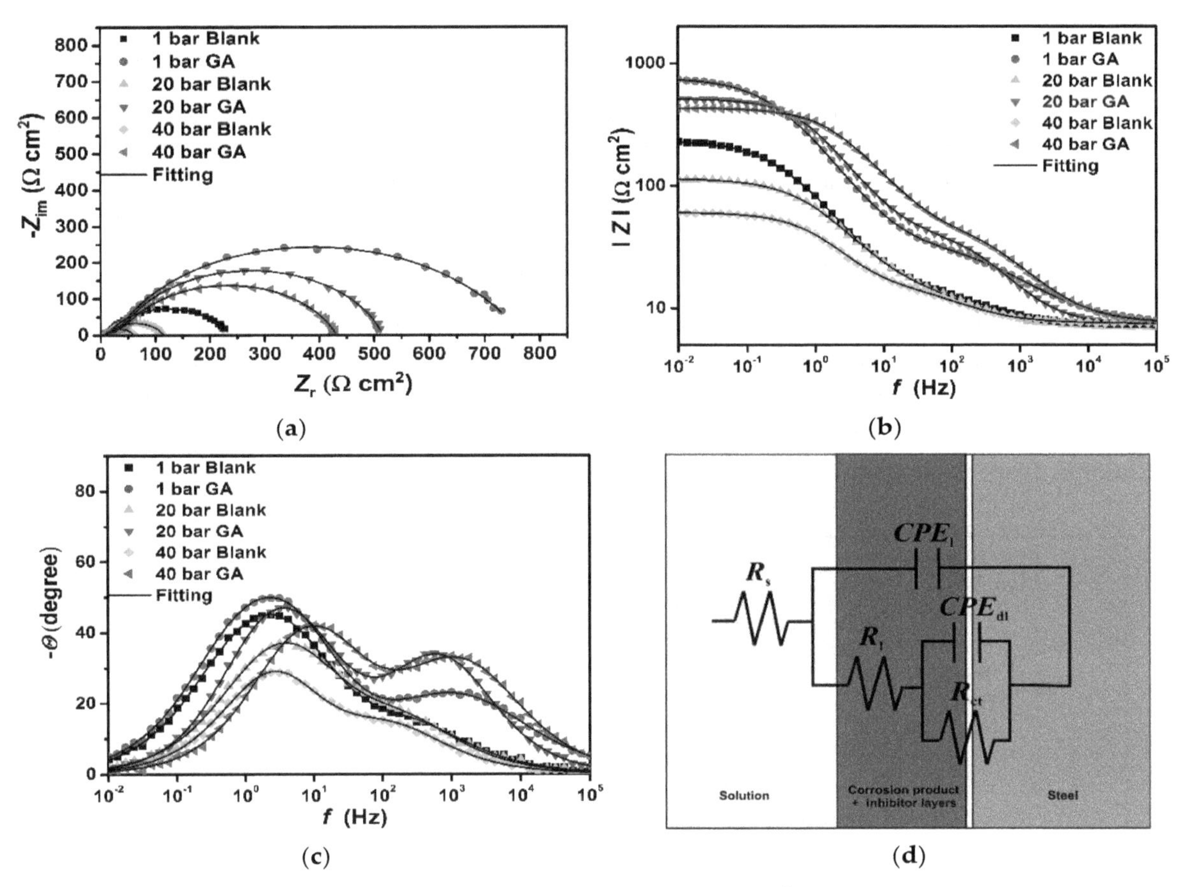

Figure 3. EIS plot recorded in the presence and absence of 1 g L^{-1} of GA after 24 h of immersion at different CO_2-partial pressures. (**a**) Nyquist; (**b**) Bode; (**c**) phase angle; (**d**) equivalent circuit

Table 1. Electrochemical impedance parameters with and without the presence of 1.0 g L^{-1} concentrations of GA after 24 h of immersion.

C (g L^{-1})	R_s (Ω cm^2)	CPE_f		R_f (Ω cm^2)	CPE_{dl}		R_{ct} (Ω cm^2)	$R_p = R_f + R_{ct}$ (Ω cm^2)	χ^2 ($\times 10^{-3}$)	IE (%)
		Y_f (mΩ^{-1} s^n cm^{-2})	n_f		Y_{dl} (mΩ^{-1} s^n cm^{-2})	n_{dl}				
1 bar Blank	7.13	1.24	0.66	11.59	1.68	0.79	217.90	229.49	1.12	-
1 bar Ga	7.39	0.34	0.61	30.29	0.68	0.81	730.60	760.89	1.50	69.83
20 bar Blank	6.92	1.28	0.68	12.18	1.85	0.76	95.07	107.25	1.34	-
20 bar GA	7.32	0.10	0.78	38.89	0.55	0.82	469.90	507.79	1.28	78.68
40 bar Blank	7.49	0.15	0.70	9.44	3.99	0.78	43.87	53.31	1.11	-
40 bar GA	7.56	0.01	0.71	50.11	0.28	0.81	374.50	424.61	1.99	87.44

The presence of a time constant at HF is reported in several studies [34,35] and it is often observed in a Fe/water system. This time constant may be due to the capacity of a porous thin layer formed onto the metal surface. In this study, and without the inhibitor, the presence of this time constant at HF is due to the formation of a thin layer of Fe_3C onto the metal surface. As mentioned before, the microstructure of the tested carbon steel is composed of circa 41% of a ferritic phase and the remaining of a perlitic phase (Figure 1). The ferritic phase is more active than the Fe_3C contained in the perlitic phase [7], in this case, the former phase will act as an anode and the latter one as a cathode. This will generate a micro-galvanic effect, which will eventually lead to the formation of a thin layer of Fe_3C onto the metal surface. However, it follows from the data that both the values of R_f and R_{ct} greatly increased in the presence of the inhibitor, which indicated that the GA molecules were adsorbed onto the metal surface leading to the formation of a protective layer that covers the surface, as confirmed also from the morphological analysis (e.g., SEM-EDS and XPS). Moreover, the difference between these two values obtained in the absence and the presence of GA increased even more with increasing CO_2 partial pressure, suggesting that this protective layer becomes more stable and compact, with the corrosion inhibition efficiency going from 69.83% up to 87.44% at P_{CO_2} = 1 bar and P_{CO_2} = 40 bar, respectively. The increase in *IE* observed with an increase in CO_2 partial pressure agrees with the results obtained with the gravimetric measurements and is in agreement with the ones reported in the literature [4]. It is evident that the addition of GA had a remarkable effect on the corrosion process of the metal and that its inhibition not only depends on the concentration of GA but also from CO_2 partial pressure. The results show that the coverage and thickness of the formed protective layer increased with CO_2 partial pressure, acting both as a barrier against the charge and the mass transfer processes that occur onto the metal surface owing to the corrosive attack of the aggressive electrolyte.

Figure 4 and Table 2 show the potentiodynamic polarization measurements and the corrosion kinetic parameters obtained from the polarization plots in the presence and absence of GA at different CO_2 partial pressures, respectively. As can be seen from Figure 4, the anodic polarization curve of the blank solution does not show the typical Tafel behavior consequently, the corrosion current densities were calculated from the extrapolation of the cathodic Tafel region.

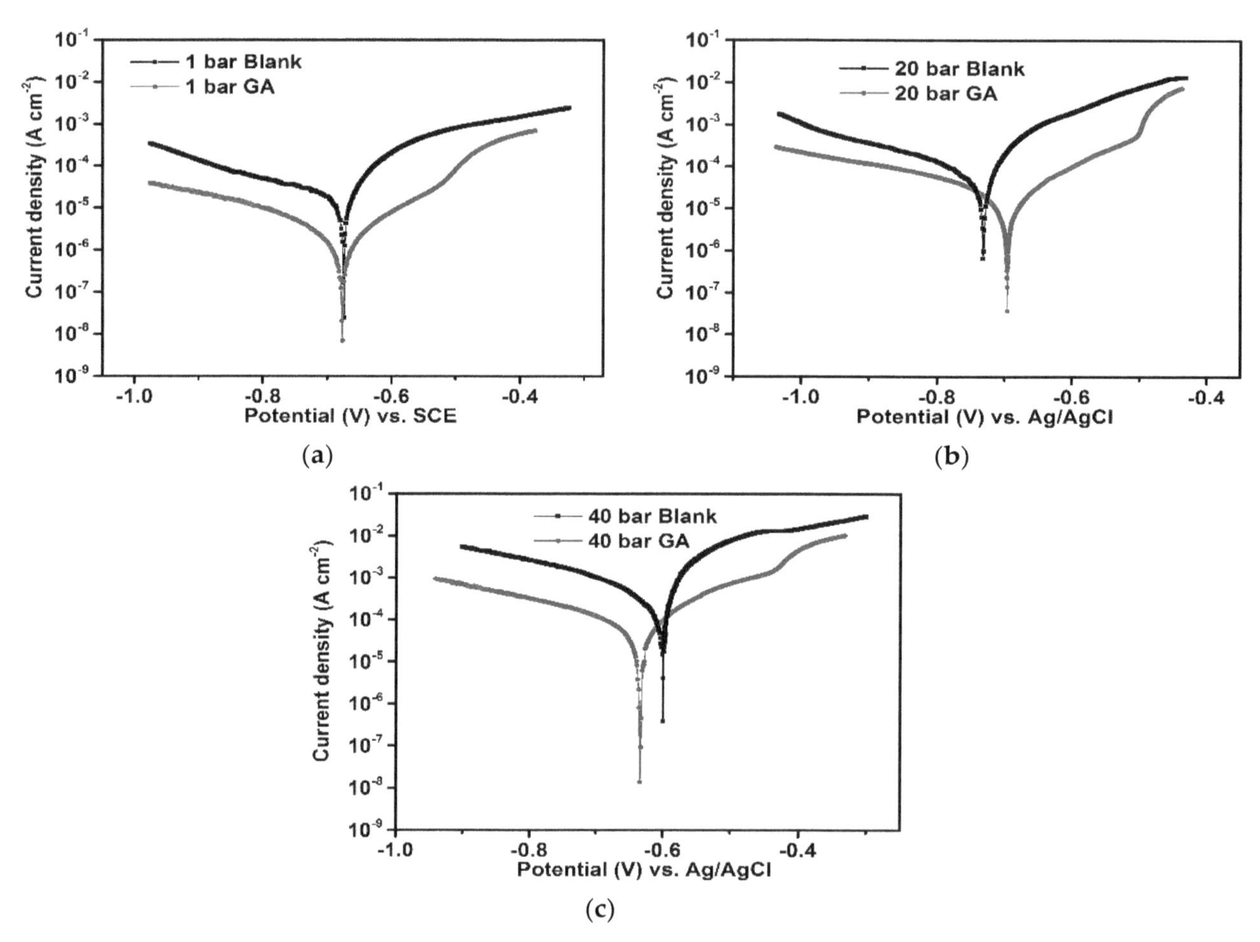

Figure 4. Potentiodynamic polarization parameters obtained in the absence and presence of 1.0 g L^{-1} of GA at different CO_2-partial pressures, after 24 h of immersion. (**a**) P_{CO_2} = 1 bar, (**b**) P_{CO_2} = 20 bar and (**c**) P_{CO_2} = 40 bar.

Table 2. Potentiodynamic polarization parameters obtained after 24 h of immersion without and with 1.0 g L^{-1} of GA.

C (g L^{-1})	E_{corr} (V)	i_{corr} (μA cm^{-2})	β_c (V dec^{-1})	*IE* (%)
1 bar Blank	−0.673	17.98	0.286	-
1 bar GA	−0.676	5.54	0.334	69.23
20 bar Blank	−0.696	99.90	0.311	-
20 bar GA	−0.736	24.09	0.391	75.88
40 bar Blank	−0.600	647.05	0.316	-
40 bar GA	−0.634	84.90	0.312	86.76

The data shows that in absence of GA, the corrosion current density of the steel increased with an increase in CO_2 partial pressure, which is linked with the increased acidity of the solution, in agreement with the gravimetric experiments. However, it is evident from the data that the corrosion current density of the steel was prominently reduced after the addition of GA to the solution. Furthermore, both the cathodic and anodic curves of the polarization curves were shifted towards lower current densities after the addition of GA. The result suggests that the inhibitor impeded both the rate of the anodic dissolution (Equation (8)) and the cathodic reactions (Equations (9)–(11)), by either covering part of the metal surface and/or blocking the active corrosion sites on the steel surface. The dominant cathodic reaction depends on the pH value of the solution. At lower pH (e.g., less than 4) the reduction of H^+ ions would be the dominant cathodic reaction (Equation (11)). At pH > 4 the dominant cathodic reaction will be the reduction of HCO_3^- ions and H_2CO_3 (Equations (9) and (10)). At higher values of CO_2 partial pressure, GA suppresses the Equation (11) (e.g., the pH of the solution is circa 3.5 at P_{CO_2} = 40 bar), through the formation of H-bonding between the hydroxyl groups of the inhibitor units and the H^+ ions, adsorbed onto the steel surface, as discussed in more detail in Section 3.5.2.

Moreover, after the addition of GA, the E_{corr} can be seen to shift with no definite trend toward both the anodic and cathodic regions. This result suggested that GA behaves as a mixed type inhibitor as also reported by other studies for this inhibitor [7,21,27,28].

3.2. *Effect of Time*

The effect of immersion time on the corrosion inhibition efficiency of the tested inhibitor was also assessed in this paper. Figure 5 and Table S3 show the corrosion rate and the corrosion inhibition efficiency after 168 h of immersion in the presence and absence of 1.0 g L^{-1} of GA at different CO_2 partial pressures at 25 °C. It follows from the table that GA still shows a very high *IE* even after a longer immersion time. However, it should be noted that *IE* slightly decreases after 168 h of immersion, compared to the one observed after 24 h of immersion.

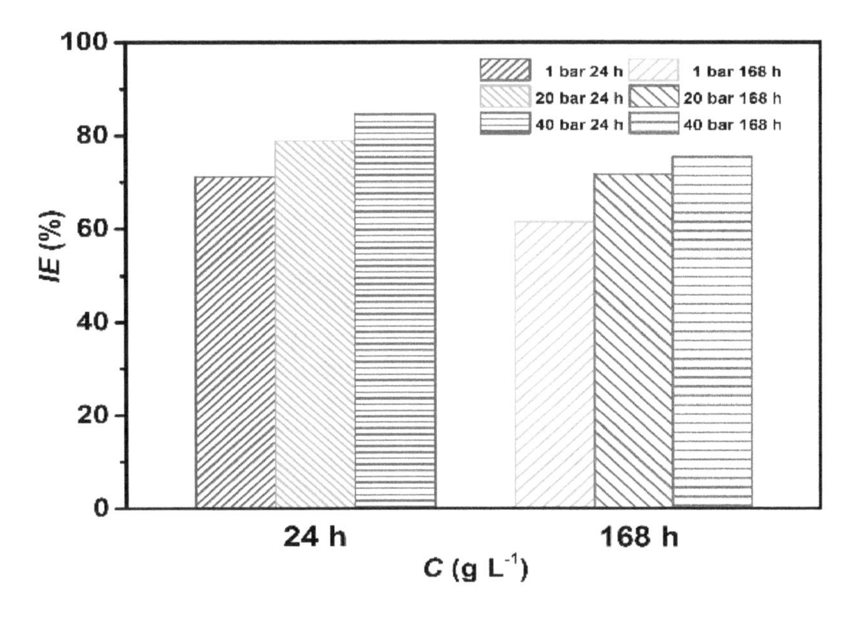

Figure 5. Corrosion inhibitor efficiency obtained at different CO_2 partial pressures after 24 and 168 h of immersion at 25 °C.

This behavior has also been reported by several studies [36,37]. The decrease in *IE* may be due to the desorption of the inhibitor from the metal surface, which makes the protective layer unstable. In this study, the desorption of GA is likely ascribed to its deprotonation due to the consumption of CO_2 from the tested solution because of the electrochemical reactions occurring into the system [5,38]. This leads to a decrease in the acidity of the solution and shifting the Equation (12) towards the deprotonation of the inhibitor.

These results confirm that GA is effectively able to protect the steel surface from sweet corrosion at high CO_2 partial pressures even after a prolonged immersion time, reflecting a strong molecular adsorption of GA on the metal surface and the formation of a stable protective layer.

3.3. Adsorption Study and Standard Adsorption Free Energy

The corrosion inhibition adsorption process of the tested inhibitor on the N80 carbon steel surface was carried out by several adsorption isotherms, such as Temkin's, Frumkin's, Langmuir's, and El-Awady's adsorption isotherms. The Temkin's adsorption isotherm was found to give the best description of the adsorption behavior of the studied inhibitor. The Temkin's adsorption isotherm is defined by the following equations:

$$\theta = \frac{-2.303\ \log K_{\text{ads}}}{2a} - \frac{2.303\ \log C}{2a} \tag{13}$$

where θ is the surface coverage ($\theta = IE\%/100$), K_{ads} the adsorption-desorption equilibrium constant, C is the inhibitor concentration, a is the molecules interaction parameter. Positive values of a imply attractive forces between the inhibitor molecules, while negative values indicate repulsive forces between them.

K_{ads} is related to the free energy of adsorption by the following equation:

$$\Delta G^{\circ}_{\text{ads}} = -RT\ \text{Ln}(K_{\text{ads}}) \tag{14}$$

where R is the gas constant (8.314 J K^{-1} mol^{-1}), T is the absolute temperature (K). The plot of surface coverage (θ) as a function of the logarithm of the inhibitor concentration at different CO_2 partial pressures is shown in Figure 6.

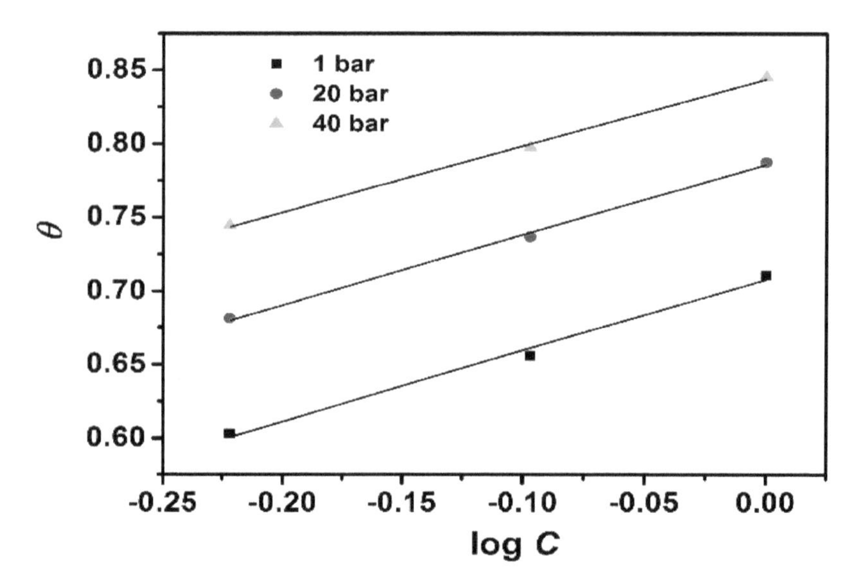

Figure 6. Temkin's adsorption isotherm for carbon steel (N80) pipeline steel in CO_2-saturated chloride at different pressures.

The plot of θ vs. log C yields a straight line and the regression coefficient ranges from 0.985 to 0.996. The calculated values of adsorption parameters $\Delta G^{\circ}_{\text{ads}}$, a and K at different CO_2 partial pressures are presented in Table 3 and the following notes can be written: (i) The values of $\Delta G^{\circ}_{\text{ads}}$ are negative for all three pressures, indicating that the adsorption of GA on the steel surface in the tested solution is a spontaneous process [7,21,22,28]. Furthermore, the value of $\Delta G^{\circ}_{\text{ads}}$ ranges between

−10.64 to −8.37 kJ mol^{-1} indicating that the adsorption of GA on the steel occurs through a physical adsorption process [7,21,22,28]; (ii) The values of "*a*" are negative for all three pressures, indicating that repulsion forces exist between the adsorbed inhibitor molecules in the adsorption layer, as also reported by other studies for the same tested inhibitor [7,22]; (iii) The values of K_{ads} increases with an increase in CO_2 partial pressure. It should be noted that K_{ads} denotes the strength between adsorbate and adsorbent. It can be inferred that a large value of K_{ads} implies a more efficient adsorption process and thus, a better corrosion inhibition efficiency [21,22]. The results suggest that the adsorption of GA increases with an increase of the environment pressure, leading to a greater surface coverage and consequently, a better protection performance.

Table 3. Parameters of the Temkin's adsorption isotherm calculated from weight loss measurements after 24 h of immersion time.

Pressure (bar)	R^2	Slope	Intercept	*a*	K_{ads}	ΔG_{ads} (kJ mol^{-1})
1	0.985	0.483	0.708	−2.38	29.10	−8.37
20	0.995	0.478	0.786	−2.41	44.09	−9.39
40	0.996	0.453	0.844	−2.50	72.97	−10.64

3.4. *Surface Analysis*

The surface morphology of the samples exposed for 24 h at different CO_2 partial pressures in the absence and presence of 1.0 g L^{-1} of GA are presented in Figures 7–9. For instance, it can be seen that the surface morphology of the samples exposed to the blank and inhibited solution differs significantly. For the blank solution, at P_{CO_2} = 1 bar, the microstructure of the sample is clearly visible (Figure 7a). The metal surface appears corroded resulting from the selective dissolution of the ferritic phase over the cementite contained in the perlitic phase. By contrast, Figure 7c,d show that after the addition of the inhibitor the metal surface becomes much smoother. It is clear from the image that the metal surface was partially covered by a protective layer, although some areas of the surface still show signs of corrosion attacks.

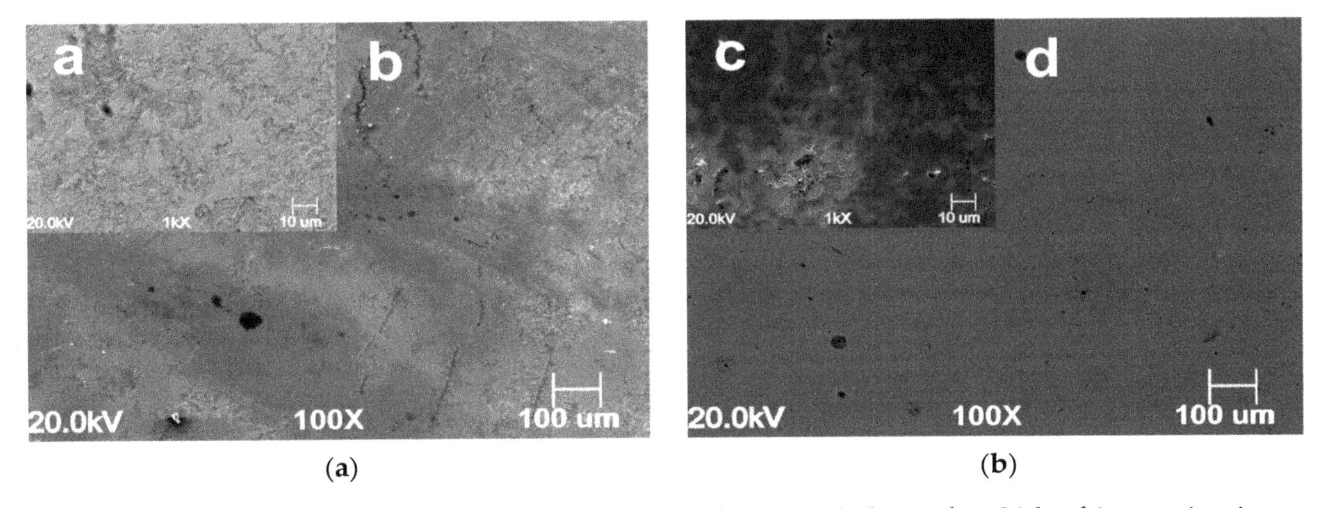

Figure 7. SEM images of the N80 carbon steel surface morphology after 24 h of immersion in the uninhibited ((**a**) **a** lower and **b** higher magnification) and inhibited ((**b**) **c** lower and **d** higher magnification) solution at 25 °C and P_{CO_2} = 1 bar CO_2.

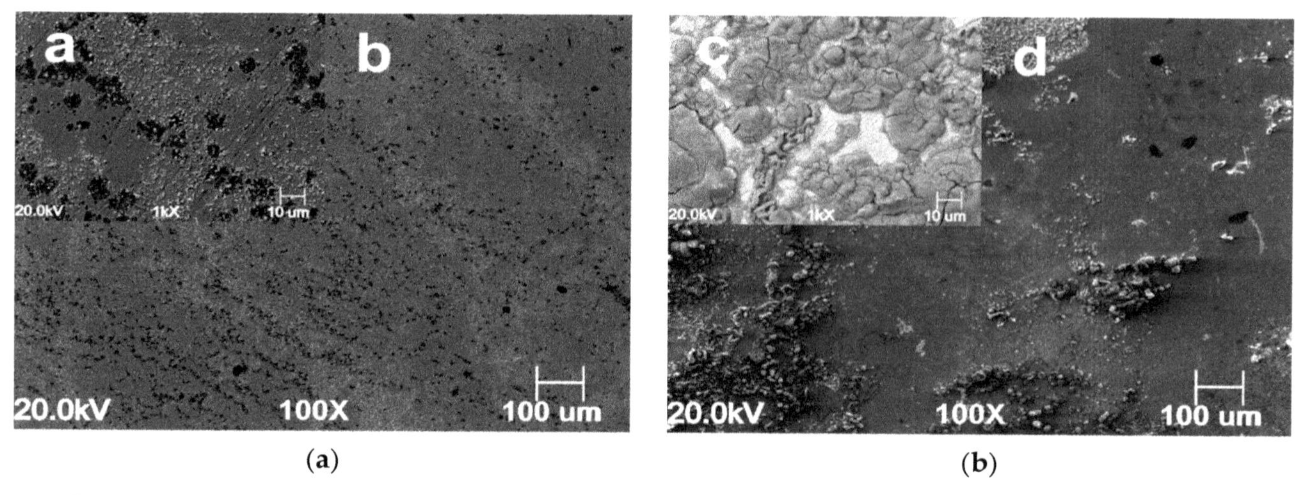

(a) (b)

Figure 8. SEM images of the N80 carbon steel surface morphology after 24 h of immersion in the uninhibited ((**a**) **a** lower and **b** higher magnification) and inhibited ((**b**) **c** lower and **d** higher magnification) solution at 25 °C and P_{CO_2} = 20 bar CO_2.

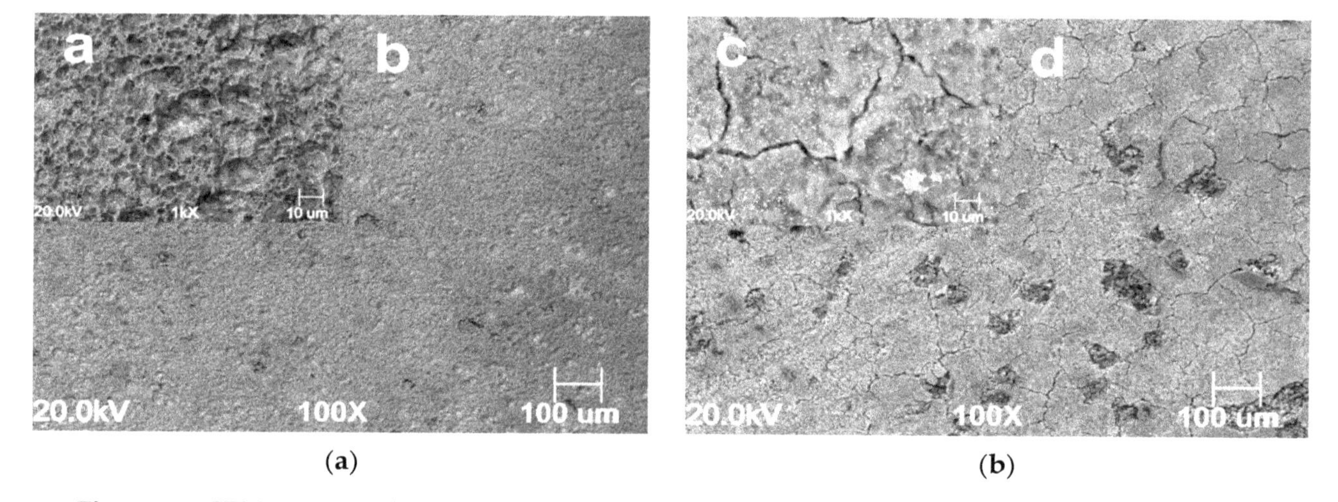

(a) (b)

Figure 9. SEM images of the N80 carbon steel surface morphology after 24 h of immersion in the uninhibited ((**a**) **a** lower and **b** higher magnification) and inhibited ((**b**) **c** lower and **d** higher magnification) solution at 25 °C and P_{CO_2} = 40 bar CO_2.

The severity of the corrosion attack increases with an increase in CO_2 partial pressure in the blank solution, as shown in Figure 8a,b and Figure 9a,b respectively carried out at P_{CO_2} = 20 bar and P_{CO_2} = 40 bar. However, it can be seen that in the presence of GA, an increase in CO_2 partial pressure led to a gradual increase in the surface coverage on the metal surface, as a result of an increase of GA molecules adsorbed onto the metal surface (Figure 8c,d and Figure 9c,d). At higher CO_2 partial pressure (e.g., P_{CO_2} = 40 bar, Figure 9) the protective action of the inhibitor is even more evident. The images show that for the uninhibited solution, the surface of the metal appears severely corroded, while the one obtained in the presence of the inhibitor shows the formation of a uniform protective layer over its entire metal surface. The results indicate that in the presence of GA and with a gradual increase in CO_2 partial pressure, the protective layer gradually becomes more compact and thicker [4]. As discussed in Section 3.1, the solubility of CO_2 increases with its partial pressure, and as a result of this, the concentration of H^+ ions into the solution also increases, hence the number of the inhibitor molecules that can be protonated and adsorbed onto the metal surface also increases according to the Equation (12), leading to a substantial reduction of the corrosion rate of the metal.

The morphology of the metal surface was also analyzed with the help of an energy-dispersive spectroscopy with the result listed in Table 4. In the absence of GA, the metal surface was characterized by a corrosion product layer mainly consisting of carbon, iron, and a small amount of oxygen elements, indicating that this corrosion layer is mainly composed of Fe_3C. These results are in agreement with that previously observed in the literature [5,7,39,40]. Other researchers reported that at a temperature below 40 °C, the corrosion product layer is generally composed of Fe_3C, and only little traces of $FeCO_3$ were observed on the metal surface [4,7,39,40], as also confirmed by the GIXRD measurements shown in Figure 10. The presence of Fe_3C on the metal surface is due to the anodic dissolution of the ferrite phase over the cementite in the perlitic phase, which leads to an accumulation of the cementite on the metal surface.

Table 4. Weight percentage of the elements calculated from EDS analyses.

Element	Weight%			
	C	O	Fe	Total
Polished	0.70	-	99.30	100
Blank (1 bar)	1.18	-	98.82	100
1.0 g L^{-1} (1 bar)	4.06	3.51	92.43	100
Blank (20 bar)	7.28	0.83	91.89	100
1.0 g L^{-1} (20 bar)	8.00	21.98	70.02	100
Blank (40 bar)	4.99	2.05	92.96	100
1.0 g L^{-1} (40 bar)	9.90	16.21	73.89	100

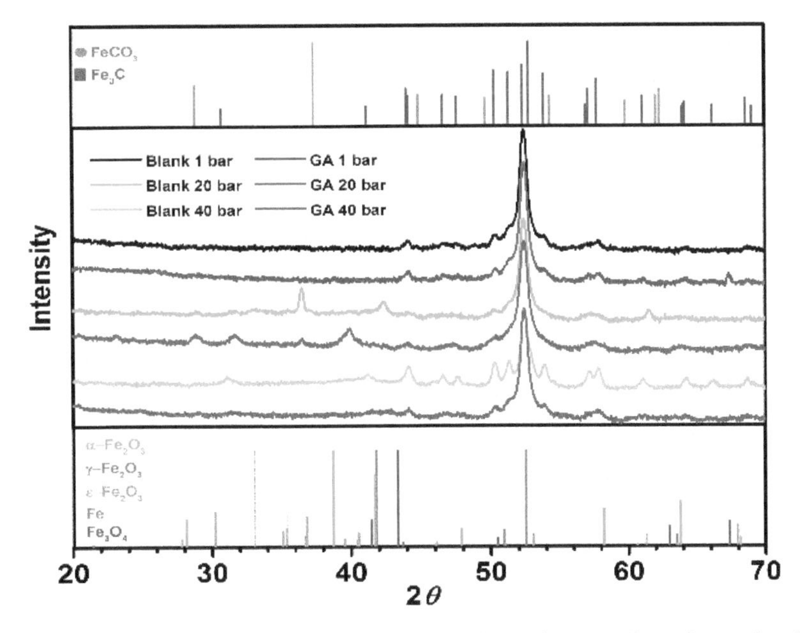

Figure 10. XRD spectra of corrosion product film formed on the metal surface after been exposed for 24 h without and with the presence of 1.0 g L^{-1} of GA at different CO_2 partial pressures at 25 °C.

It is worth mentioning that in the presence of GA the content of carbon and oxygen was found to be higher than those observed for the blank solution. It should be noted that carbon and oxygen are also the main constituents of the tested inhibitor and therefore, their higher concentration on the protective layer formed in the presence of the inhibitor can be attributed to its adsorption onto the metal surface, as also reported by other studies [4,7,20]. Moreover, it can be seen from the table that the percentage of Fe decreased in the presence of GA, likely due to the overlying effect of the inhibitor layer.

The GIXRD analysis for the samples corroded in an inhibited and uninhibited solution at P_{CO_2} = 40 bar and at 25 °C (Figure 10) shows the presence of cementite on the metal surface, although in the presence of GA the intensity of these peaks is much weaker. This result can be explained as follows: Fe_3C accumulates on the metal surface after the dissolution of the ferritic phase. However, in the presence of the inhibitor, it only accumulates in small amounts on the bare metal surface at the early stage of the experiment, since the dissolution of the ferritic phase is quickly suppressed by the absorption of the inhibitor on the surface of the metal.

Figure 11a,b show the surface morphology for specimens corroded in the blank and inhibited solution carried out at 60 °C and P_{CO_2} = 40 bar, after immersion the samples for 24 h in the tested solution, without and with the presence of GA, respectively. The corrosion product layer appears to be different for the inhibited solution compared to one observed in the presence of GA. Figure 11a shows the presence of a porous corrosion product layer formed onto the metal surface corroded in a free-inhibitor solution, pores which create paths for the solution to penetrate it and thereby leading to the dissolution of the underlying metal. On the other hand, the surface of the metal corroded in the presence of GA (Figure 11b) shows the formation of a more compact layer, which forms a better protective barrier and thereby greatly reducing the corrosion rate of the metal.

(**a**) (**b**)

Figure 11. SEM images of the N80 carbon steel morphology after 24 h of immersion in the tested solution at P_{CO_2} = 40 bar and at 60 °C, without (**a**) and with (**b**) the presence of GA.

EDS analysis reports high content of carbon, oxygen, and iron elements in both layers (C:11.11%, O:6.02% and C:13.69%, O:11.0%, in the blank and inhibited solution, respectively). The GIXRD measurements presented in Figure 12 show the characteristic XRD diffraction patterns associated with $FeCO_3$. By contrast, the intensity of the iron carbonate peaks observed in the presence of GA is almost negligible. These results suggest that the layer observed for the uninhibited solution is mainly composed of Fe_3C and $FeCO_3$, while in the presence of GA is mainly composed of Fe_3C with little traces of $FeCO_3$ [4]. Similar behavior was also reported by Ding et al. [41] related to the study of the effect of an imidazoline-type inhibitor against CO_2 corrosion of mild steel. The authors suggested that the formation of the corrosion inhibitor layer was able to suppress the formation of the iron carbonate. The precipitation of $FeCO_3$ depends on the concentration of the Fe^{2+} and CO_3^{2-} ions, pH, and temperature. When the concentrations of Fe^{2+} and CO_3^{2-} ions exceed the solubility limit, $FeCO_3$ will precipitate on the surface [9,40,42]. At higher temperatures, its solubility decreases, and therefore the likelihood of its precipitation will be also higher. In a free-inhibitor solution, the dissolution of the ferrite phase may lead to an increase in the concentration of Fe^{2+} ions in the bulk solution and thereby favoring the precipitation of $FeCO_3$ onto the surface of the metal. Conversely, in the presence of the inhibitor, the protective layer formed onto the surface of the metal slows down the corrosion processes, and thereby reducing the concentrations of Fe^{2+} ions available for the formation of $FeCO_3$.

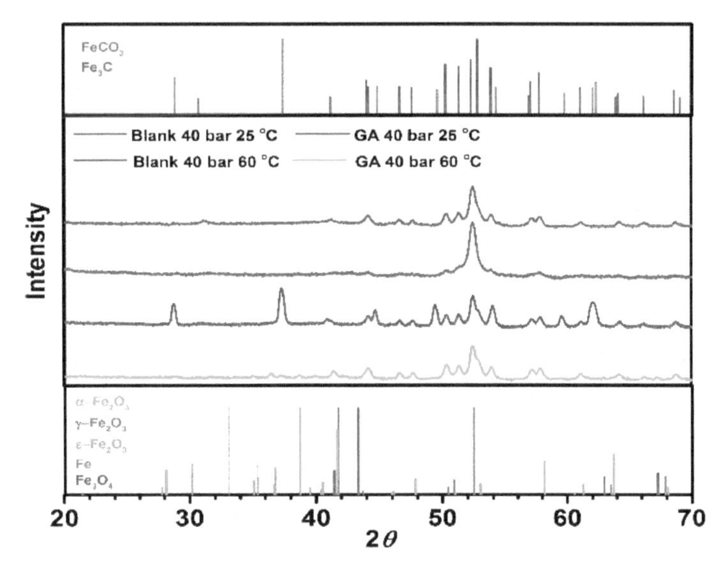

Figure 12. XRD spectra of corrosion product film formed on the metal surface after been exposed for 24 h without and with the presence of 1.0 g L^{-1} of GA at P_{CO_2} = 40 bar and at 60 °C.

Figure 13a–d show the SEM analysis of the metal surface after 168 h of immersion in the absence and presence of 1.0 g L^{-1} GA at P_{CO_2} = 40 bar, respectively. It is apparent from the figures that a thick porous layer covers both surface samples; although it seems that in the presence of the GA, this layer appears denser, thus providing a higher level of protection. To analyze the condition of the metal surface, these porous layers were removed with the help of Clark's solution. It can be seen that both surfaces show clear signs of corrosion attacks (Figure 13b; however, it is also clear from the figures that in the presence of the inhibitor (Figure 13d) the surface of the metal appears to be less damaged and smoother, with the ground scratches still visible on the surface. This result was also confirmed by the atomic force microscopy experiments performed by Azzaoui et al. [28] concerning the use of GA as a corrosion inhibitor in a 1 M HCl solution. The authors reported that in the uninhibited solution the surface of the metal was found to be more corroded with an average roughness of 1.3 μm, while in the presence of GA the average roughness was reduced to 500 nm. The authors justified this behavior due to the formation of a more compact protective layer on the metal surface that strongly reduced the diffusion of the aggressive substances to the metal, and thereby reducing the corrosion rate of the metal.

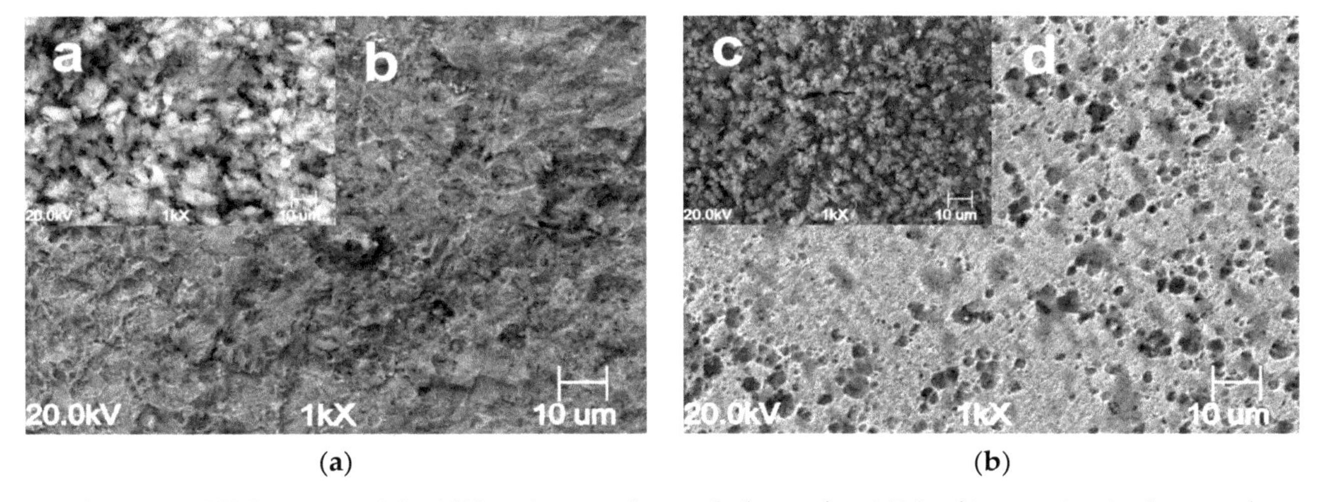

Figure 13. SEM images of the N80 carbon steel morphology after 168 h of immersion in the tested solution in the presence of 1.0 g L^{-1} of GA at P_{CO_2} = 40 bar. Without (**a**,**b**) and with the inhibitor (**c**,**d**) at 25 °C.

The SEM-EDS and GIXRD result confirm that GA provides adequate protection to the metal surface from sweet corrosion even at high CO_2 partial pressures and after long immersion times. The results are in agreement with the findings obtained with the weight loss measurements, confirming the high inhibition efficiency value observed after a long immersion time.

X-ray photoelectron spectroscopy analysis was employed as a means to confirm the adsorption of the tested inhibitor on the carbon steel surface. The analysis was carried out on the native inhibitor and the steel surface after 24 h of immersion in the tested solution in the presence of 1.0 g L^{-1} of GA at P_{CO_2} = 40 bar and at 25 °C. The XPS results presented in Figure 14a showed evidence of the presence of O, C, N, and Fe on the carbon steel surface, where the O and C contents displayed the highest amount, while the signal of N was detected with small intensity. The high-resolution peaks core levels were analyzed through a deconvolution fitting of the complex spectra. The binding energies and the corresponding quantification (%) of each peak component are presented in Table S4.

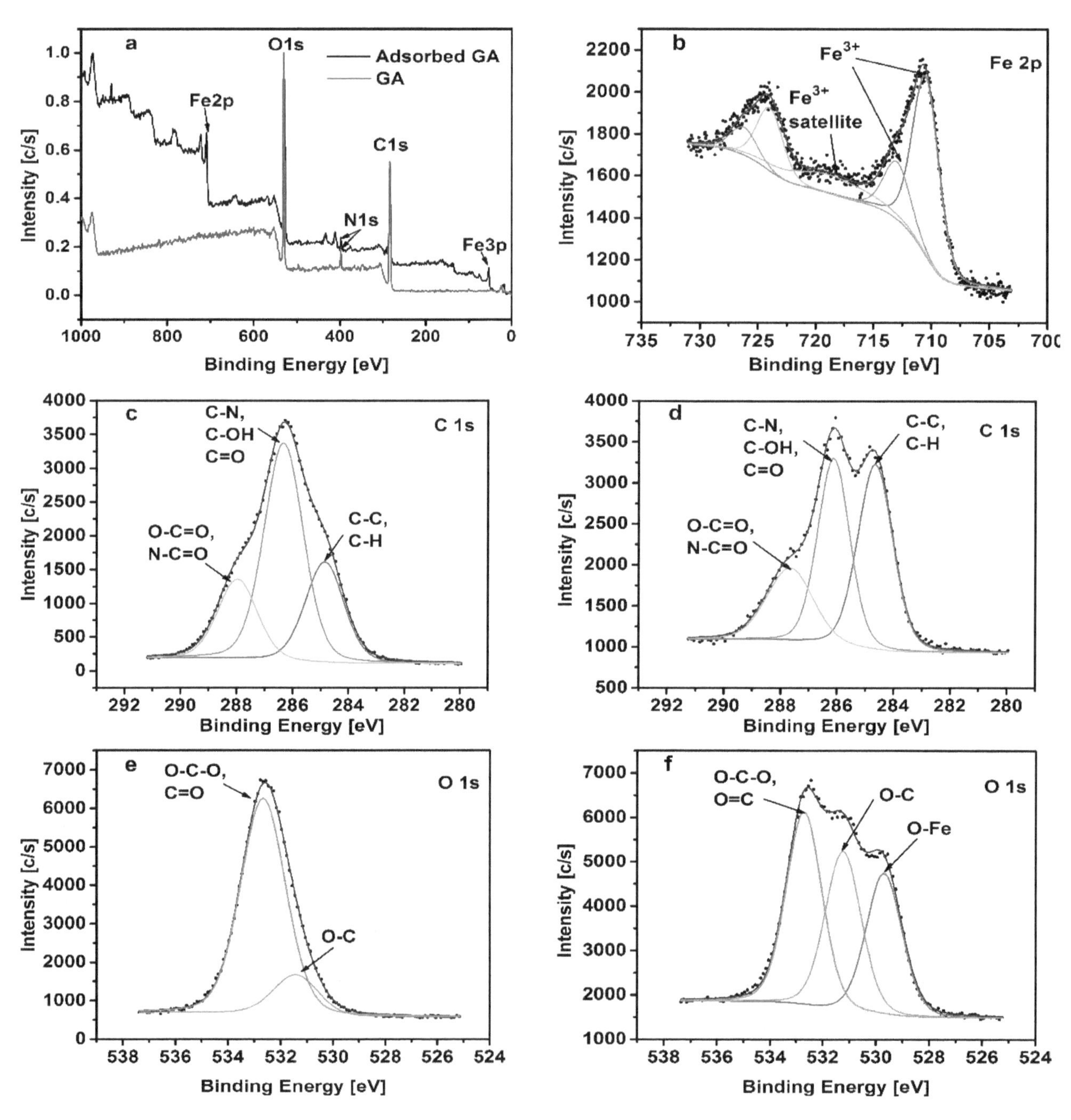

Figure 14. *Cont.*

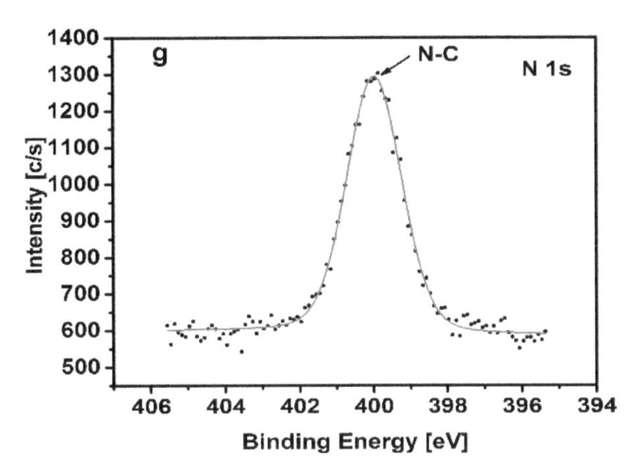

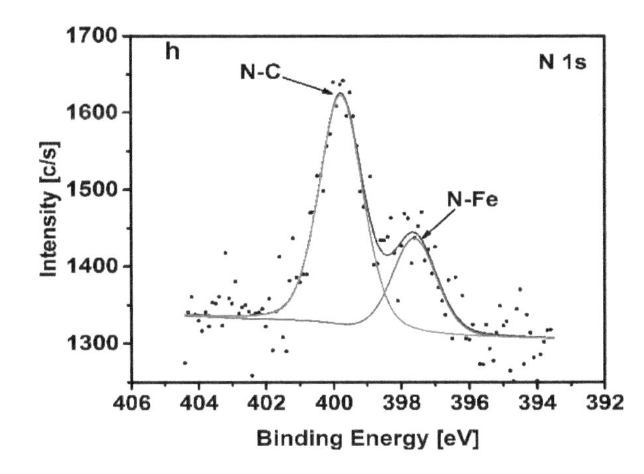

Figure 14. XPS spectra of the native gum Arabic: (**a**,**c**,**e**,**g**). XPS spectra of the film formed on the N80 carbon steel after 24 h exposure in CO_2 at P_{CO_2} = 40 bar in the presence of 1.0 g L^{-1} of GA at 25 °C: (**b**,**d**,**f**,**h**).

The deconvoluted $Fe2p_{3/2}$ peaks (Figure 14b) at 710.5 and 713.0 eV could be associated with the α-Fe_2O_3 or/and γ- Fe_2O_3 [13]. The presence of these species is likely due to the partial decomposition of iron carbonate. The literature reported that $FeCO_3$ begins decomposing at temperatures below 100 °C according to the following reaction [5,42]:

$$FeCO_3 \rightarrow FeO + CO_2 \tag{15}$$

In the presence of CO_2 or water vapor, FeO transforms into Fe_3O_4 [5,42].

$$3FeO + CO_2 \rightarrow Fe_3O_4 + CO \tag{16}$$

$$3FeO + H_2O \rightarrow Fe_3O_4 + H_2 \tag{17}$$

However, in the presence of oxygen, FeO and Fe_3O_4 transform into Fe_2O_3 [5,42].

$$4FeO{+}O_2 \rightarrow 2Fe_2O_3 \tag{18}$$

$$\text{In the air}: \ 4Fe_3O_4 + O_2 \rightarrow 6Fe_2O_3 \tag{19}$$

The C1s spectra of the native gum arabic and the adsorbed one (Figure 14c,d, respectively) show three main peaks. The C1s peak with binding energy at 284.8 eV could be attributed to the C–C/C–H bonds [26,28]. The C1s peak at 286.2 eV could be attributed to the C–OH/C=O bonds related to the different groups of GA [26,43]. This peak may also be assigned to the carbon atom bonded to nitrogen in C–N bond [13,44] and could be related to the glycoprotein and/or to the arabinogalactan-protein fractions of the inhibitor (Figure 15b,c, respectively). The last C1s peak with a binding energy of 287.7 eV could be associated with the presence of carbonyl type groups O–C=O/N–C=O that result from the protonation of the GA molecule in the acid environment [28].

It is worth mentioning that no peaks assigned to Fe_3C were found with the XPS analysis in contrast to the results reported from the GIXRD analysis, where the characteristic peaks assigned to this compound can be seen in the presence of GA (Figure 10). Fe_3C cannot be detected since the average depth of analysis for an XPS measurement is approximately 5 nm however, the cementite formed on the metal surface at the early stage of the experiment is covered by a thicker layer of inhibitor (Figure 9c,d).

The deconvoluted O1s spectra of the native and adsorbed inhibitor are displayed in Figure 14e,f, respectively. The peaks at 531.2 and 532.7 eV could be attributed to the single bonded oxygen in C–O and the double bonded oxygen C=O and/or to the single bonded oxygen in O–C–O respectively [4,13,26,28]. The latter peak may correspond to the carbonyl type groups and/or to the glycosidic C(1)-O-C(4)/C(1)-O-C(6) linkages of the GA molecules (Figure 15a), as well as, in the case of the sample exposed to the tested solution, to $FeCO_3$ formed on the metals surface, respectively [4,26,28]. Moreover, some authors reported that the peak at 231.2 eV could also be attributed to the oxygen of the hydroxyl groups (–OH) [5,43], likely due to the hydroxyl groups of the tested polysaccharide. The O1s spectrum of the adsorbed inhibitor (Figure 14f) displays an extra peak at 529.7 eV corresponding to O^{2-} related to the oxygen atoms bonded with Fe^{3+} in the Fe_2O_3 oxide [4,43,44]. The O1s results are in good agreement with the findings of the Fe2p spectrum.

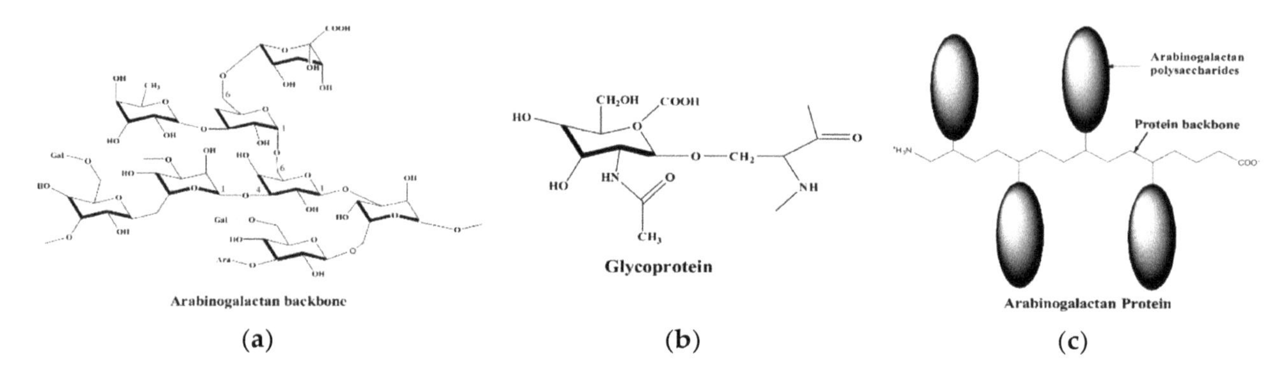

Figure 15. Structure of gum arabic: (**a**) arabinogalactan; (**b**) glycoprotein; (**c**) arabinogalactan-protein.

The presence of N1s peak in the survey for the adsorbed GA on the carbon steel surface (Figure 14a) provides evidence that gum arabic was effectively adsorbed on the tested substrate surface since the N80 carbon steel substrate does not contain nitrogen in its chemical composition. The N1s spectra of the native and adsorbed inhibitor are presented in Figure 14g,h. Both images show the presence of a peak at 400 and 399.8 eV attributed to the nitrogen atom bonded with the carbon atom, for the native and adsorbed inhibitor. However, as it can be seen that the high-resolution N1s spectrum of the tested substrate sample after the addition of GA depicts an extra peak at 397.6 eV. This extra peak can be ascribed to the coordinated nitrogen atom of the amino group with the metal surface (N–Fe bond) [44]. Other authors also suggested that this peak could be attributed to the bond between the nitrogen of the amino groups and the oxide layer on the metal surface (FeO_x) [45].

3.5. Mechanism of Inhibition

Given all the observed results, it can be inferred that the GA was effectively adsorbed on the metal surface, providing good protection to the metal surface against sweet corrosion. However, the complex chemical structure of this inhibitor makes it difficult to determine the exact adsorption mechanism involved. Gum arabic is a heterogeneous mixture of different compounds consisting of three main fractions: 80% of arabinogalactan (AG), 10.4% of arabinogalactan-protein (AGP) and 1.2% glycoprotein (GP) (Figure 15). Each of these fractions contains a range of different molecular weight components and different protein contents. Therefore, some of these compounds can be physically and others chemically adsorbed. Nevertheless, based on the results reported in this study, it can be assumed that the following three types of adsorption mechanisms or likely a combination of them may take place in the inhibiting phenomena involving GA on the steel surface.

3.5.1. Adsorption via Electrostatic Interaction

The functional groups such as hydroxyl, carboxyl, and amino present in the GA molecules, by virtue of the presence of lone pair of electrons, can be easily protonated in acid solutions such that the newly formed polycations are in equilibrium with their neutral counterpart according to the Equation (12). The high corrosion inhibition activity showed by GA is likely due to a synergistic electrostatic interaction between the protonated GA molecules with the adsorbed chloride ions, as shown in Figure 16a. As reported by several studies [7,20–22,28] chloride ions are strongly adsorbed on the positively charged metal surface, thereby creating an excess of electrons so that the metal will be negatively charged. These adsorbed chloride ions can act as an intermediate bridge between the surface and the protonated inhibitor molecules and therefore, assisting the adsorption of GA on the metal surface. This type of adsorption mechanism is likely the one that accounts for the most inhibition action of the inhibitor. In fact, the results presented in this manuscript have demonstrated clearly that the corrosion inhibition action of GA was strongly influenced by both the concentration of the inhibitor, CO_2 partial pressure, and temperature. A change in one of these two factors has a great effect on the equilibrium reaction (Equation (12)), shifting the equilibrium towards the protonated or the deprotonated form of the inhibitor. A shift to the right implies an increase in the number of protonated molecules of GA available to interact with the chloride ions adsorbed on the surface and thus, an increase in *IE* of the system.

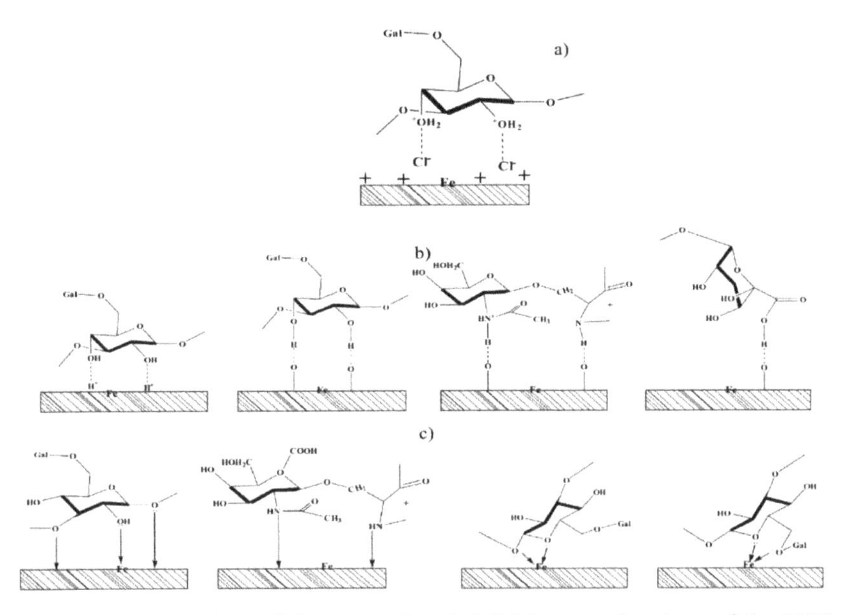

Figure 16. Schematic representation of the corrosion inhibition mechanism of the N80 carbon steel by GA. (**a**) electrostatic; (**b**) H-bond formation; (**c**) chemical adsorption.

3.5.2. Adsorption via Hydrogen Bond Formation Interaction

At higher CO_2 partial pressure (i.e., 40 bar) the pH of the solution is around 3 [32], and among the three possible occurring cathodic reactions (Equations (9)–(11)), the reduction of hydrogen ions to hydrogen gas is the dominant cathodic reaction. It is generally accepted that this reaction can be described using three steps [46]. The first step is the electrochemical adsorption of the H^+ ions (Equation (20)) followed by either the electrochemical desorption (Equation (21)) or the chemical desorption (Equation (22)).

$$H^+_{(aq)} + e^- \rightarrow H_{ads} \tag{20}$$

$$H_{ads} + H^+_{(aq)} \rightarrow H_{2(g)} \tag{21}$$

$$H_{ads} + H_{ads} \rightarrow H_{2(g)} \tag{22}$$

The potentiodynamic measurements presented in Figure 5 showed that the cathodic current density of the system was greatly reduced after the addition of GA in the solution, suggesting that GA was able to suppress the hydrogen evolution reaction (Equation (11)) to some extent. Similar results were also confirmed by other authors [21,26–28]. This assumption was also confirmed by FT-IR and Raman measurements performed on GA [7,26] and other gum-like [20,36,47] compounds. The results showed that the characteristic peak assigned to the hydroxyl groups of the carbohydrate units narrowed down and/or shifted after its adsorption on the metal surface. The authors agreed that this change in shape was likely due to a possible interaction of the hydroxyl groups of the GA molecules with the H adsorbed on the cathodic sites of the metal surface via H-bond formation (Figure 16b). Therefore, the high value of IE observed in this study at different CO_2 partial pressure can be also ascribed to the ability of GA to suppress one of these reactions (Equations (20)–(22)) via H-bonds formation, thus suppressing Equation (11) and consequently the dissolution of the steel (Equation (8)).

The adsorption of GA may also be promoted by the presence of the oxide layer on the metal surface via hydrogen bonding (Figure 16b). Studies concerning the adsorption of GA on oxide nanoparticles (i.e., iron oxide nanoparticles [48] and zinc or aluminum oxide nanoparticles [49]) reported that GA showed a strong affinity toward these oxide nanoparticles. The authors suggested that the adsorption of GA on these oxide nanoparticles surface might be due to the formation of hydrogen bonds between the functional groups of the GA molecules (e.g., hydroxyl, carboxylate, and amino) with the oxidized surface. The XPS analysis presented in this study showed that the metal surface after 24 h of exposure is covered by different oxide species such as Fe_2O_3 and/or Fe_3O_4, (e.g., Equations (15)–(19)). Therefore, the adsorption of GA assisted by the presence of oxide species formed on the metal surface via H-bonds formation is an adsorption mechanism that must be also taken into account.

3.5.3. Chemical Adsorption

The heteroatoms (i.e., O, N) present on the GA molecules by virtue of the presence of lone pair of electrons may promote the adsorption of the inhibitor via the formation of coordinate bonds with the iron from the metal surface and/or with iron from the oxide species formed on the surface [45] (Figure 16c). The XPS measurements observed in this study showed a peak at 397.6 eV likely ascribed to the coordinated nitrogen atom of the amino group with the Fe (N–Fe bond) [44]. This result suggests that although the inhibitor is mainly physically adsorbed on the surface of the metal, a small contribution of the chemical adsorption process cannot be ignored.

4. Conclusions

The corrosion inhibition effect of gum Arabic on the corrosion of carbon steel (N80) exposed in a high-pressure CO_2-saline environment has been studied and the following conclusion can be drawn:

- The weight loss results showed that the thickening agent gum arabic was found to be an efficient corrosion inhibitor for carbon steel in a high-pressure CO_2-saline environment. The Inhibition efficiency increased with an increase in inhibitor concentration and CO_2 partial pressure with the maximum value of IE found to be 84.53% at P_{CO_2} = 40 bar after 24 h of immersion. Moreover, the weight loss results also showed that GA was effectively able to protect the steel surface from sweet corrosion at high CO_2 partial pressures (i.e., 40 bar) even after a prolonged immersion time (i.e., 168 h) with a corrosion inhibition efficiency found to be 74.41%.
- The adsorption of GA on the carbon steel surface follows the Temkin's adsorption isotherm model. The negative free energy of adsorption ΔG°_{ads} indicates a strong and spontaneous adsorption of GA on the carbon steel surface. Furthermore, the value of ΔG°_{ads} indicates that the GA adsorbs mainly via physical adsorption on the metal surface.
- The SEM analysis revealed that in the presence of GA the protective layer on the metal surface becomes more compact and dense with an increase in CO_2 partial pressure. Also, the SEM analysis revealed that after 168 h of immersion, in the presence of GA, the metal surface appeared to be less damaged and smother.
- The XPS results confirmed the formation of a protective layer containing GA molecules and iron oxides on the metal surface.

Supplementary Materials:
Table S1: Corrosion rate and inhibition efficiency obtained from weight loss measurements for the N80 carbon steel at various concentrations of GA and CO_2 partial pressures after 24 h of immersion time, Table S2: Comparison of reported inhibition efficiency of some other corrosion inhibitors used in a CO_2 saturated saline solution (3.5 wt.% NaCl), Table S3: Corrosion rate and inhibition efficiency obtained from weight loss measurements for the carbon steel (N80) carried out at 1.0 g L−1 o f G A a n d C O 2 partial pressures after 168 h of immersion time at 25 °C, Table S4: XPS analysis of sample steel surface after 24 h of immersion in test solution at P_{CO_2} = 40 bar and at 25 °C in the presence of 1.0 g L^{-1} of GA.

Author Contributions: G.P. conceived, designed, and performed the measurements, analyzed the experimental data, wrote and edited the manuscript; K.K. performed the XRD analysis; R.W. and A.B. performed the XPS. analysis; M.G. performed the SEM-EDS analysis. All authors have read and agreed to the published version of the manuscript.

Acknowledgments: RW has been partly supported by the EU Project POWR.03.02.00-00-I004/16.

References

1. Sheng, J.J. Enhanced oil recovery in shale reservoirs by gas injection. *J. Nat. Gas Sci. Eng.* **2015**, *22*, 252–259. [CrossRef]
2. Bai, H.; Wang, Y.; Ma, Y.; Zhang, Q.; Zhang, N. Effect of CO_2 Partial Pressure on the Corrosion Behavior of J55 Carbon Steel in 30% Crude Oil/Brine Mixture. *Materials* **2018**, *11*, 1765. [CrossRef] [PubMed]
3. Bai, H.; Wang, Y.; Ma, Y.; Ren, P.; Zhang, N. Pitting Corrosion and Microstructure of J55 Carbon Steel Exposed to CO_2/Crude Oil/Brine Solution under 2–15 MPa at 30–80 °C. *Materials* **2018**, *11*, 2374. [CrossRef] [PubMed]
4. Mustafa, A.H.; Ari-Wahjoedi, B.; Ismail, M.C. Inhibition of CO_2 Corrosion of X52 Steel by Imidazoline-Based Inhibitor in High Pressure CO_2-Water Environment. *J. Mater. Eng. Perform.* **2012**, *22*, 1748–1755. [CrossRef]
5. Islam, A.; Farhat, Z.N. Characterization of the Corrosion Layer on Pipeline Steel in Sweet Environment. *J. Mater. Eng. Perform.* **2015**, *24*, 3142–3158. [CrossRef]
6. Aristia, G.; Hoa, L.Q.; Bäßler, R. Corrosion of Carbon Steel in Artificial Geothermal Brine: Influence of Carbon Dioxide at 70 °C and 150 °C. *Materials* **2019**, *12*, 3801. [CrossRef]
7. Palumbo, G.; Górny, M.; Banaś, J. Corrosion Inhibition of Pipeline Carbon Steel (N80) in CO_2-Saturated Chloride (0.5 M of KCl) Solution Using Gum Arabic as a Possible Environmentally Friendly Corrosion Inhibitor for Shale Gas Industry. *J. Mater. Eng. Perform.* **2019**, *28*, 6458–6470. [CrossRef]
8. Palumbo, G.; Banas, J.; Bałkowiec, A.; Mizera, J.; Lelek-Borkowska, U. Electrochemical study of the corrosion behaviour of carbon steel in fracturing fluid. *J. Solid State Electrochem.* **2014**, *18*, 2933–2945. [CrossRef]
9. Linter, B.; Burstein, G. Reactions of pipeline steels in carbon dioxide solutions. *Corros. Sci.* **1999**, *41*, 117–139. [CrossRef]
10. Palumbo, G.; Banaś, J. Inhibition effect of guar gum on the corrosion behaviour of carbon steel (K-55) in fracturing fluid. *Solid State Phenom.* **2015**, *227*, 59–62. [CrossRef]
11. Tang, J.; Hu, Y.; Han, Z.; Wang, H.; Zhu, Y.; Wang, Y.; Nie, Z.; Wang, Y. Experimental and Theoretical Study on the Synergistic Inhibition Effect of Pyridine Derivatives and Sulfur-Containing Compounds on the Corrosion of Carbon Steel in CO_2-Saturated 3.5 wt.% NaCl Solution. *Molecules* **2018**, *23*, 3270. [CrossRef]
12. Ortega-Toledo, D.; Gonzalez-Rodriguez, J.G.; Casales, M.; Martinez, L.; Martinez-Villafañe, A. Co2 corrosion inhibition of X-120 pipeline steel by a modified imidazoline under flow conditions. *Corros. Sci.* **2011**, *53*, 3780–3787. [CrossRef]
13. Singh, A.; Ansari, K.R.; Quraishi, M.A.; Lgaz, H. Effect of Electron Donating Functional Groups on Corrosion Inhibition of J55 Steel in a Sweet Corrosive Environment: Experimental, Density Functional Theory, and Molecular Dynamic Simulation. *Materials* **2018**, *12*, 17. [CrossRef] [PubMed]
14. Ghareba, S.; Omanovic, S. The effect of electrolyte flow on the performance of 12-aminododecanoic acid as a carbon steel corrosion inhibitor in CO_2-saturated hydrochloric acid. *Corros. Sci.* **2011**, *53*, 3805–3812. [CrossRef]
15. Umoren, S.; Alahmary, A.A.; Gasem, Z.M.; Solomon, M. Evaluation of chitosan and carboxymethyl cellulose as ecofriendly corrosion inhibitors for steel. *Int. J. Boil. Macromol.* **2018**, *117*, 1017–1028. [CrossRef] [PubMed]
16. Ansari, K.; Chauhan, D.S.; Quraishi, M.; Mazumder, M.A.; Singh, A. Chitosan Schiff base: An environmentally benign biological macromolecule as a new corrosion inhibitor for oil & gas industries. *Int. J. Boil. Macromol.* **2020**, *144*, 305–315. [CrossRef]
17. Lin, Y.; Singh, A.; Ebenso, E.E.; Quraishi, M.A.; Zhou, Y.; Huang, Y. Use of HPHT Autoclave to Determine Corrosion Inhibition by Berberine extract on Carbon Steels in 3.5% NaCl Solution Saturated with CO_2. *Int. J. Electrochem. Sci.* **2015**, *10*, 194–208.
18. Singh, A.; Lin, Y.; Liu, W.; Ebenso, E.E.; Pan, J. Extract of Momordica charantia (Karela) Seeds as Corrosion Inhibitor for P110SS Steel in CO_2 Saturated 3.5% NaCl Solution. *Int. J. Electrochem. Sci.* **2013**, *8*, 12884–12893.
19. Singh, A.; Lin, Y.; Ebenso, E.E.; Liu, W.; Pan, J.; Huang, B. Gingko biloba fruit extract as an eco-friendly corrosion inhibitor for J55 steel in CO_2 saturated 3.5% NaCl solution. *J. Ind. Eng. Chem.* **2015**, *24*, 219–228. [CrossRef]

20. Palumbo, G.; Berent, K.; Proniewicz, E.; Banaś, J. Guar Gum as an Eco-Friendly Corrosion Inhibitor for Pure Aluminium in 1-M HCl Solution. *Materials* **2019**, *12*, 2620. [CrossRef]
21. Bentrah, H.; Rahali, Y.; Chala, A. Gum Arabic as an eco-friendly inhibitor for API 5L X42 pipeline steel in HCl medium. *Corros. Sci.* **2014**, *82*, 426–431. [CrossRef]
22. Umoren, S. Inhibition of aluminium and mild steel corrosion in acidic medium using Gum Arabic. *Cellulose* **2008**, *15*, 751–761. [CrossRef]
23. Singh, A.; Ansari, K.; Quraishi, M. Inhibition effect of natural polysaccharide composite on hydrogen evolution and P110 steel corrosion in 3.5 wt% NaCl solution saturated with CO_2: Combination of experimental and surface analysis. *Int. J. Hydrogen Energy* **2020**, *45*, 25398–25408. [CrossRef]
24. Umoren, S.; Ogbobe, O.; Igwe, I.; Ebenso, E. Inhibition of mild steel corrosion in acidic medium using synthetic and naturally occurring polymers and synergistic halide additives. *Corros. Sci.* **2008**, *50*, 1998–2006. [CrossRef]
25. Mobin, M.; Alam Khan, M. Investigation on the Adsorption and Corrosion Inhibition Behavior of Gum Acacia and Synergistic Surfactants Additives on Mild Steel in 0.1 MH_2SO_4. *J. Dispers. Sci. Technol.* **2013**, *34*, 1496–1506. [CrossRef]
26. Abu-Dalo, M.A.; Othman, A.A.; Al-Rawashdeh, N.A.F. Exudate gum from acacia trees as green corrosion inhibitor for mild steel in acidic media. *Int. J. Electrochem. Sci.* **2012**, *7*, 9303–9324.
27. Shen, C.; Alvarez, V.; Koenig, J.D.B.; Luo, J.-L. Gum Arabic as corrosion inhibitor in the oil industry: Experimental and theoretical studies. *Corros. Eng. Sci. Technol.* **2019**, *54*, 444–454. [CrossRef]
28. Azzaoui, K.; Mejdoubi, E.; Jodeh, S.; Lamhamdi, A.; Rodríguez-Castellón, E.; Algarra, M.; Zarrouk, A.; Errich, A.; Salghi, R.; Lgaz, H. Eco friendly green inhibitor Gum Arabic (GA) for the corrosion control of mild steel in hydrochloric acid medium. *Corros. Sci.* **2017**, *129*, 70–81. [CrossRef]
29. Spellman, F.R. *Environmental Impacts of Hydraulic Fracturing*; Informa UK Limited: London, UK, 2012.
30. *ASTM-G1-90, Standard Practice for Preparing, Cleaning, and Evaluation Corrosion Test Specimens*; ASTM International: West Conshohocken, PA, USA, 1999.
31. Choi, Y.-S.; Nešić, S. Determining the corrosive potential of CO_2 transport pipeline in high pCO_2–water environments. *Int. J. Greenh. Gas. Control* **2011**, *5*, 788–797. [CrossRef]
32. Li, X.; Peng, C.; Crawshaw, J.; Maitland, G.; Trusler, J.M. The pH of CO_2-saturated aqueous NaCl and NaHCO3 solutions at temperatures between 308 K and 373 K at pressures up to 15 MPa. *Fluid Phase Equilibria* **2018**, *458*, 253–263. [CrossRef]
33. Dong, B.; Liu, W.; Zhang, Y.; Banthukul, W.; Zhao, Y.; Zhang, T.; Fan, Y.; Li, X.; Wei, L.; Yonggang, Z.; et al. Comparison of the characteristics of corrosion scales covering 3Cr steel and X60 steel in CO_2-H2S coexistence environment. *J. Nat. Gas Sci. Eng.* **2020**, *80*, 103371. [CrossRef]
34. Bousselmi, L.; Fiaud, C.; Tribollet, B.; Triki, E. Impedance spectroscopic study of a steel electrode in condition of scaling and corrosion. *Electrochim. Acta* **1999**, *44*, 4357–4363. [CrossRef]
35. Bousselmi, L.; Fiaud, C.; Tribollet, B.; Triki, E. The characterisation of the coated layer at the interface carbon steel-natural salt water by impedance spectroscopy. *Corros. Sci.* **1997**, *39*, 1711–1724. [CrossRef]
36. Roy, P.; Karfa, P.; Adhikari, U.; Sukul, D. Corrosion inhibition of mild steel in acidic medium by polyacrylamide grafted Guar gum with various grafting percentage: Effect of intramolecular synergism. *Corros. Sci.* **2014**, *88*, 246–253. [CrossRef]
37. Saha, S.K.; Dutta, A.; Sukul, D.; Ghosh, P.; Banerjee, P. Adsorption and corrosion inhibition effect of Schiff base molecules on the mild steel surface in 1 M HCl medium: A combined experimental and theoretical approach. *Phys. Chem. Chem. Phys.* **2015**, *17*, 5679–5690. [CrossRef]
38. Outirite, M.; Lagrenée, M.; Lebrini, M.; Traisnel, M.; Jama, C.; Vezin, H.; Bentiss, F. Ac impedance, X-ray photoelectron spectroscopy and density functional theory studies of 3,5-bis(n-pyridyl)-1,2,4-oxadiazoles as efficient corrosion inhibitors for carbon steel surface in hydrochloric acid solution. *Electrochim. Acta* **2010**, *55*, 1670–1681. [CrossRef]
39. Paolinelli, L.; Perez, T.; Simison, S. The effect of pre-corrosion and steel microstructure on inhibitor performance in CO_2 corrosion. *Corros. Sci.* **2008**, *50*, 2456–2464. [CrossRef]
40. Mora-Mendoza, J.; Turgoose, S. Fe3C influence on the corrosion rate of mild steel in aqueous CO_2 systems under turbulent flow conditions. *Corros. Sci.* **2002**, *44*, 1223–1246. [CrossRef]

41. Ding, Y.; Brown, B.; Young, D.; Singer, M. Effectiveness of an Imidazoline-Type Inhibitor Against CO_2 Corrosion of Mild Steel at Elevated Temperatures (120 °C–150 °C). In Proceedings of the CORROSION 2018, Phoenix, AZ, USA, 15–19 April 2018; p. 22.
42. Heuer, J.; Stubbins, J. An XPS characterization of $FeCO_3$ films from CO_2 corrosion. *Corros. Sci.* **1999**, *41*, 1231–1243. [CrossRef]
43. Boumhara, K.; Tabyaoui, M.; Jama, C.; Bentiss, F. Artemisia Mesatlantica essential oil as green inhibitor for carbon steel corrosion in 1M HCl solution: Electrochemical and XPS investigations. *J. Ind. Eng. Chem.* **2015**, *29*, 146–155. [CrossRef]
44. Bouanis, M.; Tourabi, M.; Nyassi, A.; Zarrouk, A.; Jama, C.; Bentiss, F. Corrosion inhibition performance of 2,5-bis(4-dimethylaminophenyl)-1,3,4-oxadiazole for carbon steel in HCl solution: Gravimetric, electrochemical and XPS studies. *Appl. Surf. Sci.* **2016**, *389*, 952–966. [CrossRef]
45. Hashim, N.Z.N.; Anouar, E.H.; Kassim, K.; Zaki, H.M.; Alharthi, A.I.; Embong, Z. XPS and DFT investigations of corrosion inhibition of substituted benzylidene Schiff bases on mild steel in hydrochloric acid. *Appl. Surf. Sci.* **2019**, *476*, 861–877. [CrossRef]
46. Barker, R.; Burkle, D.; Charpentier, T.; Thompson, H.; Neville, A. A review of iron carbonate ($FeCO_3$) formation in the oil and gas industry. *Corros. Sci.* **2018**, *142*, 312–341. [CrossRef]
47. Messali, M.; Lgaz, H.; Dassanayake, R.; Salghi, R.; Jodeh, S.; Abidi, N.; Hamed, O. Guar gum as efficient non-toxic inhibitor of carbon steel corrosion in phosphoric acid medium: Electrochemical, surface, DFT and MD simulations studies. *J. Mol. Struct.* **2017**, *1145*, 43–54. [CrossRef]
48. Williams, D.N.; Gold, K.A.; Holoman, T.R.P.; Ehrman, S.H.; Wilson, O.C. Surface Modification of Magnetic Nanoparticles Using Gum Arabic. *J. Nanopart. Res.* **2006**, *8*, 749–753. [CrossRef]
49. Leong, Y.; Seah, U.; Chu, S.; Ong, B. Effects of Gum Arabic macromolecules on surface forces in oxide dispersions. *Colloids Surf. A Physicochem. Eng. Asp.* **2001**, *182*, 263–268. [CrossRef]

6

Cyclic Stress Response Behavior of Near β Titanium Alloy and Deformation Mechanism Associated with Precipitated Phase

Siqian Zhang [1], Haoyu Zhang [1,*], Junhong Hao [1], Jing Liu [1], Jie Sun [2] and Lijia Chen [1]

[1] School of Materials Science and Engineering, Shenyang University of Technology, Shenyang 110870, China; sqzhang@alum.imr.ac.cn (S.Z.); wangronnie2016@163.com (J.H.); zwtsysmm@163.com (J.L.); chenlijia@sut.edu.cn (L.C.)

[2] State Key Laboratory of Rolling and Automation, Northeastern University, Shenyang 110004, China; sunjie@ral.neu.edu.cn

* Correspondence: zhanghaoyu@sut.edu.cn

Abstract: The cyclic stress response behavior of Ti-3Al-8V-6Cr-4Mo-4Zr alloy with three different microstructures has been systematically studied. The cyclic stress response was highly related to the applied strain amplitude and precipitated phase. At low strain amplitude, the plastic deformation was mainly restricted to soft α phase, and a significant cyclic saturation stage was shown until fracture for all three alloys. At high strain amplitude, three alloys all displayed an initial striking cyclic softening. However, the softening mechanism was obviously difference. Interestingly, a significant cyclic saturation stage was noticed after an initial cyclic softening for alloy aging for 12 h, which could be attributed to the deformation of {332}<113> twin and precipitation of α″ martensite.

Keywords: cyclic stress response; cyclic softening; cyclic saturation; {332}<113> twinning; stress induced α″ martensite

---

1. Introduction

Near β titanium alloys are extensively used in industry due to their high specific strength, adequate ductility, fracture toughness, biocompatibility and excellent corrosion resistance [1–3]. It is generally known that low cycle fatigue behavior is of utmost importance in the selection of engineering materials, and cyclic stress–strain data have been connected with fatigue crack initiation and fatigue crack propagation rate [4–7]. Therefore, in order to ensure engineering applications, the cyclic stress response behavior must be comprehensively characterized.

As titanium alloys are cyclically strained during fatigue, the response stresses generally show an increase or a decrease as the number of cycles increases, which is termed cyclic hardening or cyclic softening, respectively. In practical applications, the cyclic stress response behavior of titanium alloys tends to be remarkably complex, which is highly dependent on the test temperature, applied plastic strain amplitude, strain rate, microstructure and so on [8–11]. For example, it may display an initial cyclic hardening, then an obvious softening, and finally a saturation stage is reached. Or, an obvious cyclic saturation stage is noticed after an initial cyclic softening until fracture.

Generally speaking, the cyclic stress response behavior is influenced by the competitive effect of the back stress and friction stress during fatigue deformation, which is closely related to dislocation motion [12]. For two phase titanium alloys, the softening behavior is due to a decrease in the kinematic component of stress at room temperature. It is controlled by dislocation configuration in the α phase changing from the heterogeneous distribution to the gradual homogenization [13]. On the contrary, at high temperature, the observed cyclic softening behavior is related to the second phase shearing process because of the homogeneously distributed dislocations and the occurrence of cross-slip. During fatigue deformation, the interaction between the dislocation-dislocation or mobile dislocation and the precipitation phase results in hardening [14].

For near β titanium alloys, different heat treatment processes often result in obviously different α/β morphology [15–17]. However, little information is available on the intrinsic connection between the cyclic deformation behavior and the transition of precipitation phases. In addition, it should be emphasized that the formation of α″ is often induced during fatigue deformation [18,19]. Additionally, for titanium alloys, twinning has great effect on maintaining the homogeneous plastic deformation [20–24]. Therefore, it is necessary to study the effect of α″ and twins on cyclic stress response behavior.

2. Materials and Methods

A 300 mm diameter ingot of the Ti-3Al-8V-6Cr-4Mo-4Zr alloy was produced by vacuum arc melting using pure Ti, V-Al alloy, Cr, Zr and Mo as raw materials. The ingot was forged at 850 °C to 55 mm diameter cylindrical bars, and then hot rolled at 800 °C to 16 mm in diameter rods. The chemical compositions results are given in Table 1. According to the previous studies [25], after solution treated at 800 °C for 0.5 h (AC)+aging at 500 °C, the alloys have a good resistance/ductility combination. So, samples were heat treated at 800 °C for 0.5 h and cooled in air. Then some of these specimens were aged at 500 °C for 4, 8, 12 and 24 h, respectively, and cooled in air.

Table 1. Chemical composition of alloys (wt.%).

Alloy	Al	V	Cr	Mo	Zr	Ti
Ti-3Al-8V-6Cr-4Mo-4Zr	2.9	7.8	6.1	3.9	4.0	Bal.

Uniaxial tensile properties were tested at room temperature (~25 °C) and atmosphere using a rectangular specimen with cross section 2.0 mm × 3.0 mm and a gage length 13 mm. The LCF specimens with a gage length 20 mm and diameter 5 mm were processed, ground and polished. The LCF tests under total strain controlled were conducted at room temperature (~25 °C) and atmosphere using an MTS landmark 370.10 servohydraulic test system (MTS, Eden Prairie, Minnesota, USA) with the strain ratios (R) of −1 and a frequency of 0.5 Hz.

The TEM slices were mechanically thinned down to a thickness of approximately 50 μm. Discs of 3 mm in diameter were punched out of the thin sheets and electro-polished with 60 mL perchloric acid, 85 mL n-butanol and 150 mL methanol at temperature −33 °C. Microstructural observations were conducted with SEM (Zeiss Gemini 500, Heidenheim, Germany) and TEM (JEOL JEM-2100, Tokyo, Japan).

3. Results

3.1. Microstructure before Fatigue Deformation

Figure 1 shows TEM microstructure of the alloy with three different heat treatment processes. It can be found that α phase was precipitated from β grains after heat treatment. With increasing the ageing time, the quantity of α phase decreased (volume fraction), but the size increased gradually. Moreover, the distance between α phase was also increasing.

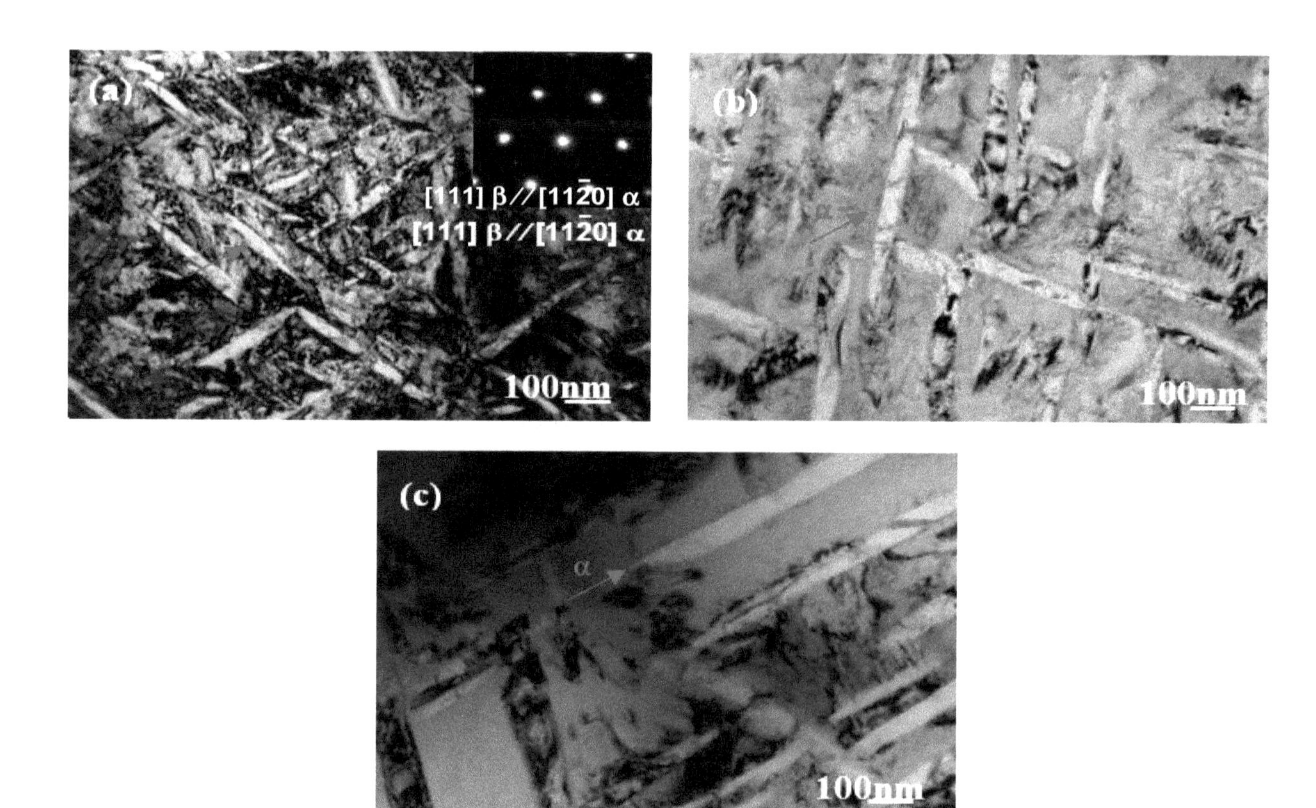

Figure 1. Microstructures of the alloy in different heat treatment conditions, (**a**) 800 °C/30 min + 500 °C/4 h (**b**) 800 °C/30 min + 500 °C/12 h (**c**) 800 °C/30 min + 500 °C/24 h.

3.2. Tensile Properties

The mechanical properties of the alloy with different heat treatment microstructures were evaluated first by tensile tests, and the results are shown in Table 2. As the aging time increased, both the ultimate tensile strength (σ_b) and yield strength ($\sigma_{0.2}$) increased significantly and then decreased rapidly, but the ductility almost stayed the same. Therefore, we can confirm that the alloy possessed the best tensile property after 12 h aging.

Table 2. Tensile properties of the alloy at room temperature.

Heat Treatment States	σ_b/MPa	$\sigma_{0.2}$/MPa	φ(Elongation at Maximum Tensile Strength)	ψ(Reduction of Area)
800 °C/30 min + 500 °C/4 h	1205.2	1137.8	12.98%	16.94%
800 °C/30 min + 500 °C/12 h	1452.1	1436.9	12.75%	15.85%
800 °C/30 min + 500 °C/24 h	1115.6	1088.6	12.67%	15.18%

3.3. Strain Amplitude in Fatigue

It can be seen from Figure 2 that the three curves (total, elastic and plastic strains) are linear on log–log scale. There is an intersection of elastic and plastic strain-life curves, called as the transition fatigue life 2Nt, the life at which elastic and plastic regions of strain are equal. For the alloy after aging, at the lower cycle region when 2Nf ≤ 2Nt, the plastic strain plays a main role and fatigue properties are dominated by strength, while at the higher cycle region when 2Nf ≥ 2Nt, the elastic strain plays a main role and fatigue properties are dominated by ductility.

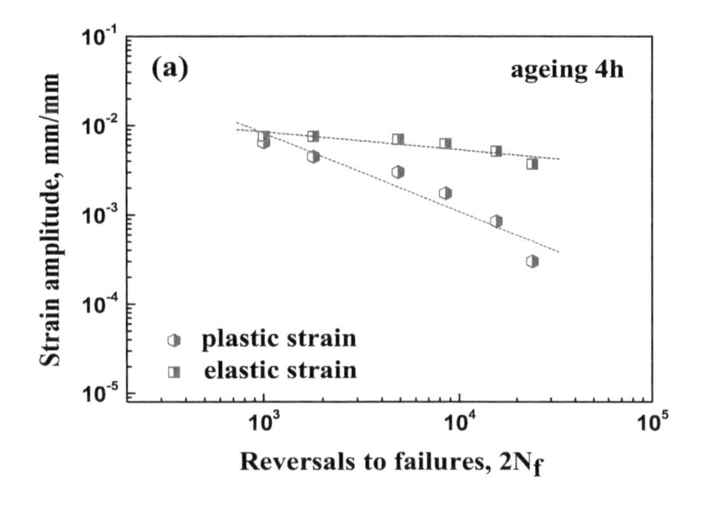

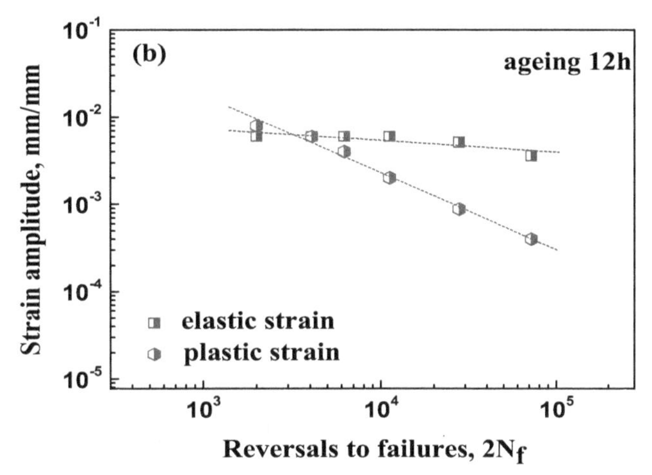

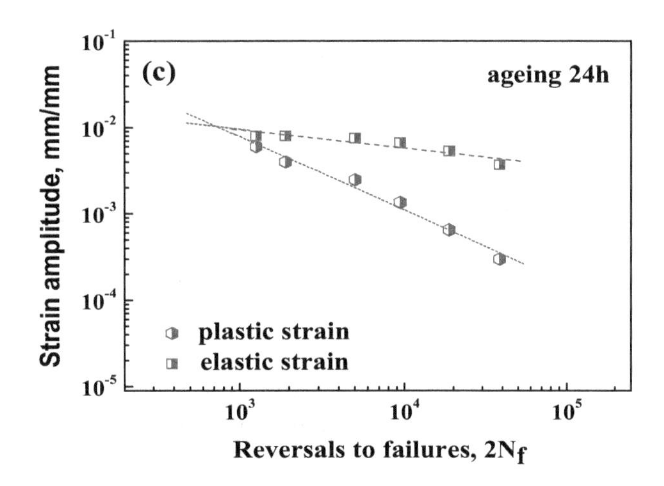

Figure 2. Strain amplitude versus reversals to failure curves for Ti-3Al-8V-6Cr-4Mo-4Zr alloys in different state. (**a**) Aging after 4 h, (**b**) aging after 12 h, (**c**) aging after 24 h.

3.4. Cyclic Stress Response Behaviour

Strain-controlled low cycle fatigue behaviors of the alloys with different heat treatment microstructures were studied at $R = -1$ and the frequency of 0.5 Hz at room temperature and atmosphere. The cyclic stress response curves of alloy with different aging times are showed in Figure 3. It can be found that fatigue life for each alloy decreased as the strain amplitude increases. The cyclic stress response behavior was highly dependent on aging time and strain amplitude.

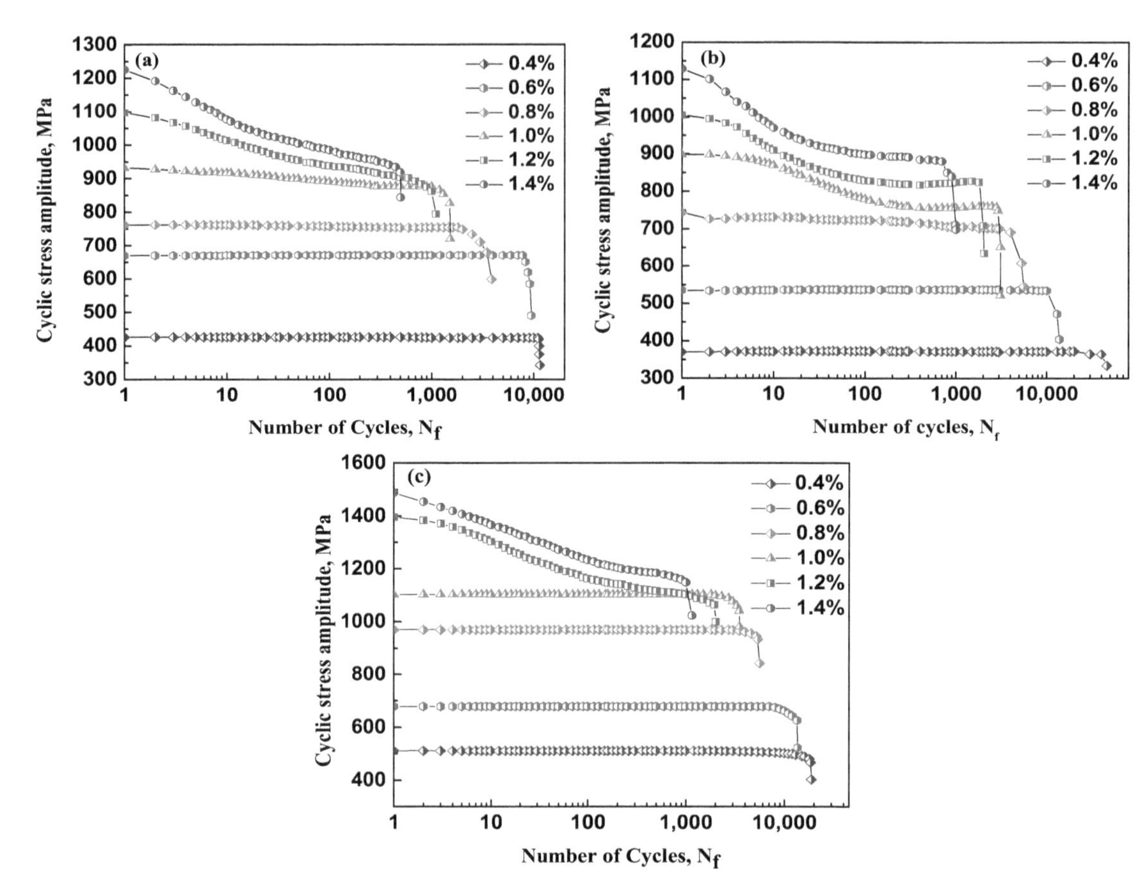

Figure 3. Cyclic stress response curves of the alloy after aging treatment. (**a**) 4 h, (**b**) 12 h (**c**) 24 h.

(a) At low total strain amplitude (alloys aging for 4 h and 24 h: $\Delta\varepsilon/2 \leq 1.0\%$, alloy aging for 12 h: $\Delta\varepsilon/2 \leq 0.8\%$), cyclic saturation was exhibited, followed by a short stage of softening.

(b) At high total strain amplitude (alloys aging for 4 h and 24 h: $\Delta\varepsilon/2 > 1.0\%$, alloy aging for 12 h: $\Delta\varepsilon/2 > 0.8\%$), the alloy generally exhibited a rapid cyclic softening. It is worth mentioning that, for the alloy aging for 12 h at $\Delta\varepsilon/2 \geq 1.0\%$, cyclic saturation stage after the initial softening was reached and extended through the last part of the fatigue life.

3.5. Microstructure after Fatigue Deformation

Microstructures corresponding to the different heat treatment alloy after the low cycle fatigue (LCF) were characterized by TEM analysis. Figure 4 is a TEM microstructure of alloy after LCF under 0.4% total strain amplitude. From the picture, it can be noticed that the dislocation density increased obviously after LCF deformation, compared to heat treatment microstructure. Figure 5 shows the TEM microstructure of alloy after LCF under 1.4% total strain amplitude. Compared with the microstructure under 0.4% total strain amplitude, the dislocation density increased further, especially for alloy aging for 4 h (Figure 5a). In addition, through a lower magnification overview of the microstructure (Figure 5b), it can be also found that the alloy aging for 12 h consists of nano-size twins which were locally distributed. The selected-area taken from the red circle region marked in Figure 5b is presented in Figure 5c, from which we can determined that the ~5 nm twin plate is a typical {332}<113> twin pattern. Interestingly, some extra reflections which can be inferred as the α'' martensite structure were also visible (Figure 5d), in addition to the formation of {332}<113> twin. It can be seen that the stress induced the formation of α'' martensite is located on the one side of the twin boundary. Moreover, a detwinning process also took place after LCF (Figure 5e). Some parts of the twins decreased obviously as the size of α'' martensite was increased. Interestingly, the twins might transform back into the matrix, thus significantly decreasing the area of the twin interface i.e., decreasing of the total interfacial energy, which might be the driving force of the process of detwinning. The {332}<113> twins partly transformed into the matrix and the α″ martensite at the shared interface took place of the detwinning area (Figure 5e). These observations indicated that the formation of stress induced α″ martensite might have undergone a complex process, and the nature of the phenomenon needs to be investigated deeply later. As to the alloy aging for 4 h and 24 h under 1.4% total strain amplitude, With the increase of dislocation density, it is easy to produce multi-system slip and reduce the effective stress, thus showing the trend of cyclic softening(Figure 5a,f).

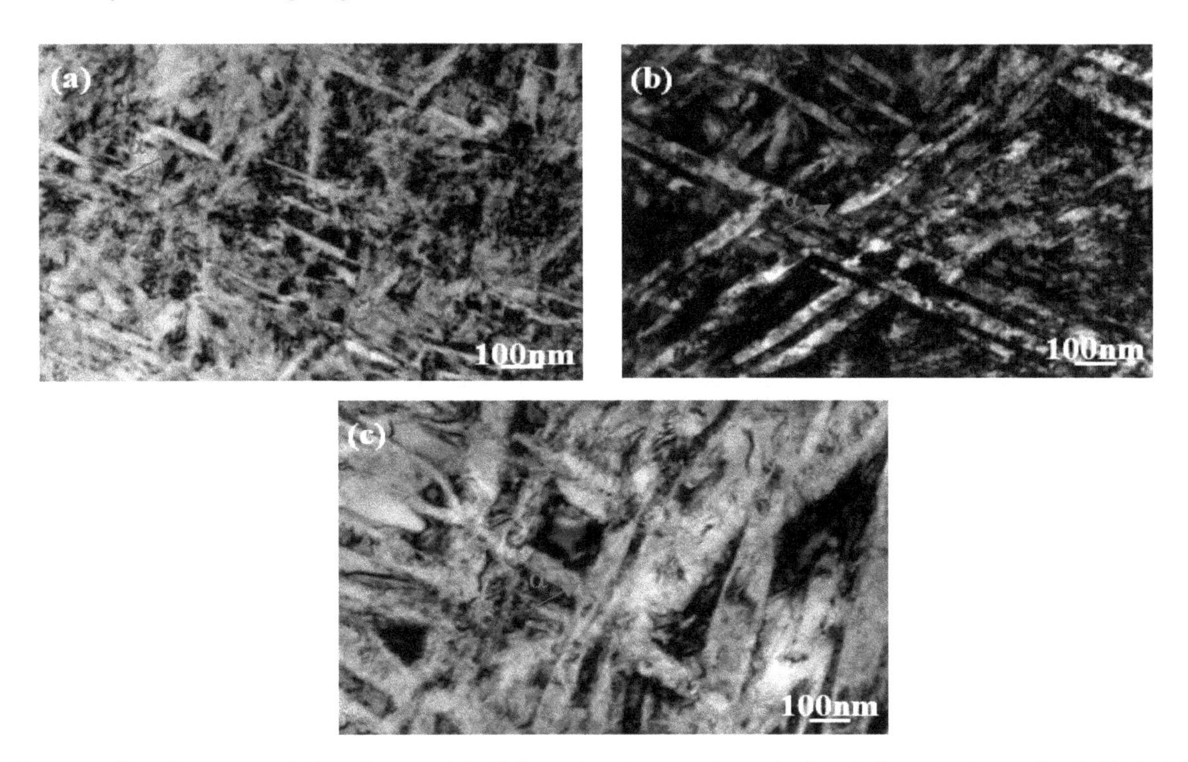

Figure 4. TEM images of the alloys with different aging treatment after fatigue tests under 0.4% total strain range. (**a**) 4 h, (**b**) 12 h and (**c**) 24 h.

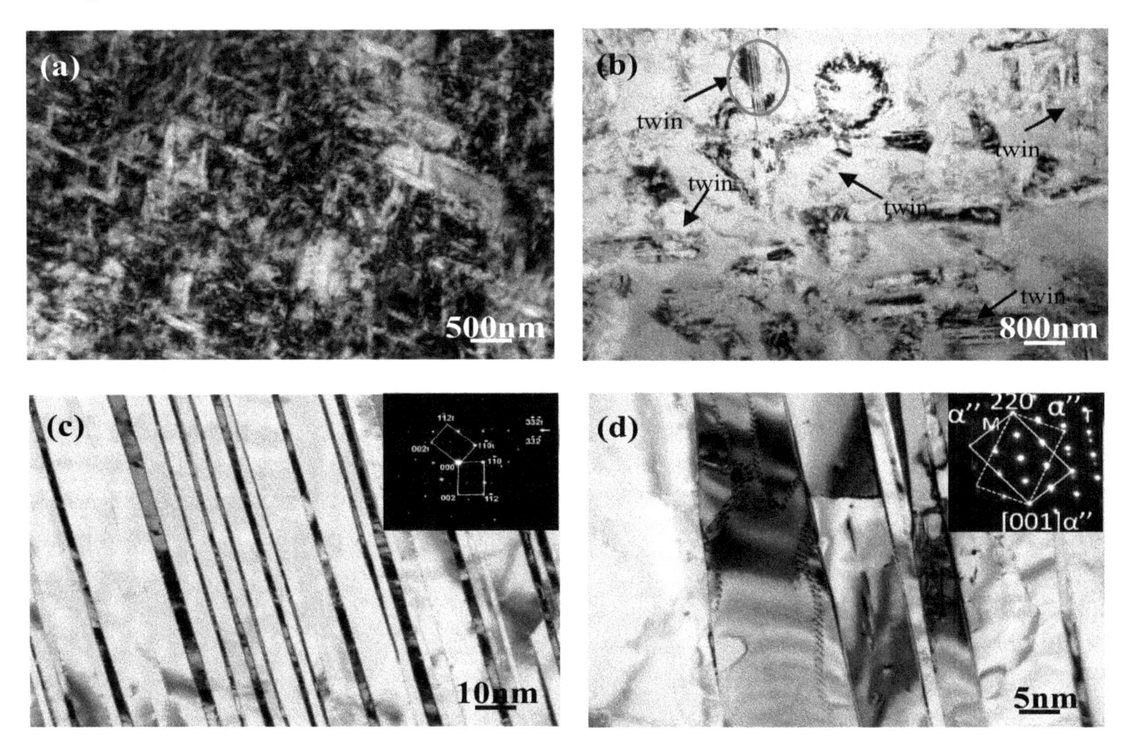

Figure 5. *Cont.*

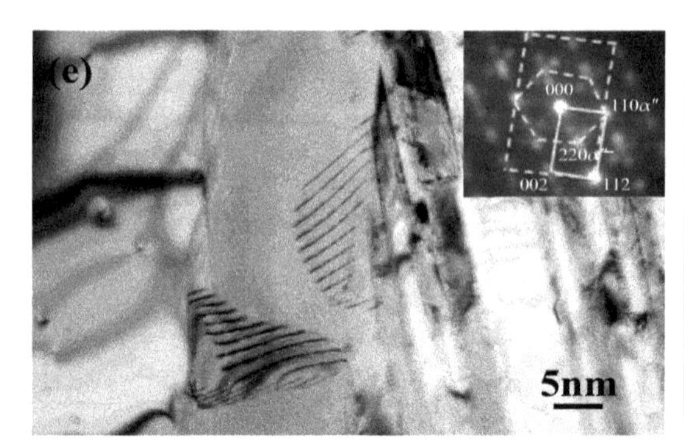

Figure 5. TEM images of the alloys with different aging treatment after fatigue tests under 1.4% total strain range. (**a**) 4 h, (**b–e**) 12 h and (**f**) 24 h.

4. Discussion

According to the above results, we can infer that the cyclic stress response behavior of near β titanium alloy was highly dependent on the applied plastic strain amplitude and precipitated phase.

At low strain amplitude, an apparently cyclic saturation stage was exhibited until fracture for all three alloys. It appeared that the plastic deformation was restricted to the soft α phase, with the hard β-phase staying in its elastic domain. Therefore, it can be considered that the dislocations movement in matrix was negligible due to the large quantity of α/β interfaces which hinder gliding. It is interesting, for alloys aged for 24 h possess a perfect cyclic stability up to a total strain amplitude of $\Delta\varepsilon/2 = 1.0\%$ and stress amplitude up to 1100 MPa. It could be because the distance between α phase was relatively large, and α″ martensite and twins were not easily precipitated for alloys aged for 24 h.

At high strain amplitude, three alloys all displayed an initial striking cyclic softening. However, the softening mechanism may be obviously different. For alloys aging for 4 h and 24 h, the softening behavior was closely related to mobile dislocation. It was well known that prismatic slip could be activated at high cyclic strain amplitude. The continuous softening was controlled by the gradual homogenization of the initially heterogeneous distribution of the dislocations in the α phase in association with cross-slip. The cross-slip activation at room temperature may be related to the presence of additional elements in the α phase and to a high stacking fault energy associated with the prismatic slip [11]. In addition, for alloys aging for 12 h, the formation of the twin due to strain incompatibility might also have effect on the initial softening response of the samples during cycling. It was noticed that the microstructure evolution and cyclic stress response were connected with the {332}<113> twins in this study. As the twins formed and grew, the trend of the cyclic stress curve, shown in Figure 4, might be well explained by softening the mechanism in terms of a crystallographic orientation of a deforming crystal structure. Earlier reports indicated that deformation twinning can induce the structural softening in a twinned region, since texture evolution was associated with the lattice reorientation [26]. TEM studies have indicated that slip precedes twin formation [27].

Interestingly, for alloy aging for 12 h, a significantly cyclic saturation stage was noticed after an initial cyclic softening. This would imply that there was a balance between the counteracting effects of the softening and hardening mechanism. According to the TEM microstructure, we can infer that the precipitation of α″ martensite was the main reason for the hardening. As the cycling proceeds, the formation of α″ pined the dislocations and further increased the stress. It can be said that the irregular twin boundary assists the transformation of α″ martensite through lattice reorientation [28,29]. From Figure 4, it can be found that some of the twins partly transformed into the matrix and α″ martensite at the shared interface took the place of the detwinning area due to the decrease in strain energy. During the LCF process, the residual stress near the twin region might be partially released and the lattice strain around the twin would gradually relax. Furthermore, the twins make opposite contributions to the macroscopic strains. As deformation progresses, the stress can be released by local compatible deformation. Under an external applied stress, a crystal with a twinned part has to make a positive contribution with respect to the macroscopic strain field to be compatible with its surroundings, which gives rise to the occurrence of detwinning [30]. Then, the SIM at the Shared interface took place of the detwinning area. The cyclic saturation stress in the alloy aging for 12 h ($\Delta\varepsilon/2 > 0.8\%$) which associates with the increased strain hardening caused by SIM can be clarified by dynamic Hall–Petch theory [31]. The formation of α'' martensite rather adjoined with the twin boundaries obviously decreased the effective dislocation glide distance, and β grains could be segmented into smaller zones by the deformation of the SIM phase and the phase boundaries acted as impenetrable barriers for the dislocation glide [32]. Figure 4 shows that many dislocations slip into α'' martensite and form an immovable dislocation array, and the movable dislocation decreases with the cyclic stress, which provides support for the enhanced strain hardening, and the cooperation between α'' martensite and twins causes the formation of cyclic stress stability.

5. Conclusions

In the present work, the cyclic stress response behavior of Ti-3Al-8V-6Cr-4Mo-4Zr alloy with three different microstructures has been studied. The following main conclusions can be drawn in this work:

(1) The alloy aging for 12 h possessed the optimal α/β morphology and exhibited the best tensile and LCF properties.
(2) At low strain amplitude, an obvious cyclic saturation stage was revealed until fracture for all three alloys.
(3) At high strain amplitude, three alloys all displayed an initial striking cyclic softening. However, for alloy aging for 12 h, a cyclic saturation stage was obtained after an initial cyclic softening.
(4) The deformation of {332}<113> twin and precipitation of α'' martensite were found to have a crucial influence on the cyclic stress response behavior.

Author Contributions: Contributed materials and performed the experiments, J.L., H.Z., J.H. and L.C.; designed the experiments, J.S.; analyzed the data and wrote the paper, S.Z. All authors have read and agreed to the published version of the manuscript.

References

1. Koizumi, H.; Takeuchi, Y.; Imai, H.; Kawai, T.; Yoneyama, T. Application of titanium and titanium alloys to fixed dental prostheses. *J. Prosthodont. Res.* **2019**, *63*, 266–270. [CrossRef] [PubMed]
2. Hong, S.H.; Hwang, Y.J.; Park, S.W.; Park, C.H.; Yeom, J.-T.; Park, J.M.; Kim, K. Low-cost beta titanium cast alloys with good tensile properties developed with addition of commercial material. *J. Alloys Compd.* **2019**, *793*, 271–276. [CrossRef]
3. Bin Asim, U.; Siddiq, A.; Kartal, M.E. A CPFEM based study to understand the void growth in high strength dual-phase titanium alloy (Ti-10V-2Fe-3Al). *Int. J. Plast.* **2019**, *122*, 188–211. [CrossRef]
4. Srivatsan, T.; Soboyejo, W.; Lederich, R. The cyclic fatigue and fracture behavior of a titanium alloy metal matrix composite. *Eng. Fract. Mech.* **1995**, *52*, 467–491. [CrossRef]
5. Grosdidier, T.; Combress, Y.; Gautier, E.; Philippe, M.J. Effect of microstructure variations on the formation of deformation-induced martensite and associated tensile properties in a b metastable Ti alloy. *Metall. Mater. Trans. A* **2000**, *31A*, 1095–1106. [CrossRef]
6. Hua, K.; Zhang, Y.; Gan, W.; Kou, H.; Beausir, B.; Li, J.; Esling, C. Hot deformation behavior originated from dislocation activity and β to α phase transformation in a metastable β titanium alloy. *Int. J. Plast.* **2019**, *119*, 200–214. [CrossRef]
7. Sun, Q.Y.; Song, S.J.; Zhu, R.H.; Gu, H.C. Toughening of titanium alloys by twinning and martensite transformation. *J. Mater. Sci.* **2002**, *37*, 2543–2547. [CrossRef]
8. Gouthama, N.S.; Singh, V. Low cycle fatigue behaviour of Ti alloy Timetal 834 at 873K. *Int. J. Fatigue* **2007**, *29*, 843–851. [CrossRef]
9. Zhang, Y.; Chen, Z.; Qu, S.J.; Feng, A.; Mi, G.; Shen, J.; Huang, X.; Chen, D. Microstructure and cyclic deformation behavior of a 3D-printed Ti–6Al–4V alloy. *J. Alloys Compd.* **2020**, *825*, 153971. [CrossRef]

10. Ibrahim, A.M.H.; Balog, M.; Krizik, P.; Novy, F.; Cetin, Y.; Svec, P.; Bajana, O.; Drienovsky, M. Partially biodegradable Ti-based composites for biomedical applications subjected to intense and cyclic loading. *J. Alloys Compd.* **2020**, *839*, 155663. [CrossRef]
11. Luquiau, D.; Feaugas, X.; Clavel, M. Cyclic softenimg of the Ti-10V-2Fe-3Al titanium alloy. *Mater. Sci. Eng. A.* **1997**, *224*, 146–156. [CrossRef]
12. Cottrell, A.H. *Dislocations and Plastic Flow in Crystals*; Oxford University Press: Oxford, London, 1953.
13. Béranger, A.; Feaugas, X.; Clavel, M. Low cycle fatigue behavior of an α + β titanium alloy: Ti6246. *Mater. Sci. Eng. A* **1993**, *172*, 31–41. [CrossRef]
14. Giugliano, D.; Cho, N.-K.; Chen, H.; Gentile, L. Cyclic plasticity and creep-cyclic plasticity behaviours of the SiC/Ti-6242 particulate reinforced titanium matrix composites under thermo-mechanical loadings. *Compos. Struct.* **2019**, *218*, 204–216. [CrossRef]
15. Santos, P.F.; Niinomi, M.; Cho, K.; Liu, H.; Nakai, M.; Narushima, T.; Ueda, K.; Itoh, Y. Effects of Mo Addition on the Mechanical Properties and Microstructures of Ti-Mn Alloys Fabricated by Metal Injection Molding for Biomedical Applications. *Mater. Trans.* **2017**, *58*, 271–279. [CrossRef]
16. Du, Z.; Ma, Y.; Liu, F.; Zhao, X.; Chen, Y.; Li, G.; Liu, G.; Chen, Y. Improving mechanical properties of near beta titanium alloy by high-low duplex aging. *Mater. Sci. Eng. A* **2019**, *754*, 702–707. [CrossRef]
17. Bertrand, E.; Castany, P.; Péron, I.; Gloriant, T. Twinning system selection in a metastable β-titanium alloy by Schmid factor analysis. *Scr. Mater.* **2011**, *64*, 1110–1113. [CrossRef]
18. Lee, B.-S.; Im, Y.-D.; Kim, H.G.; Kim, K.; Kim, W.-Y.; Lim, S.-H. Stress-Induced α″ Martensitic Transformation Mechanism in Deformation Twinning of Metastable β-Type Ti-27Nb-0.5Ge Alloy under Tension. *Mater. Trans.* **2016**, *57*, 1868–1871. [CrossRef]
19. Blackburn, M.J.; Feeny, J.A. Stress-induced transformations in Ti-Mo alloys. *J. Inst. Met.* **1971**, *99*, 132–134.
20. Lai, M.J.; Tasan, C.C.; Raabe, D. On the mechanism of {332} twinning in metastable β titanium alloys. *Acta Mater.* **2016**, *111*, 173–186. [CrossRef]
21. Gao, Y.; Zheng, Y.; Fraser, H.; Wang, Y. Intrinsic coupling between twinning plasticity and transformation plasticity in metastable β Ti-alloys: A symmetry and pathway analysis. *Acta Mater.* **2020**, *196*, 488–504. [CrossRef]
22. Gong, M.Y.; Xu, S.; Capolungo, L.; Tomé, C.N.; Wang, J. Interactions between <a>dislocations and three-dimensional {1122} twin in Ti. *Acta Mater.* **2020**, *195*, 597–610. [CrossRef]
23. Chen, N.; Aldareguia, J.M.M.; Kou, H.C.; Qiang, F.M.; Wu, Z.H.; Li, J.S. Reversion martensitic phase transformation induced {3 3 2}<1 1 3> twinning in metastable β-Ti alloys. *Mater. Lett.* **2020**, *272*, 127883. [CrossRef]
24. Ren, L.; Xiao, W.; Kent, D.; Wan, M.; Ma, C.; Zhou, L. Simultaneously enhanced strength and ductility in a metastable β-Ti alloy by stress-induced hierarchical twin structure. *Scr. Mater.* **2020**, *184*, 6–11. [CrossRef]
25. Cao, S.; Zhou, X.; Lim, C.V.S.; Boyer, R.R.; Williams, J.C.; Wu, X. A strong and ductile Ti-3Al-8V-6Cr-4Mo-4Zr (Beta-C) alloy achieved by introducing trace carbon addition and cold work. *Scr. Mater.* **2020**, *178*, 124–128. [CrossRef]
26. Shin, S.; Zhu, C.Y.; Vecchio, K.S. Observations on {332}<113> twinning- induced softening in Ti-Nb Gum Metal. *Mater. Sci. Eng. A.* **2018**, *724*, 189–198. [CrossRef]
27. De Cooman, B.C.; Estrin, Y.; Kim, S.K. Twinning-induced plasticity (TWIP) steels. *Acta Mater.* **2018**, *142*, 283–362. [CrossRef]
28. Chen, B.; Sun, W. Transitional structure of {332}<113> β twin boundary in a deformed metastable β-type Ti-Nb-based alloy, revealed by atomic resolution electron microscopy. *Scr. Mater.* **2018**, *150*, 115–119. [CrossRef]
29. Castany, P.; Yang, Y.; Bertrand, E.; Gloriant, T. Reversion of a parent {130}<310>α″ martensitic Twinning system at the origin of {332}<113> β twins observed in metastable β titanium alloys. *Phys. Rev. Lett.* **2016**, *117*, 245501. [CrossRef] [PubMed]
30. Qu, L.; Yang, Y.; Lu, Y.F.; Feng, L.; Ju, J.H.; Ge, P.; Zhou, W.; Han, D.; Ping, D.H. A detwinning process of {332}<113> twins in beta titanium alloys. *Scr. Mater.* **2013**, *69*, 389–392. [CrossRef]
31. Idrissi, H.; Renard, K.; Schryvers, D.; Jacques, P. On the relationship between the twin internal structure and the work-hardening rate of TWIP steels. *Scr. Mater.* **2010**, *63*, 961–964. [CrossRef]
32. Cho, K.; Morioka, R.; Harjo, S.; Kawasaki, T.; Yasuda, H.Y. Study on formation mechanism of {332}<113> deformation twinning in metastable β-type Ti alloy focusing on stress-induced α″ martensite phase. *Scr. Mater.* **2020**, *177*, 106–111. [CrossRef]

7

Effect of Tungsten on Creep Behavior of 9%Cr–3%Co Martensitic Steels

Alexandra Fedoseeva, Nadezhda Dudova, Rustam Kaibyshev and Andrey Belyakov *

Laboratory of Mechanical Properties of Nanostructured Materials and Superalloys, Belgorod National Research University, Pobeda 85, Belgorod 308015, Russia; fedoseeva@bsu.edu.ru (A.F.); dudova@bsu.edu.ru (N.D.); rustam_kaibyshev@bsu.edu.ru (R.K.)

* Correspondence: belyakov@bsu.edu.ru

Abstract: The effect of increasing tungsten content from 2 to 3 wt % on the creep rupture strength of a 3 wt % Co-modified P92-type steel was studied. Creep tests were carried out at a temperature of 650 °C under applied stresses ranging from 100 to 220 MPa with a step of 20 MPa. It was found that an increase in W content from 2 to 3 wt % resulted in a +15% and +14% increase in the creep rupture strength in the short-term region (up to 10^3 h) and long-term one (up to 10^4 h), respectively. On the other hand, in the long-term creep region, the effect of W on creep strength diminished with increasing rupture time, up to complete disappearance at 10^5 h, because of depletion of excess W from the solid solution in the form of precipitation of the Laves phase particles. An increase in W content led to the increased amount of Laves phase and rapid coarsening of these particles under long-term creep. The contribution of W to the enhancement of creep resistance has short-term character.

Keywords: martensitic steels; creep; precipitation; electron microscopy

1. Introduction

The heat resistant steels with 9–12%Cr are widely used as structural materials for boilers, main steam pipes, and turbines of fossil power plants with increased thermal efficiency [1,2]. The excellent creep resistance of these steels is attributed to the tempered martensite lath structure (TMLS) consisting of prior austenite grains (PAG), packets, blocks, and laths, and containing a high density of separate dislocations and a dispersion of secondary phase particles [1–5]. Stability of TMLS is provided by $M_{23}C_6$-type carbides and MX (where M is V and/or Nb, and X is C and/or N) carbonitrides precipitated during tempering at boundaries and within ferritic matrix, respectively, and boundary Laves phase particles precipitated during creep [1,2,4–12]. MX carbonitrides, which are highly effective in pinning of lattice dislocations, play a vital role in superior long-term creep resistance of the high chromium martensitic steels, whereas boundary $M_{23}C_6$ carbides and Laves phase particles exerting a high Zener drag force stabilize the TMLS [1,2,5,7–14]. This dispersion of secondary phase particles withstands short-term creep [4,9,15]. The Laves phase provides effective stabilization of TMLS and therefore promotes creep resistance, although their effect on the rearrangement of lattice dislocation is negligibly small [9,11,15]. However, the particles of Laves phase grow with a high rate under creep condition and, at present, these boundary precipitations are considered to be responsible for the creep ductility of the P92-type steels during long-term aging [16]. The enhanced long-term creep strength could be achieved by hindering this microstructural evolution. An effective way to achieve this goal is to slow down the diffusion-controlled processes, such as the climb of dislocations, knitting reaction between dislocation and lath boundaries, particle coarsening, etc., by such substitutional additives as Co, W, and Mo [1,9,17–19].

It was recently shown that cobalt addition significantly hinders the coarsening of $M_{23}C_6$ carbides and MX carbonitrides under creep conditions, which results in the superior creep resistance of martensitic steels [9,13,18,20]. This positive effect of Co is attributed to hindering diffusion within ferrite [9,18]. Efficiency of W as an alloying element in hindering diffusion is much higher than that of Co. As a result, W and Mo are known as effective alloying additives to enhance creep resistance of high chromium martensitic steels [1,19]. These elements provide an effective solid solution strengthening [1]. It was shown [21] that addition of 1% W gives +35 MPa increase in the creep rupture strength at 600 °C for 1000 h. However, in contrast with cobalt, the tungsten and molybdenum have limited solubility within ferrite, and their excessive content leads to precipitation of such W- and Mo-rich particles as Laves phase $Fe_2(W,Mo)$ or M_6C carbides [1,7,9–12,22,23]. This depletion does not occur in the 9%Cr steel containing no or low amount of W [7]. Depletion of solid solution by these elements highly deteriorates the creep resistance [7,10,11]. It is worth noting that, at present, the most of experimental data on the effect of W on creep behavior were obtained for cobalt-free high chromium martensitic steels. The aim of the present work is to report the effect of W addition on the creep strength and microstructure evolution during creep of two 9%Cr martensitic steels containing 3%Co and distinguished by W content.

2. Materials and Methods

Two Co-modified P92-type steels with 2 and 3 wt % W denoted here as the 9Cr2W and the 9Cr3Wsteels, respectively, were produced by air melting as 40 kg ingots. Chemical compositions of these steels, measured by a FOUNDRY-MASTER UVR optical emission spectrometer (Oxford Instruments, Ambingdon, UK) are presented in Table 1.

Table 1. Chemical composition of the steels studied (wt %).

Steel	Fe	C	Cr	Co.	Mo	W	V	Nb	B	N	Si	Mn
9Cr2W	bal.	0.12	9.3	3.1	0.44	2.0	0.2	0.06	0.005	0.05	0.08	0.2
9Cr3W	bal.	0.12	9.5	3.2	0.45	3.1	0.2	0.06	0.005	0.05	0.06	0.2

Square bars with cross-section of 13 × 13 mm^2 were cast and hot-forged in the temperature interval 1150–1050 °C after homogenization annealing at 1100 °C for 1 h by the Central Research Institute for Machine-Building Technology, Moscow, Russia. Both steels were solution-treated at 1050 °C for 30 min, cooled in air, and subsequently tempered at 750 °C for 3 h. Tensile tests were carried out on specimens having a cross section of 1.4 × 3 mm^2 and a 16 mm gauge length using an Instron 5882 Universal Testing Machine (Instron, Norwood, MA, USA) at room temperature and at 650 °C with a strain rate of 2×10^{-3} s^{-1}. Flat specimens with a gauge length of 25 mm and a cross section of 7 × 3 mm^2 (for 220–140 MPa) and cylindrical specimens with a gauge length of 100 mm and a diameter of 10 mm (for 120–80 MPa) were subjected to creep tests until rupture. The creep tests were carried out in the air at 650 °C under different initial stresses ranging from 80 to 220 MPa with a step of 20 MPa. The 100,000 h creep rupture strength was estimated by extrapolation of the experimental data using the Larson–Miller equation [24]:

$$P = T(\lg\tau + 36) \times 10^{-3} \quad (1)$$

where P is the parameter of Larson–Miller; T is temperature, (K); τ is time to rupture, (h).

The structural characterization was carried out using a transmission electron microscope, JEOL-2100, (TEM) (JEOL Ltd., Tokyo, Japan) with an INCA energy dispersive X-ray spectroscope (EDS) (Oxford Instruments, Abingdon, UK) and scanning electron microscope, Quanta 600FEG, (SEM) (FEI, Hillsboro, OR, USA) on ruptured creep specimens in the gauge sections corresponding to uniform deformation in the middle between grip portion and fracture surface. The size distribution and mean radius of the secondary phase particles were estimated by counting of 150 to 250 particles per specimen on at least 15 arbitrarily selected typical TEM images for each data point. The error bars are given according to the standard deviation. Identification of the precipitates was done on the basis of the combination of EDS composition measurements of the metallic elements and indexing of electron diffraction patterns by TEM. The subgrain sizes were evaluated on TEM micrographs by the linear intercept method including all clearly visible (sub)boundaries. The dislocation densities in the grain and subgrain interiors were estimated as a number of intersections of individual dislocations with upper or down foil surfaces per unit area on at least six arbitrarily selected typical TEM images for each data point [25]. The dislocation observation was carried out under multiple-beam conditions with large excitation vectors for several diffracted planes for each TEM image. The W-rich M_6C carbides and Laves phase particles could be clearly distinguished from other precipitates by their bright contrast in the back scattered electron (BSE) image (Z-contrast) [26]. M_6C carbides and Laves phase particles were separated from each other by EDS composition measurements by TEM and particle size distribution [27]. The volume fractions of the precipitated phases were calculated by the Thermo-Calc software (Version 5.0.4 75, Thermo-Calc software AB, Stockholm, Sweden, 2010) using the TCFE7 database for the following compositions of steels (in wt %): 0.1%C-9.4%Cr-0.5%Mo-2.0 (or 3.0)%W-3.0%Co-0.2%V-0.05%Nb-0.05%N-0.005%B and Fe-balance. The following phases were chosen independently for calculation: austenite (FCC_A1), ferrite (BCC_A2), cementite, $M_{23}C_6$ carbide, M_7C_3 carbide, M_6C carbide, and Laves phase (Fe_2(W, Mo) (C14).

3. Results

3.1. Tempered Martensite Lath Structure

After tempering at 750 °C, TMLS forms in both steels. However, in the 9Cr2W steel, the additional formation of subgrains was observed (Figure 1a), whereas TMLS is dominant in the 9Cr3W steel (Figure 1b). The average sizes of PAGs were 11 and 20 μm for the 9Cr2W and 9Cr3W steels, respectively. More details about the microstructure of the steels studied after normalization at 1050 °C and tempering at 750 °C can be found elsewhere [27].

Figure 1. Mixed lath structure and subgrain one (**a**) in the 9Cr2W steel and homogeneous tempered martensite lath structure in the 9Cr3W steel (**b**) after normalization at 1050 °C and tempering at 750 °C.

The lath thickness was approximately 0.4 μm for both steels. The high dislocation density of approximately 2×10^{14} m^{-2} was observed within the lath and subgrain interiors. In the structure of both steels, $M_{23}C_6$ carbides located on the boundaries of PAGs, packets, blocks, and laths, and MX-type carbonitrides uniformly distributed within the martensitic laths were observed. The mean size of $M_{23}C_6$ carbides was about 90 nm. V-rich MX carbonitrides with a "wing" shape [8,15] have a mean longitudinal size of 20 nm. Nb-rich MX carbonitrides with a round shape have an average size of 40 nm. Dimensions of these particles in both steels were the same. The W-rich precipitates of M_6C carbide (Fe_3W_3C) and Laves phase (Fe_2W) were found in the 9Cr3W steel alongside the $M_{23}C_6$ and MX particles [27]. Therefore, the solubility of (3%W + 0.5%Mo) excesses the thermodynamically equilibrium solubility limit even at the tempering temperature of 750 °C. The 1 wt %W additive provides the precipitation of the W-rich M_6C carbides and Laves phase even under tempering. No formation of thermodynamically stable W-rich Laves phases was reported in the conventional 9%Cr martensitic steels after tempering [1,2,7–12], and the appearance of the less stable W-rich M_6C carbides was found only in a 10%Cr–2%W steel [23]. Under tempering, the partial transformation of M_6C carbides into Laves phase particles may occur if M_6C carbides are occupied by other M_6C_6 carbides and do not have access to W segregation in the vicinity of PAG–lath boundaries [27].

3.2. Tensile Test

The W effect on engineering stress-strain curves is shown in Figure 2, and yield stress (YS), ultimate tensile stress (UTS), and ductility δ are summarized in Table 2.

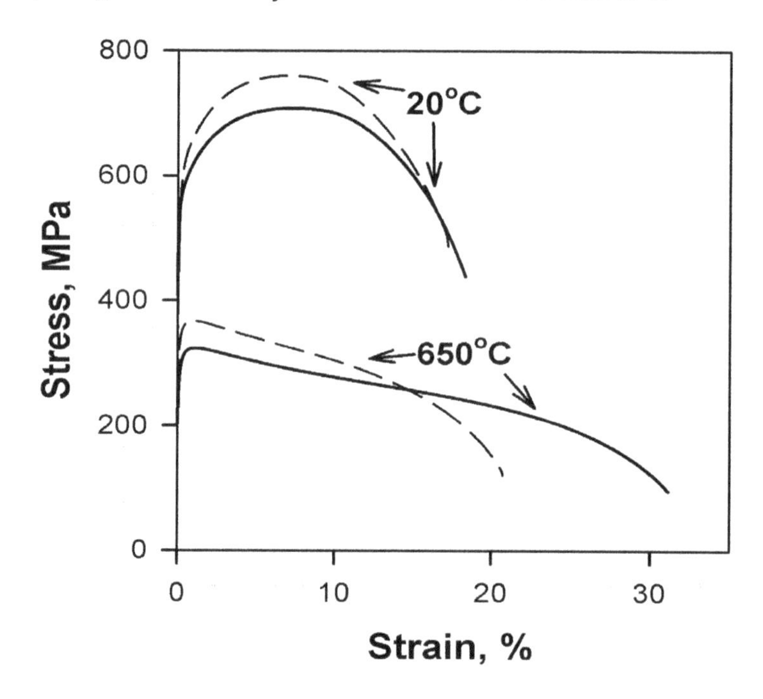

Figure 2. Tensile behavior of the 9Cr2W (solid lines) and 9Cr3W (dash lines) steels after heat treatment consisting of normalizing at 1050 °C and tempering at 750 °C. Tensile tests were carried out at room temperature and at 650 °C (creep test temperature).

Table 2. Values of yield stress (YS), ultimate tensile stress (UTS), and ductility (δ), obtained under tension, at room temperature and 650 °C for the 9Cr2W and the 9Cr3W steels.

Steel	Temperature Test	YS, MPa	UTS, MPa	δ, %
9Cr2W	20 °C	560	708	19
	650 °C	295	325	32
9Cr3W	20 °C	570	760	18
	650 °C	340	370	21

The shapes of the engineering σ–ε curves at room and elevated test temperatures for both steels were nearly the same, whereas YS and UTS are higher and δ is smaller for the 9Cr3W steel. The σ–ε curves at elevated test temperature showed the continuous yielding for both steels. After a short stage of extensive strain hardening, the apparent steady-state flow appeared and occurred up to necking. Next, post-uniform necking elongation took place up to fracture. Ductility at room test temperature was nearly the same for both steels, while at elevated temperature ductility of the 9Cr2W steel was +52% more than for 9Cr3W steel. Increments of +2% and +7% in YS and UTS at room temperature and of +15% and +14% in YS and UTS at elevated test temperature were observed for the steel with increased W content. It is obvious that these increments are provided by solid solution strengthening due to increasing W content up to 3%.

3.3. Creep Behavior

Figure 3a shows the creep rupture data of the steels at a temperature of 650 °C. In general, the creep rupture time of both Co-modified steels is significantly longer in comparison with the P92 steel, that is indicative of the positive effect of Co on the creep strength [1,5,15,28].

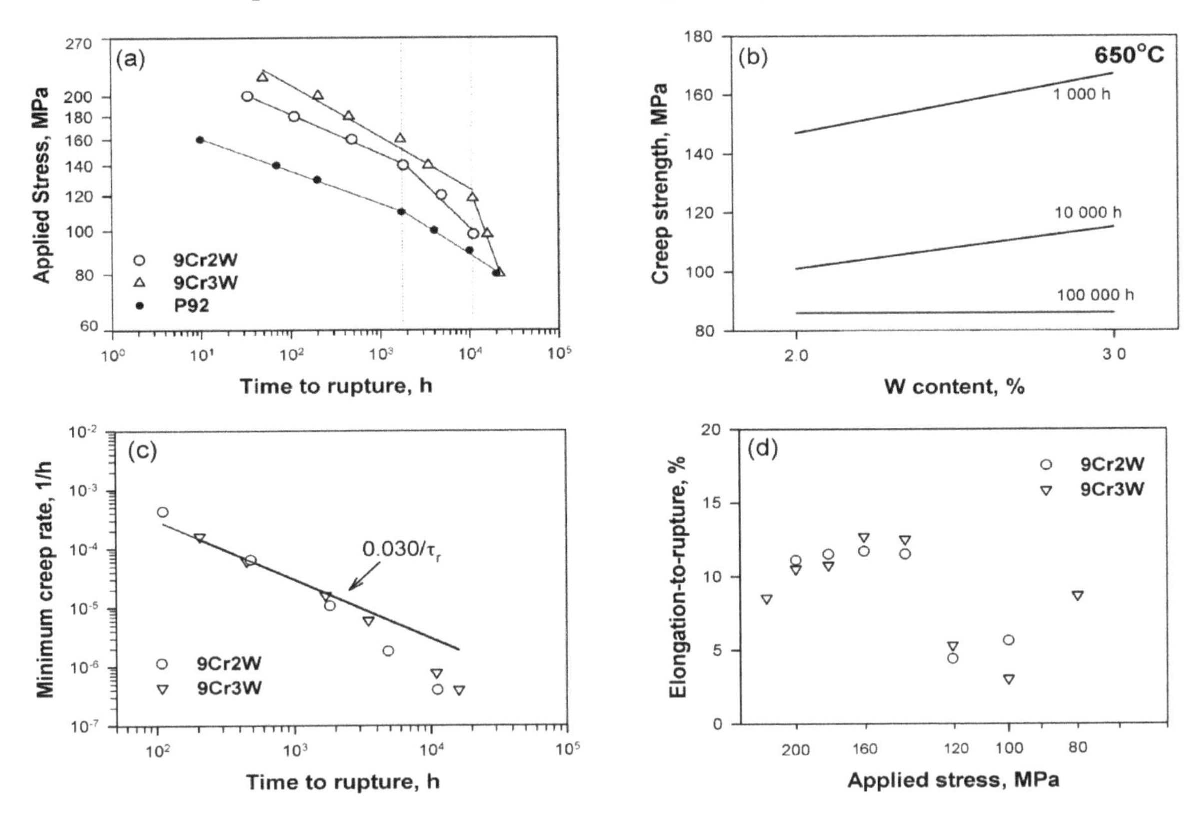

Figure 3. (**a**) Time to rupture vs. stress curves for the 9Cr2W and 9Cr3W steels in comparison with data for a P92 steel [2,7]; (**b**) the 1000 h, 10,000 h, and 100,000 h creep rupture strengths of steels at 650 °C as a function of W content; (**c**) minimum creep rate as a function of time to rupture for the 9Cr2W and 9Cr3W steels; and (**d**) applied stress vs. elongation-to-rupture for the 9Cr2W and 9Cr3W steels. The dotted lines in (**a**) indicate the time to rupture corresponding to the creep strength breakdown.

In the short-term region up to 10^3 and 10^4 h, the creep strength increase is +15% and +14% from 145 to 167 MPa and from 101 to 115 MPa, respectively, due to increased W content from 2 to 3 wt %. In the long-term creep region, the effect of W additives on the creep strength tends to diminish. For both steels the 100,000 h creep rupture strength of ~85 MPa predicted through the Larson–Miller parameter (Figure 3b) is nearly the same, and, therefore, the positive effect of W disappears. However, this value is 15% higher than the creep rupture strength of ~72 MPa for the P92 steel predicted through the Larson–Miller parameter from data published in previous works [28,29].

It was recently shown that creep strength breakdown is a tertiary creep phenomenon [7,13]. The Monkman–Grant relationship relating the rupture time, τ_r, to the minimum or steady-state creep rate is described as the Equation (2):

$$\tau_r = (c'/\dot{\varepsilon}_{\min})^{m'}, \tag{2}$$

where c' and m' are constants. This relationship is used for the prediction of creep life of heat-resistant steels [1,13]. Analysis of Equation (2) for the studied steels (Figure 3c) shows that this approach is suitable for describing the relation of rupture time with the offset strain rate, $\dot{\varepsilon}_{\min}$. For short-term conditions, which corresponds to τ_r less than approximately 2000–3500 h, τ_r is inversely proportional to $\dot{\varepsilon}_{\min}$ (Figure 3c) at $m' = 1$ [13]. The constant c' is 3.0×10^{-2}. For long-term conditions, which correspond to a τ_r greater than approximately 2000–3500 h, the relationship between τ_r and $\dot{\varepsilon}_{\min}$

deviates downward. The transition from short-term creep to a long-term one appears as the deviation from the linear dependence described by Equation (2), which indicates $m' < 1$ [13]. Loss of ductility occurs at low stresses of 120–100 MPa in both steels (Figure 3d). For the 9Cr2W steel, a decrease in the elongation-to-rupture correlates with the creep strength breakdown appearance in Figure 3a, while for the 9Cr3W steel, the changes in elongation-to-rupture have irregular character. For high stresses from 220 to 140 MPa, elongation-to-rupture increases from 8% to 13%, then remarkably reduces to 3–5% at low stresses of 120 and 100 MPa, and then increases to 8% at 80 MPa. The relation of loss of ductility and creep strength breakdown is not revealed for the 9Cr3W steel. In contrast with the dependencies of the applied stress vs. the rupture time (Figure 3a), there was no distinct inflection point for the transition from the short-term region to the long-term one by the shapes of aforementioned curves. The minimum creep rate decreased from approximately 10^{-7} to 10^{-10} s^{-1} with a decrease in the applied stress from 220 to 100 MPa. There was a linear dependence between the minimum creep rate and the applied stress (Figure 4). The experimental data obey a power law relationship throughout the whole range of the applied stress of the usual form [1,7,13]:

$$\dot{\varepsilon}_{\min} = A \times \sigma^n \exp\left(\frac{-Q}{RT}\right), \quad (3)$$

where $\dot{\varepsilon}_{\min}$ is the minimum creep rate, σ is the applied stress, Q is the activation energy for a plastic deformation, R is the gas constant, T is the absolute temperature, A is a constant, and n is the "apparent" stress exponent.

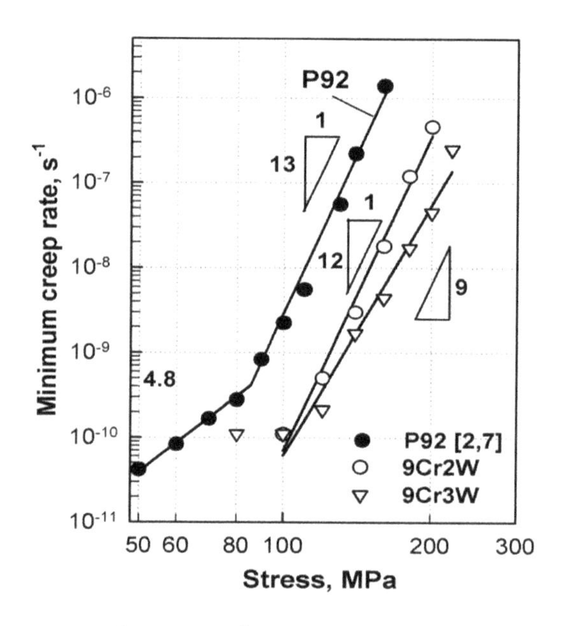

Figure 4. Minimum creep rate as a function of time to rupture for the 9Cr2W and 9Cr3W steels in comparison with data for the P92 steel [2,7].

For the applied stresses from 220 to 100 MPa, these plots provide the best linear fit with a regression coefficient of 0.98 for $n = 12$ and 9 for the 9Cr2W and 9Cr3W steels, respectively. This n value at all tested stress regimes remains constant. The steady-state creep of the 9Cr2W and 9Cr3W steels was controlled using the same process for the short- and long-term regions at a creep rate ranging from 10^{-6} to 10^{-10} s^{-1}. This clearly indicates that there is no creep strength breakdown during the steady-state creep of both steels [7,13]. However, for the 9Cr3W steel, minimum creep rate at 80 MPa is similar with that at 100 MPa, which indicates the change in the process that controls the steady-state creep.

3.4. Crept Microstructures

SEM-BSE and TEM images of both steels after creep rupture tests are shown in Figure 5.

Laves phase, enriched by W and Mo, could be distinguished as the white particles in Z-contrast, and Cr-enriched $M_{23}C_6$ carbides as the grey particles on the grey background of matrix [10,30]. The short-term creep tests (≤2000 h) did not change the lath structure in both steels (Figure 5a,b). W-rich particles exhibited nearly round shape. After the long-term creep rupture tests (≥2000 h) the lath thickness increased (Figure 5c,d), and transformation of the lath structure into subgrain structure took place in both steels. The well-defined subgrain structure evolved in the 9Cr2W steel, only. TEM studies support this conclusion (Figure 5e,f). In the 9Cr2W steel, the subgrains rapidly grew from 0.6 to 1.5 μm after 2000 h (Figure 6a), whereas in the 9Cr3W steel, this size insignificantly changed from 0.6 to 0.7 μm (Figure 6b). It should be noted that upon further creep tests from 2000 to 10,000 h, the subgrain sizes remained almost unchanged in both steels (Figure 6a,b).

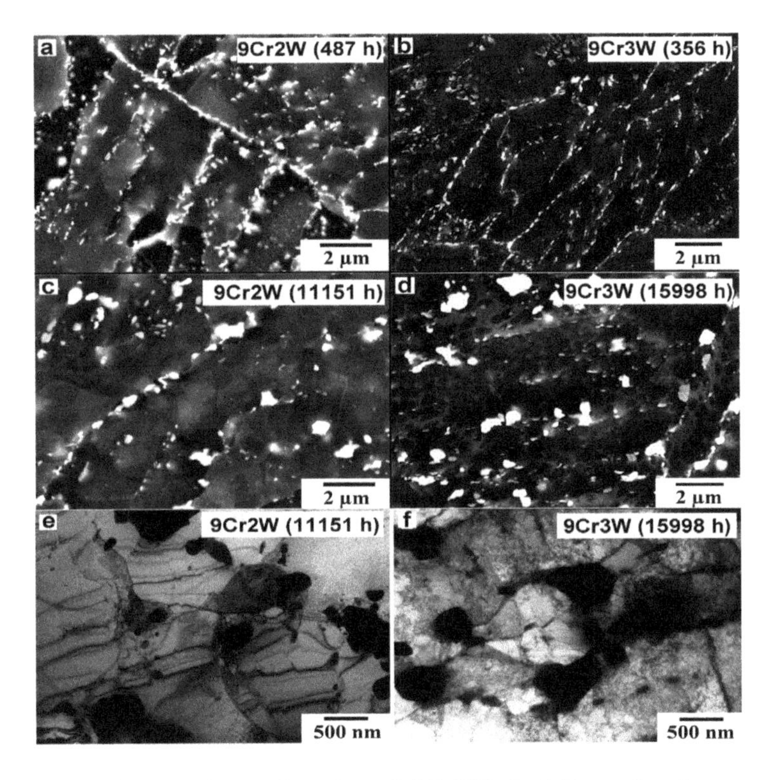

Figure 5. Microstructure of the 9Cr2W (**a**,**c**,**e**) and 9Cr3W (**b**,**d**,**f**) steels after creep rupture tests at 650 °C under the stress of: (**a**) 160 MPa, 487 h; (**b**) 180 MPa, 356 h; (**c**,**e**) 100 MPa, 11,151 h; (**d**,**f**) 100 MPa, 15,998 h; obtained by SEM (**a**–**d**) and TEM (**e**,**f**).

There is a difference in the effect of creep on the distributions of second phase particles in the two steels. In the 9Cr3W steel, the fine particles densely distributed along the boundaries can be seen after short-term creep (Figure 5b), whereas in the 9Cr2W steel, these precipitates were slightly coarser, and their density was less (Figure 5a).

$M_{23}C_6$ carbides. In the 9Cr2W steel, coarsening of $M_{23}C_6$ carbides occurred faster than in the 9Cr3W steel during creep tests (Figure 6c,d). Mean size of these carbides increased to 200–250 nm after 1000 h in the 9Cr2W steel (Figure 6c), whereas in the 9Cr3W steel, the size of these carbides remained less than 100 nm up to rupture times of ~5000 h (Figure 6d). Therefore, the W addition enhanced the coarsening resistance of $M_{23}C_6$ carbides in short-term creep conditions. At the same time, under long-term creep conditions, the average dimension of $M_{23}C_6$ carbides and their morphology became essentially the same in both steels. The average Cr and W contents in the $M_{23}C_6$ carbides were the same in both steels and tended to slightly increase with increasing rupture time (Figure 7a). In contrast, the portion of Fe decreased with increasing rupture time. Under long-term exposure, the chemical composition of $M_{23}C_6$ carbides tended to approach the thermodynamically equilibrium composition (Table 3).

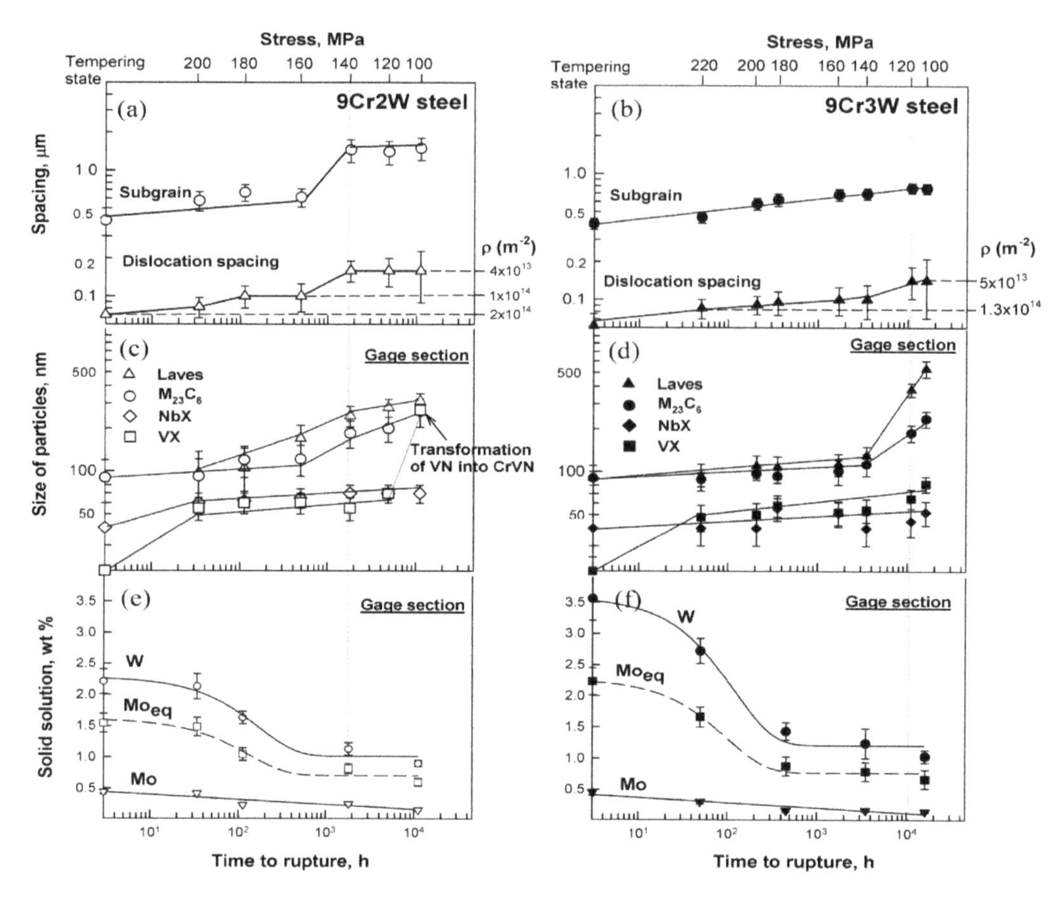

Figure 6. Mean size of spacing (**a**,**b**), mean size of particles of different phases (**c**,**d**), and change in W and Mo content in the solid solution (**e**,**f**) as function of time in the 9Cr2W (**a**,**c**,**e**) and 9Cr3W (**b**,**d**,**f**) steels during creep tests at 650 °C under the stresses of 100–220 MPa. The vertical dotted lines indicate the time to rupture corresponding to the creep strength breakdown. Mo_{eq} = Mo + 0.5W.

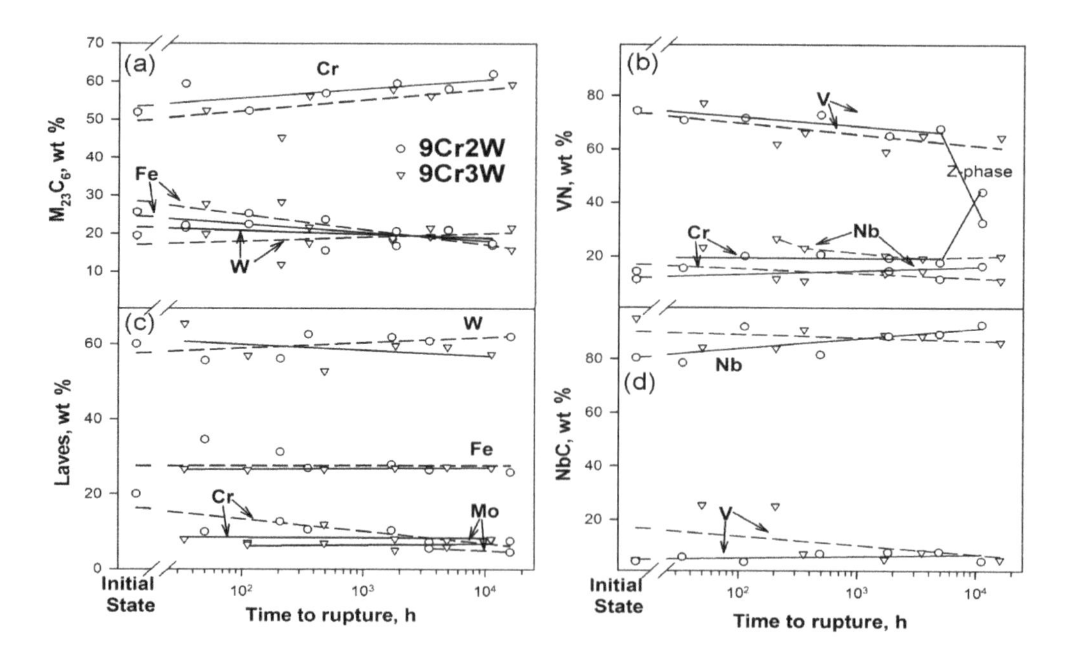

Figure 7. Change in the chemical compositions of (**a**) $M_{23}C_6$ carbides; (**b**) V-rich MX; (**c**) Laves phase particles; and (**d**) Nb-rich MX during creep tests at 650 °C under the stresses of 100–220 MPa.

Table 3. Weight fraction of elements in the different phases in the steels studied at 650 and 750 °C as calculated by Thermo-Calc.

Element		Cr	Fe	W	V	Nb	N	C	Mo	Co.
					650 °C (Creep)					
MX	9Cr2W	-	-	-	61.98	17.67	18.25	0.51	-	-
	9Cr3W	-	-	-	61.96	18.59	17.78	0.44	-	-
$M_{23}C_6$	9Cr2W	60.54	14.08	7.50	0.44	-	-	5.00	12.25	-
	9Cr3W	60.32	13.95	8.59	-	-	-	5.00	11.24	-
Laves	9Cr2W	6.26	32.57	56.99	-	-	-	-	3.99	-
	9Cr3W	6.42	32.20	57.77	-	-	-	-	3.42	-
Solid solution	9Cr2W	8.39	87.59	0.68	0.02	-	-	-	0.18	3.13
	9Cr3W	8.47	87.16	0.68	0.04	-	-	-	0.16	3.13
					750 °C (Tempering)					
MX	9Cr2W	-	-	-	61.29	17.99	17.73	0.77	-	-
	9Cr3W	-	-	-	59.50	19.38	17.78	0.70	-	-
$M_{23}C_6$	9Cr2W	51.78	20.61	13.92	-	-	-	4.83	8.03	-
	9Cr3W	51.32	20.30	16.83				4.78	5.80	-
Laves	9Cr2W	5.35	33.62	56.49	-	-	-	-	4.31	-
	9Cr3W	5.50	33.27	57.22	-	-	-	-	3.76	-
Solid solution	9Cr2W	8.51	86.42	1.60	0.03	-	-	-	0.37	3.00
	9Cr3W	8.60	85.98	1.63	0.04	-	-	-	0.32	3.00

Laves phase. The precipitation behavior of Laves phase in the two steels was quite different (Figure 6c,d). In the 9Cr2W steel, the precipitation of Laves phase particles started to occur with a high rate under creep condition. Next, these particles gradually grew with a high rate up to 250 nm at creep rupture time of 11,151 h (Figure 6c). In contrast, in the 9Cr3W steel, the particles of Laves phase precipitated under tempering remained their size of ~100 nm up to 5000 h of creep tests. Then, the rapid coarsening of Laves phase started to occur, and their average size attained ~550 nm at creep rupture time of ~16,000 h (Figure 6d). Thus, high W content slowed down the coarsening of Laves phase and provided the improved coarsening resistance of Laves phase up to about 10,000 h. However, upon further creep, the extensive precipitation of Laves phase induced their growth with an increased rate. The final average size of Laves phase particles in the 9Cr3W steel became even higher than that in the 9Cr2W steel (530 nm and 312 nm, respectively). In the 9Cr3W steel, the volume fraction of Laves phase was significantly higher (2.394%) in comparison with the 9Cr2W steel (1.315%) as calculated by Thermo-Calc software (Table 4). It is indicative of the full depletion of excess content of W at 650 °C from the solid solution under long-term creep.

Table 4. The volume fraction of second phases in the steels at 650 °C as calculated by Thermo-Calc.

Phase	Volume Fraction (%)	
	9Cr2W Steel	9Cr3W Steel
$M_{23}C_6$	1.958	1.966
MX	0.246	0.238
Laves phase	1.315	2.394

In both steels, Fe and Cr contents insignificantly decreased in the Laves phase, and Mo content slightly increased during creep (Figure 7c). W content decreased in the 9Cr2W steel, whereas in the 3 wt % W steel it increased. However, these chemical composition changes of the Laves phase particles were insignificant. Therefore, the Laves phase particles initially precipitated in accordance with the thermodynamically equilibrium content (Table 3), and their coarsening was not associated with the changes in their chemical composition.

No evidence for M_6C carbides presence in the 9Cr3W steel was found after creep tests. These carbides were replaced by more stable Laves phase under short-term creep as in other high-chromium martensitic steels [22].

Nb-rich MX carbonitrides. Thermo-Calc calculation predicted the existence of a unified (V,Nb)N nitride with a very low C content at 650 °C only in both steels (Table 4), whereas Nb-rich and V-rich separation of MX phase remained under creep condition at this temperature. In the 9Cr3W steel, Nb-rich MX carbonitrides grew from 40 to 50 nm under long-term creep condition (Figure 6d). In contrast, in the 9Cr2W steel, they rapidly grew from 40 to 60 nm under short-term creep conditions and then to 70 nm under long-term creep conditions (Figure 6c). Thus, W additives hindered the coarsening of Nb-rich MX carbonitrides. Effect of W additions is more pronounced under short-term creep. Nb and V content in the Nb-rich MX particles slightly increased and decreased, respectively, with increasing rupture time (Figure 7d) in opposition to thermodynamically equilibrium values (Table 3).

V-rich MX carbonitrides. The average size of V-rich MX particles attained 60 nm after long-term creep tests in both steels. In the 9Cr2W steel, the well-known replacement of V-rich MX carbonitrides by Z-phase (CrVN nitride), which is in fact the most thermodynamically stable nitride [31,32], started to occur after ~5000 h. The full transformation of all V-rich MX carbonitrides was found after 10,000 h of creep tests. The mean size of Z-phase was about 300 nm. Therefore, the size of V-rich carbonitrides increased by a factor of ~4.5 and they could not more contribute to the creep resistance of the 9Cr2W steel (Figure 6c). +1 wt %W additive shifted the onset of this transformation from 5000 h to 16,000 h; the separate Z-phase particles in the 9Cr3W steel were revealed after 16,000 h [33]. Nb content in the V-rich MX particles slightly increased with rupture time in both steel (Figure 7b). V content in the V-rich MX particles tended to decrease in the 9Cr3W steel with rupture time. At rupture time ≤5000 h, in the 9Cr2W steel, the V and Cr contents were essentially independent on the rupture time. Therefore, no gradual increase in Cr content within the V-rich MX particles resulting in transformation of their cubic lattice into tetragonal lattice of Z-phase [20,32,34] was detected.

Solid solution. Solid solution of the 9Cr3W steel (Figure 6f) was more enriched by W than that of the 9Cr2W steel (Figure 6e) during short-term creep up to time t of ~500–700 h. After this exposure, there is no significant difference in W content in the solid solution of two steels.

Depletion of W and Mo from the solid solution took place with increasing rupture time. However, the rate of the W depletion was higher in the 9Cr3W steel under short-term creep conditions. Depletion of W, $f(\mathrm{W})$, from the solid solution could be described as:

$$f(\mathrm{W}) \sim 1.3exp(-0.0059t) \quad \text{for the 9Cr2W steel,} \tag{4}$$

$$\text{and } f(\mathrm{W}) \sim 2.4exp(-0.0083t) \quad \text{for the 9Cr3W steel.} \tag{5}$$

Depletion of Mo, $f(\mathrm{Mo})$, from the solid solution could be described as:

$$f(\mathrm{Mo}) \sim -0.08t \quad \text{for both steels.} \tag{6}$$

This finding indicates the strong dependence of W content on the rupture time up to an achievement of the thermodynamically equilibrium value of W in the ferritic matrix (Table 3). At rupture time ≥700 h, the $Mo_{eq} = \Sigma(Mo + 0.5W)$ [35] was the same for both steels since no significant difference in W and Mo content was found in the ferritic matrix of both steels by TEM.

4. Discussion

4.1. Contribution of W to the Solid Solution Strengthening

The experimental data showed that increasing W content from 2 to 3 wt % improves the creep resistance of the 0.1C-9Cr-3Co-0.5Mo-VNbBN steel at 650 °C under short-term creep conditions due to solid solution strengthening as main hardening mechanism. As a result, the TMLS essentially remains in the 9Cr3W steel, whereas a well-defined subgrain structure evolves in the 9Cr2W steel. Despite this fact, the creep strength of the 9Cr3W steel approaches to the level of the 9Cr2W steel with increasing rupture time up to 10^5 h according to estimation by Larson–Miller parameter. It is worth noting that effect of Co on the microstructural evolution under creep is nearly the same [18]. However, both Co and W additions could not prevent the breakdown of creep strength taking place at a rupture time of ~2000 h for the P92 steel and the 9Cr2W steel, and at a rupture time of ~10,000 for the 9Cr3W steel (Figure 3a,c).

Mo_{eq} content in the solid solution of both steels is significantly higher than the solubility limit for these elements at 650 °C and the precipitation of Laves phase provides approaching of W and Mo contents to their thermodynamically equilibrium levels under creep condition. It is known [1,7,10–12,15,22] that the precipitation of Laves phase during creep is hindered by low rate of these substitutional elements diffusion. The continuous decrease in W content in the solid solution indicates the continuous decomposition of the supersaturated solid solution accompanied by the precipitation of the fine Laves phase particles during creep tests. Under long-term condition, the volume fraction of Laves phase attains the thermodynamically equilibrium value (Table 4). The same Mo_{eq} value in the solid solution of both steels is attained after 700 h of short-term creep exposure. This indicates no advantage in solid solution strengthening for the 9Cr3W steel after 700 h of creep test.

However, the 9Cr3W steel demonstrates +14% increments in creep strength even after 700 h of creep testing. Increasing W content up to 3% contributes to creep strength not only by solid solution strengthening but by creation the preconditions for improved creep strength, such as homogeneous TMLS, increased dislocation density, and narrow size distributions of boundary particles after tempering, which provide advanced creep strength upon creep time more than 700 h. Therefore, under short-term creep, the W addition slows down the rearrangement of lattice dislocations by climb that may prevent the aforementioned transformation of lath boundaries to sub-boundaries. However, under long-term conditions, a saturation of the solid solution by W and Mo is the same in both steels and the retardation of dislocation climb by W addition could not provide the stabilization of TMLS. It is obvious that the influence of W content on the stability of TMLS under long-term creep conditions is attributed to its effect on a dispersion of secondary phase.

4.2. Contribution of W to the Particle Strengthening

The experimental data showed that increasing W content from 2 to 3 wt % improves the creep resistance of the steel in the short-term conditions due to sluggish kinetics of the coarsening of $M_{23}C_6$ carbides, precipitation of fine Laves phase particles, and preventing transformation of TMLS into subgrain structure. The effect of W on the coarsening behavior of $M_{23}C_6$ carbides and Laves phase was considered in the companion work [36] in sufficient details. In that study, we briefly summarize that 1 wt % W addition affects the size distribution of these two types of particles after tempering that leads to the difference in the coarsening behavior between two steels [36] during creep. In the work [37], it was found that the coarsening of $M_{23}C_6$ carbides correlates with the changes in their chemical composition. It is worth noting that the coarsening of $M_{23}C_6$ carbides in both steels is accompanied by an increase in Cr and W contents and a decrease in Fe content (Figure 6a). In work [37], it has been observed that $M_{23}C_6$ carbides grow during 50,000 h up to the chemical equilibrium. In both steels, the character of evolution of chemical composition of $M_{23}C_6$ carbides is essentially the same. The Cr and (W + Mo) contents in $M_{23}C_6$ carbides increase at the expense of Fe and reach the thermodynamically equilibrium values (Table 3). It is worth noting that Thermo-Calc calculation at 650 °C predicts the Cr and (W + Mo) content in $M_{23}C_6$ carbides higher and lower, respectively, than that at 750 °C. Under tempering condition, only Cr content attained thermodynamically equilibrium value in $M_{23}C_6$ carbides. Cr atoms diffuse slightly slower than Fe atoms in ferrite, and the diffusion rate of W and Mo atoms is the lowest in comparison with Cr [38]. This is the reason for approaching the thermodynamically equilibrium chemical composition of $M_{23}C_6$ carbides during a long-term exposure.

In both steels, the various second phase particles pin the lath boundaries. MX particles homogeneously distributed within the lath exert Zener drag pressure, which can be evaluated as [9,15,23,39,40]:

$$P_Z = \frac{3\gamma F_V}{d}, \tag{7}$$

where γ is the boundary surface energy per unit area (0.153 J m^{-2}) [15,40], F_v is the volume fraction of particles calculated by Thermo-Calc, and d is a mean size of particles, m.

The boundary particles of $M_{23}C_6$ and Laves phases exert Zener drag pressure estimated as [9,15,23,39,40]:

$$P_B = \frac{\gamma F_{vB} D}{d^2}, \tag{8}$$

where D is subgrain size or lath width, μm, and F_{vB} is the fraction of particles located at the boundaries. The pinning pressures were calculated separately for $M_{23}C_6$ carbides and Laves phase according with:

$$P_B = \frac{\gamma F_{vB} D_0}{d_0^2} \cdot \frac{\beta_i}{\beta_0}, \tag{9}$$

where β_i is density of $M_{23}C_6$ carbides or Laves phase, located along boundaries, for each applied stress. Calculation of Zener drag pressure for the different kinds of particles was considered in the previous work [9] in details.

+1 wt % W addition as well as an applied stress affect the particles sizes and their volume fractions (Figure 6 and Table 3). As a result, there is a difference in the pinning pressure between the two steels, and Zener drag pressure depends on the creep test time (Figure 8). Both steels contain essentially the same volume fraction of MX carbonitrides, and the pinning pressure from these particles (P_Z) is the same for both steels. At high applied stress, MX carbonitrides give a minor contribution to the total Zener pressure due to the low volume fraction of these particles and random distribution within subgrain and laths.

In the 9Cr2W steel, the $M_{23}C_6$ carbides gives the main contribution to overall Zener drag force at any applied stress [23], whereas in the 9Cr3W steel the highest Zener drag is exerted by Laves phase particles at the applied stresses more than 100 MPa [39]. Only at an applied stress of 100 MPa the Zener drag exerted by $M_{23}C_6$ carbides in this steel (9Cr3W) becomes higher than that exerted by Laves phase particles, and the P_B and P_Z values originated from Laves phase and MX carbonitrides, respectively, are the same at this condition (Figure 8b). In contrast, in the 9Cr2W steel, the P_Z value drops at an applied stress of 100 MPa, owing to transformation of V-rich MX carbonitrides to Z-phase, and the pinning pressures from the boundary $M_{23}C_6$ carbides (P_B) and MX carbonitrides with a random particle distribution (P_Z) are similar at an applied stress of 120 MPa (Figure 8a).

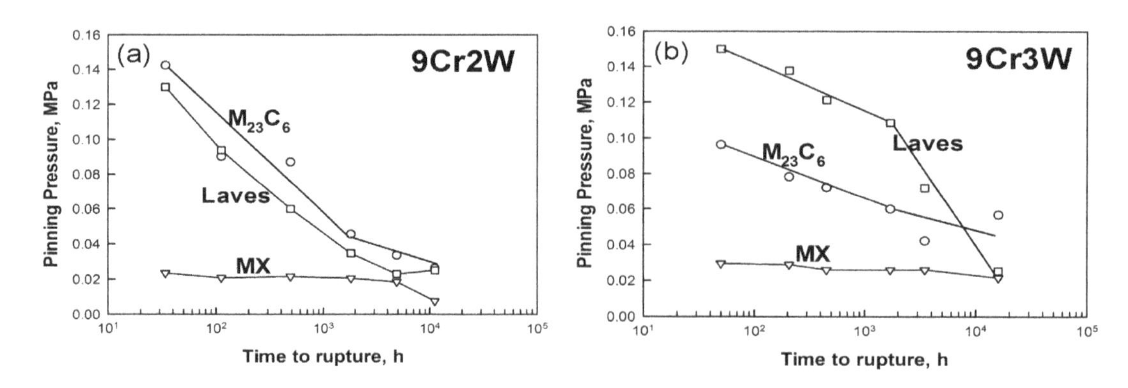

Figure 8. Change in the pinning pressures from different kinds of particles on the grain and lath boundaries of the (**a**) 9Cr2W steel and (**b**) 9Cr3W steel during creep tests at 650 °C under the stresses of 100–220 MPa.

Thus, the Zener pressure exerted by boundary particles remains for a long time until the particle coarsening occurs. Size stabilization of the Laves phase particles in the 9Cr3W steel provides a significant contribution to high Zener pressure, whereas their rapid coarsening removes this effect. Thus, Laves phase particles give short-term contribution to the precipitation strengthening. The decrease in Zener pressure exerted by Laves phase does not lead to the transformation of TMLS into subgrain structure as in the 9Cr2W steel because $M_{23}C_6$ carbides are able to provide the high level of Zener pressure [23] (more than 0.05 MPa) at the low stresses in the steel with 3 wt % W.

Therefore, there is no positive effect of increased W content on the creep resistance of the Co-modified P92 steel under long-term creep conditions. An alloying of 0.1C-9Cr-3Co-0.5Mo-VNbBN martensitic steel by 3%W does not seem justified for the applying at long-term creep condition (for 10^5 h) at 650 °C. Therefore, the W content of ~2 wt % is optimal for the Co-containing high chromium martensitic steels.

5. Conclusions

The microstructures of two 0.1C-9Cr-3Co-0.5Mo-VNbBN martensitic steels with different content of W (2 wt % and 3 wt %) in the tempered and crept at 650 °C under stresses of 100–220 MPa conditions were studied. The main results can be summarized as follows:

1. The structure of both steels tempered at 750 °C for 3 h is the tempered martensite lath structure with the lath thickness of 0.4 μm. An increased W content leads to the formation of W-rich Laves phase particles and M_6C carbides on the boundaries in addition to the $M_{23}C_6$ carbides located also on the boundaries and MX carbonitrides distributed uniformly within the ferritic matrix.
2. The steel with 3 wt % W content demonstrates a +15% increase in the 10,000 h creep rupture strength at 650 °C due to hindering the coarsening of the $M_{23}C_6$ carbides, MX carbonitrides and Laves phase particles. Tungsten also slows down the transformation of V-rich MX carbonitrides into Z-phase particles.
3. An increase in W content in the steels provides an increase in the amount of Laves phase. Under long-term conditions, the depletion of excess W from the solid solution leads to the rapid coarsening of Laves phase particles; the contribution of this phase to Zener drag has a short-term character.
4. The predicted long-term creep rupture strength for 100,000 h is about 85 MPa for both steels. This value is independent on W content due to depletion of its excess from the solid solution up to thermodynamically equilibrium value due to Laves phase precipitation. Therefore, there is no positive effect of increased W content on the creep resistance of the Co-modified P92 steel under long-term creep conditions. An alloying of 0.1C-9Cr-3Co-0.5Mo-VNbBN martensitic steel by 3%W does not seem justified for long-term creep condition (for 10^5 h) at 650 °C. Therefore, the W content of ~2 wt % is optimal for the Co-containing high chromium martensitic steels.

Acknowledgments: This study was financially supported by the Ministry of Education and Science of Russian Federation, under project of Government Task No. 11.2868.2017/PCh. The authors are grateful to V. Skorobogatykh and I. Shchenkova, Central Research Institute for Machine-Building Technology, for supplying the test material and to the staff of the Joint Research Center, "Technology and Materials", Belgorod National Research University, for their assistance with instrumental analysis.

Author Contributions: A.F., N.D., A.B. and R.K. formulated the original problem, designed the study, developed the methodology, and wrote the manuscript. A.F. and N.D. performed the experiment, collected data, and assisted with data analysis. A.B. and R.K. provided direction, guidance, and interpretation of data.

References

1. Abe, F.; Kern, T.U.; Viswanathan, R. *Creep Resistant Steels*; Part I; Woodhead Publishing in Materials: Cambridge, UK, 2008; p. 800.
2. Kaybyshev, R.O.; Skorobogatykh, V.N.; Shchenkova, I.A. New martensitic steels for fossil power plant: Creep resistance. *Phys. Met. Metallogr.* **2010**, *109*, 186–200. [CrossRef]

3. Kitahara, H.; Ueji, R.; Tsuji, N.; Minamino, Y. Crystallographic features of lath martensite in low-carbon steel. *Acta Mater.* **2006**, *54*, 1279–1288. [CrossRef]
4. Ghassemi-Armaki, H.; Chen, R.; Maruyama, K.; Igarashi, M. Premature creep failure in strength enhanced high Cr ferritic steels caused by static recovery of tempered martensite lath structures. *Mater. Sci. Eng. A* **2010**, *527*, 6581–6588. [CrossRef]
5. Abe, F. Analysis of creep rates of tempered martensitic 9% Cr steel based on microstructure evolution. *Mater. Sci. Eng. A* **2009**, *510*, 64–69. [CrossRef]
6. Kostka, A.; Tak, K.-G.; Hellmig, R.J.; Estrin, Y.; Eggeler, G. On the contribution of carbides and micrograin boundaries to the creep strength of tempered martensite ferritic steels. *Acta Mater.* **2007**, *55*, 539–550. [CrossRef]
7. Abe, F. Effect of fine precipitation and subsequent coarsening of Fe2W laves phase on the creep deformation behavior of tempered martensitic 9Cr-W steels. *Metall. Mater. Trans. A* **2005**, *36*, 321–331. [CrossRef]
8. Taneike, M.; Sawada, K.; Abe, F. Effect of carbon concentration on precipitation behavior of $M_{23}C_6$ carbides and MX carbonitrides in martensitic 9Cr steel during heat treatment. *Metall. Mater. Trans. A* **2004**, *35*, 1255–1261. [CrossRef]
9. Dudova, N.; Plotnikova, A.; Molodov, D.; Belyakov, A.; Kaibyshev, R. Structural changes of tempered martensitic 9%Cr-2%W-3%Co steel during creep at 650 °C. *Mater. Sci. Eng. A* **2012**, *534*, 632–639. [CrossRef]
10. Kipelova, A.; Belyakov, A.; Kaibyshev, R. Laves phase evolution in a modified P911 heat resistant steel during creep at 923K. *Mater. Sci. Eng. A* **2012**, *532*, 71–77. [CrossRef]
11. Fedorova, I.; Belyakov, A.; Kozlov, P.; Skorobogatykh, V.; Shenkova, I.; Kaibyshev, R. Laves-phase precipitates in a low-carbon 9% Cr martensitic steel during aging and creep at 923K. *Mater. Sci. Eng. A* **2014**, *615*, 153–163. [CrossRef]
12. Isik, M.I.; Kostka, A.; Yardley, V.A.; Pradeep, K.G.; Duarte, M.J.; Choi, P.P.; Raabe, D.; Eggeler, G. The nucleation of Mo-rich Laves phase particles adjacent to $M_{23}C_6$ micrograin boundary carbides in 12% Cr tempered martensite ferritic steels. *Acta Mater.* **2015**, *90*, 94–104. [CrossRef]
13. Ghassemi-Armaki, H.; Chen, R.; Maruyama, K.; Igarashi, M. Creep behavior and degradation of subgrain structures pinned by nanoscale precipitates in strength-enhanced 5 to 12 Pct Cr ferritic steels. *Metall. Mater. Trans. A* **2011**, *42*, 3084–3094. [CrossRef]
14. Yin, F.-S.; Tian, L.-Q.; Xue, B.; Jiang, X.-B.; Zhou, L. Effect of Carbon Content on Microstructure and Mechanical Properties of 9 to 12 pct Cr Ferritic/Martensitic Heat-Resistant Steels. *Metall. Mater. Trans. A* **2012**, *43*, 2203–2209. [CrossRef]
15. Dudko, V.; Belyakov, A.; Molodov, D.; Kaibyshev, R. Microstructure evolution and pinning of boundaries by precipitates in a 9 pct. Cr heat resistant steel during creep. *Metall. Mater. Trans. A* **2013**, *44*, S162–S172. [CrossRef]
16. Zhong, W.; Wang, W.; Yang, X.; Li, W.; Yan, W.; Sha, W.; Wang, W.; Shan, Y.; Yang, K. Relationship between Laves phase and the impact brittleness of P92 steel reevaluated. *Mater. Sci. Eng. A* **2015**, *639*, 252–258. [CrossRef]
17. Helis, L.; Toda, Y.; Hara, T.; Miyazaki, H.; Abe, F. Effect of cobalt on the microstructure of tempered martensitic 9Cr steel for ultra-supercritical power plants. *Mater. Sci. Eng. A* **2009**, *510*, 88–94. [CrossRef]
18. Kipelova, A.; Odnobokova, M.; Belyakov, A.; Kaibyshev, R. Effect of Co on creep behavior of a P911 steel. *Metall. Mater. Trans. A* **2013**, *44*, 577–583. [CrossRef]
19. Sawada, K.; Takeda, M.; Maruyama, K.; Ishii, R.; Yamada, M.; Nagae, Y.; Komine, R. Effect of W on recovery of lath structure during creep of high chromium martensitic steels. *Mater. Sci. Eng. A* **1999**, *267*, 19–25. [CrossRef]
20. Fedoseeva, A.; Dudova, N.; Kaibyshev, R. Creep strength breakdown and microstructure evolution in a 3%Co modified P92 steel. *Mater. Sci. Eng. A* **2016**, *654*, 1–12. [CrossRef]
21. Tsuchida, Y.; Okamoto, K. Improvement of creep rupture strength of high Cr ferritic steel by addition of W. *ISIJ Int.* **1995**, *35*, 317–323. [CrossRef]
22. Li, Q. Precipitation of Fe2W laves phase and modeling of its direct influence on the strength of a 12Cr-2W steel. *Metall. Mater. Trans. A* **2006**, *37*, 89–97. [CrossRef]
23. Dudova, N.; Kaibyshev, R. On the precipitation sequence in a 10% Cr steel under tempering. *ISIJ Int.* **2011**, *51*, 826–831. [CrossRef]

24. Wilshire, B.; Scharning, P. Prediction of long term creep data for forged 1Cr-1Mo-0.25V steel. *Mater. Sci. Technol.* **2008**, *24*, 1–9. [CrossRef]
25. Hirsch, P.B.; Howie, A.; Nicholson, R.B.; Pashley, D.W.; Whelan, M.J. *Electron Microscopy of Thin Crystals*, 2nd ed.; Krieger: New York, NY, USA, 1977; p. 563.
26. Dimmler, G.; Weinert, P.; Kozeschnik, E.; Cerjak, H. Quantification of the Laves phase in advanced 9–12% Cr steels using a standard SEM. *Mater. Charact.* **2003**, *51*, 341–352. [CrossRef]
27. Fedoseeva, A.; Dudova, N.; Glatzel, U.; Kaibyshev, R. Effect of W on tempering behaviour of a 3% Co modified P92 steel. *J. Mater. Sci.* **2016**, *51*, 9424–9439. [CrossRef]
28. Kimura, K.; Toda, Y.; Kushima, H.; Sawada, K. Creep strength of high chromium steel with ferrite matrix. *Int. J. Press. Vessels Pip.* **2010**, *87*, 282–288. [CrossRef]
29. Yoshizawa, M.; Igarashi, M.; Moriguchi, K.; Iseda, A.; GhassemiArmaki, H.; Maruyama, K. Effect of precipitates on long-term creep deformation properties of P92 and P122 type advanced ferritic steels for USC power plants. *Mater. Sci. Eng. A* **2009**, *510*, 162–168. [CrossRef]
30. Hattestrand, A.; Andren, H.O. Evaluation of particle size distributions of precipitates in a 9% chromium steel using energy filtered transmission electron microscopy. *Micron* **2001**, *32*, 489–498. [CrossRef]
31. Cipolla, L.; Danielsen, H.K.; Venditti, D.; Di Nunzio, P.E.; Hald, J.; Somers, M.A.J. Conversion of MX nitrides to Z-phase in a martensitic 12% Cr steel. *Acta Mater.* **2010**, *58*, 669–679. [CrossRef]
32. Danielsen, H.K.; Di Nunzio, P.E.; Hald, J. Kinetics of Z-Phase Precipitation in 9 to 12 pct Cr Steels. *Metall. Mater. Trans. A* **2013**, *44*, 2445–2452. [CrossRef]
33. Fedoseeva, A.; Dudova, N.; Kaibyshev, R. Effect of Tungsten on a Dispersion of M(C,N) Carbonitrides in 9% Cr Steels Under Creep Conditions. *Trans. Indian Inst. Met.* **2016**, *69*, 211–215. [CrossRef]
34. Kaibyshev, R.O.; Skorobogatykh, V.N.; Shchenkova, I.A. Formation of the Z-phase and prospects of martensitic steels with 11% Cr for operation above 590 °C. *Met. Sci. Heat Treat.* **2010**, *52*, 90–99. [CrossRef]
35. Klueh, R.L. Elevated temperature ferritic and martensitic steels and their application to future nuclear reactors. *Int. Mater. Rev.* **2005**, *50*, 287–310. [CrossRef]
36. Fedoseeva, A.; Dudova, N.; Kaibyshev, R. Effect of stresses on the structural changes in high-chromium steel upon creep. *Phys. Met. Metall.* **2017**, *118*, 591–600. [CrossRef]
37. Ghassemi-Armaki, H.; Chen, R.; Kano, S.; Maruyama, K.; Hasegawa, Y.; Igarashi, M. Strain-induced coarsening of nanoscale precipitates in strength enhanced high Cr ferritic steels. *Mater. Sci. Eng. A* **2012**, *532*, 373–380. [CrossRef]
38. Mehrer, H.; Stolica, N.; Stolwijk, N.A. Landolt Bornstein- Numerical Data and Functional Relationships in Science and Technology, New Series, Group III: Crystals and Solid State Physics. In *Diffusion in Solid Metals and Alloys*; Springer: Berlin/Heidelberg, Germany, 1990; Volume 26, pp. 47–48. ISBN 978-3-540-50886-1.
39. Fedoseeva, A.; Dudova, N.; Kaibyshev, R. Creep behavior and microstructure of a 9Cr-3Co-3W martensitic steel. *J. Mater. Sci.* **2017**, *52*, 2974–2988. [CrossRef]
40. Humphreys, F.J.; Hatherly, M. *Recrystallization and Related Annealing Phenomena*, 2nd ed.; Elsevier: Atlanta, GA, USA, 2004; pp. 91–112.

8

The Effect of Lath Martensite Microstructures on the Strength of Medium-Carbon Low-Alloy Steel

Chen Sun [1,2], Paixian Fu [1,*], Hongwei Liu [1], Hanghang Liu [1], Ningyu Du [1,2] and Yanfei Cao [1,*]

1 Shenyang National Laboratory for Materials Science, Institute of Metal Research, Chinese Academy of Sciences, Shenyang 110016, China; csun15s@imr.ac.cn (C.S.); hwliu@imr.ac.cn (H.L.); hhliu15b@imr.ac.cn (H.L.); nydu16s@imr.ac.cn (N.D.)

2 School of Materials Science and Engineering, University of Science and Technology of China, Hefei 230026, China

* Correspondence: pxfu@imr.ac.cn (P.F.); yfcao10s@imr.ac.cn (Y.C.)

Abstract: Different austenitizing temperatures were used to obtain medium-carbon low-alloy (MCLA) martensitic steels with different lath martensite microstructures. The hierarchical microstructures of lath martensite were investigated by optical microscopy (OM), electron backscattering diffraction (EBSD), and transmission electron microscopy (TEM). The results show that with increasing the austenitizing temperature, the prior austenite grain size and block size increased, while the lath width decreased. Further, the yield strength and tensile strength increased due to the enhancement of the grain boundary strengthening. The fitting results reveal that only the relationship between lath width and strength followed the Hall–Petch formula of. Hence, we propose that lath width acts as the effective grain size (EGS) of strength in MCLA steel. In addition, the carbon content had a significant effect on the EGS of martensitic strength. In steels with lower carbon content, block size acted as the EGS, while, in steels with higher carbon content, the EGS changed to lath width. The effect of the Cottrell atmosphere around boundaries may be responsible for this change.

Keywords: medium-carbon low-alloy steel; lath martensite; effective grain size; strength; carbon content

1. Introduction

Medium-carbon low-alloy (MCLA) steel has an excellent combination of strength, toughness and hardenability and is widely used in structural components with large sections, such as generator spindles and automotive crankshafts [1–3]. Lath martensite is a typical microstructure seen in MCLA steel after quenching. There are several elements in lath martensitic microstructures. Prior austenite grains (PAGs) are divided into several packets, which consist of parallel blocks. The blocks are composed of laths, arranged parallel to each other. Low-angle grain boundaries (LAGBs) exist among laths, while high-angle grain boundaries (HAGBs) exist among packets and blocks [4–6].

An early work [7] indicated that the PAG size was the effective grain size (EGS) of strength in the Hall–Petch formula. Subsequently, the Hall–Petch relationship between yield strength and packet size was discovered by Swarr et al. in Fe–0.2C alloy [8], by Roberts in Fe–Mn alloy [9], and by Wang et al. in 17CrNiMo6 steel [10]. With the development of characterization technology, Morito et al. [11] revealed that the block width in Fe–0.2C and the Fe–0.2C–2Mn alloys is the key structural parameter controlling the strength of lath martensite by utilizing electron backscattered diffraction (EBSD). Similar results were presented by Zhang et al. [12], Long et al. [13], and Li et al. [14]. However, the martensite lath was also considered to be an effective control unit of strength in the research of Smith et al. in 42CrMo steel [15] and Kim et al. in Fe-0.55C alloy [16]. So far, there are still controversies regarding the EGS of strength in lath martensite.

Previous studies focused mainly on low-carbon steels. However, reports are scarce on the relationship between lath martensite microstructures and strength in medium-carbon steels has been. In lath martensite, carbon atoms segregate around dislocations and grain boundaries, rather than interstitial solution in the lattice [17]. The different carbon contents in steel can change the amount of segregated carbon atoms around the grain boundaries, which therefore affects its strength. In fact, research on low-carbon steel [8–14] draws the conclusion that the block/packet size is the key structural parameter in controlling the strength of lath martensite, while research on medium-carbon steel [16,18] tends to suggest that the strength depends primarily on the lath width. However, little work has been done on medium-carbon steel. Therefore, it is necessary to study the relationship between the martensite multi-level microstructure and strength in medium-carbon steel in order to verify the above finding.

In this work, austenitizing temperatures of 850–940 °C were used to obtain different sizes of PAG, block, and lath in the experimental MCLA steel. Then, optical microscopy (OM), EBSD, and transmission electron microscopy (TEM) were utilized to quantify the multi-level structural parameters at different austenitizing temperatures. The classical Hall–Petch formula of strength was assessed with PAG size, block width, and lath width respectively in order to clarify the EGS that governs the strength in the experimental MCLA steel. In addition, the influence of carbon content on the effective grain size of strength is summarized and further elucidated based on the above results as well as data from published research.

2. Experimental Procedures

2.1. Materials and Heat Treatment

The MCLA martensite steel used in this investigation was melted in a vacuum furnace and cast into a 25 kg ingot. Then, the ingot was forged into a round rod. Specimens with a dimension of 60 mm × 60 mm × 60 mm were taken from the rods. The chemical composition of the experimental steel was determined by an inductively coupled plasma emission spectrometer (ICP-6300, Thermo Fisher Scientific Company, Waltham, USA), and the result is shown in Table 1. Figure 1 shows the dilatometric curve of test steel, indicating that the Ac_1 (austenitization starting point) and Ac_3 (austenitization ending point) were respectively 735 °C and 818 °C. Based on the dilatometric curve, the heat treatment processes were as follows. First, the specimens were annealed at 860 °C for 3 h, followed by furnace cooling. Then the specimens were austenitized at different temperatures of 850, 880, 910, and 940 °C for 3 h, and quenched by water cooling.

2.2. Microstructure Observation

The PAGs were observed via OM (AxioCam MRc5, ZEISS Company, Oberkochen, Germany). The OM specimens were etched with a supersaturated picric acid solution, which was configured with 25 g of water and 0.7 g of picric acid. The martensite packets and blocks were characterized with EBSD. As the lath width is generally 0.2–0.3 μm [5,12,13], the step size was chosen to be 0.2 μm. The block widths were measured with reference to EBSD maps, as processed with the Oxford Instruments Channel 5 HKL software. The martensite laths were observed via TEM (Tecnai G220, FEI Company, Hillsboro, AL, USA). At least 500 PAGs, 300 blocks, and 200 laths were measured in order to ascertain the average PAG size, block width, and lath width. The preparation methods used for the TEM samples and EBSD samples can be found in our previous research [3]. The dislocation densities of experimental steels were determined via X-ray diffraction (XRD) (D/Max-2500PC, Rigaku Company, Tokyo, Japan) with Cu Kα radiation (λ = 1.5406 Å). The scanning angle was 20–100°, and the scanning speed was 1°/min.

Table 1. Chemical compositions of the experimental medium-carbon low-alloy (MCLA) martensite steel (wt%).

C	Si	Mn	Cr	Mo	S	P	Ni	V
0.41	0.26	0.70	1.10	0.26	0.0003	0.010	0.55	0.19

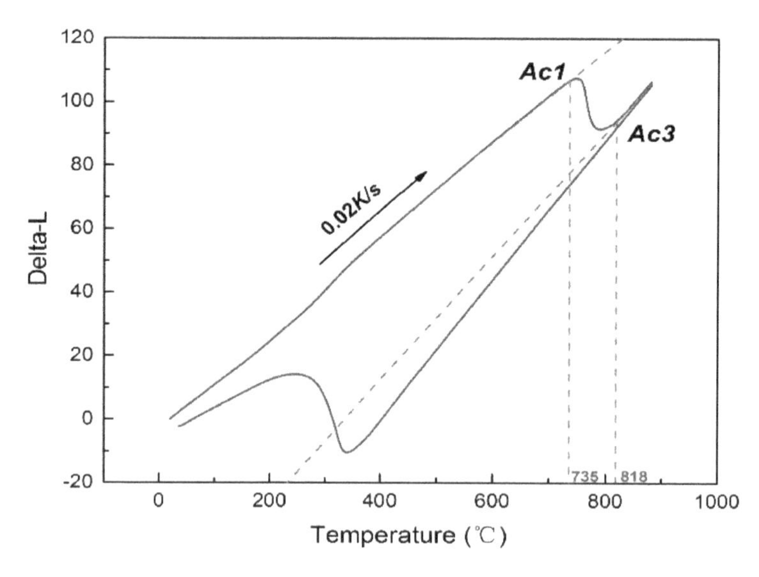

Figure 1. The dilatometric curve of the experimental MCLA steel.

2.3. Mechanical Tests

The tensile tests were performed with a Z150 tensile machine (Zwick/Roell Company, Ulm, Germany). At least three standard tensile samples with a diameter of 5 mm were used for each heat treatment condition.

3. Results and Discussion

3.1. Microstructural Characterization

Figure 2 shows the typical morphologies of PAG in MCLA martensitic steel when austenitized at different temperatures. The average PAG sizes of samples austenitized at different temperatures are shown in Figure 3a. The PAGs were uniform after austenitizing at 850 °C, as shown in Figure 2a. As the austenitizing temperature rises to 880 °C some PAGs coarsened severely, leading to an increase in the average PAG size. When the austenitizing temperature was increased further, the dissolution of fine precipitates weakened the pinning effect on the grain boundary [13], resulting in the fast growth rate of PAG size, as shown in Figure 3a.

The EBSD orientation maps of MCLA steel austenitized at different temperatures are shown in Figure 4, and the measured block width (d_B) is shown in Figure 3b. It can be seen that the PAGs are composed of several martensite packets which consist of blocks with similar extension directions. The PAG boundaries are HAGBs with a misorientation less than 45° (black lines) [13]. The HAGBs with a misorientation higher than 45° (yellow lines) are the martensite packet and block boundaries, which are distributed inside the PAGs. As the austenitizing temperature increased, the martensite blocks were arranged in a more orderly fashion. This is due to the weakening of the resistance of the lath nucleation and growth at high austenitizing temperatures [13]. At the same time, the sizes of the packets and blocks increased significantly. As shown in Figure 3b, as the austenitizing temperature increased from 850 °C to 940 °C, the average d_B increased from ~1.8 μm to ~2.5 μm.

The TEM observations of martensite laths in as-quenched MCLA steels are shown in Figure 5, and the measured lath width (d_L) is shown in Figure 3c. The laths arrange in a parallel manner in packets, and contain high-density dislocations. The d_L gradually decreased with increasing austenitizing temperature. The size distribution of d_L is shown in Figure 6. It can be seen that all distribution curves tend to show a normal distribution. The peak of the normal distribution curve moved to the left as the austenitizing temperature increased, revealing that higher austenitizing temperature resulted in a fine lath width. The high austenitizing temperature promotes the dissolution of residual carbides into austenite, which decreases the martensite starting temperature and increases the nucleation rate of martensite [19,20]. The low martensite starting temperature, coupled with the high nucleation rate, result in small d_L [13,21].

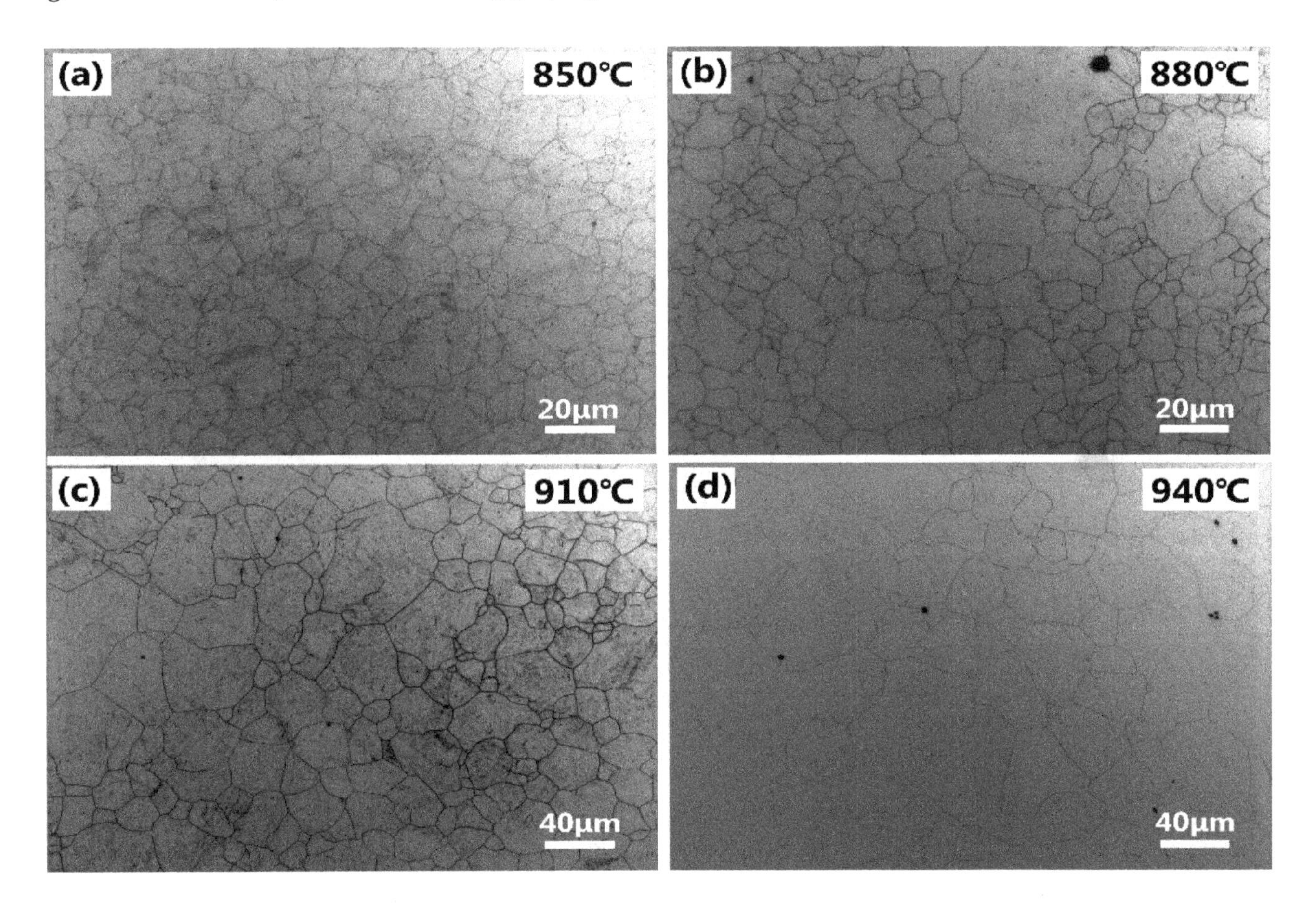

Figure 2. Prior austenite grain of MCLA martensitic steel austenitized at the different temperatures of (**a**) 850 °C, (**b**) 880 °C, (**c**) 910 °C, and (**d**) 940 °C observed by optical microscopy.

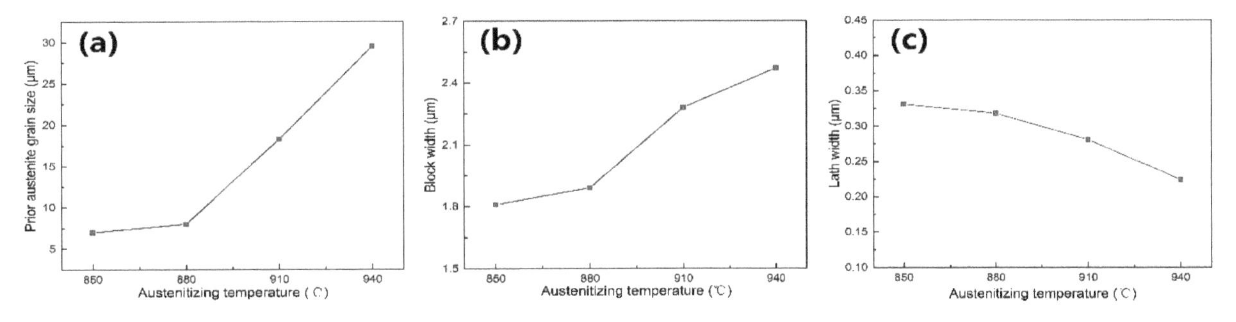

Figure 3. Relationship between austenitizing temperature and (**a**) prior austenite grain size, (**b**) block width, and (**c**) lath width.

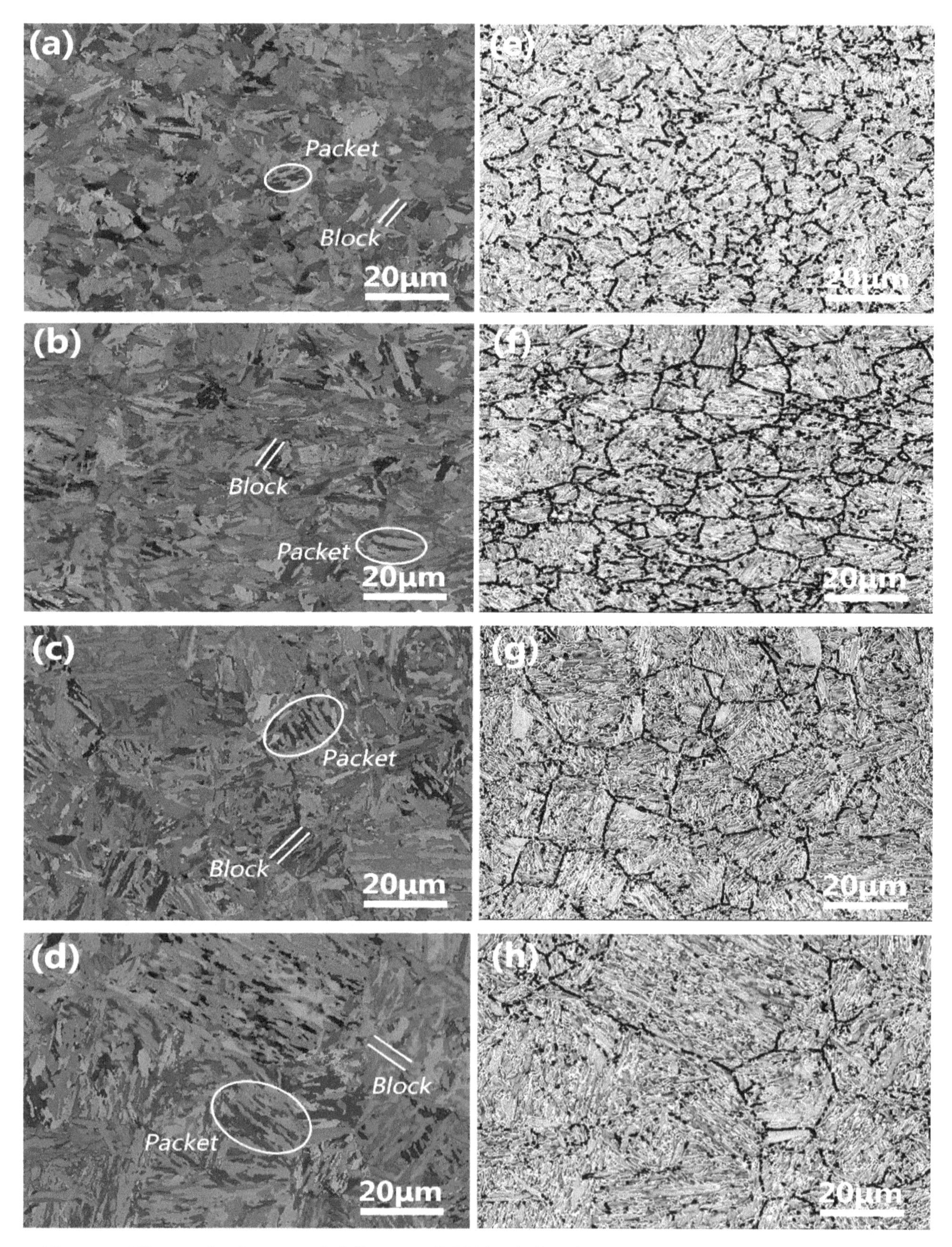

Figure 4. Electron backscattered diffraction (EBSD) of all Euler maps and the corresponding band contrast maps of MCLA martensitic steel austenitized at different temperatures of (**a**,**e**) 850 °C; (**b**,**f**) 880 °C; (**c**,**g**) 910 °C; (**d**,**h**) 940 °C (black lines: 45° > θ > 15°; and yellow lines: θ > 45°).

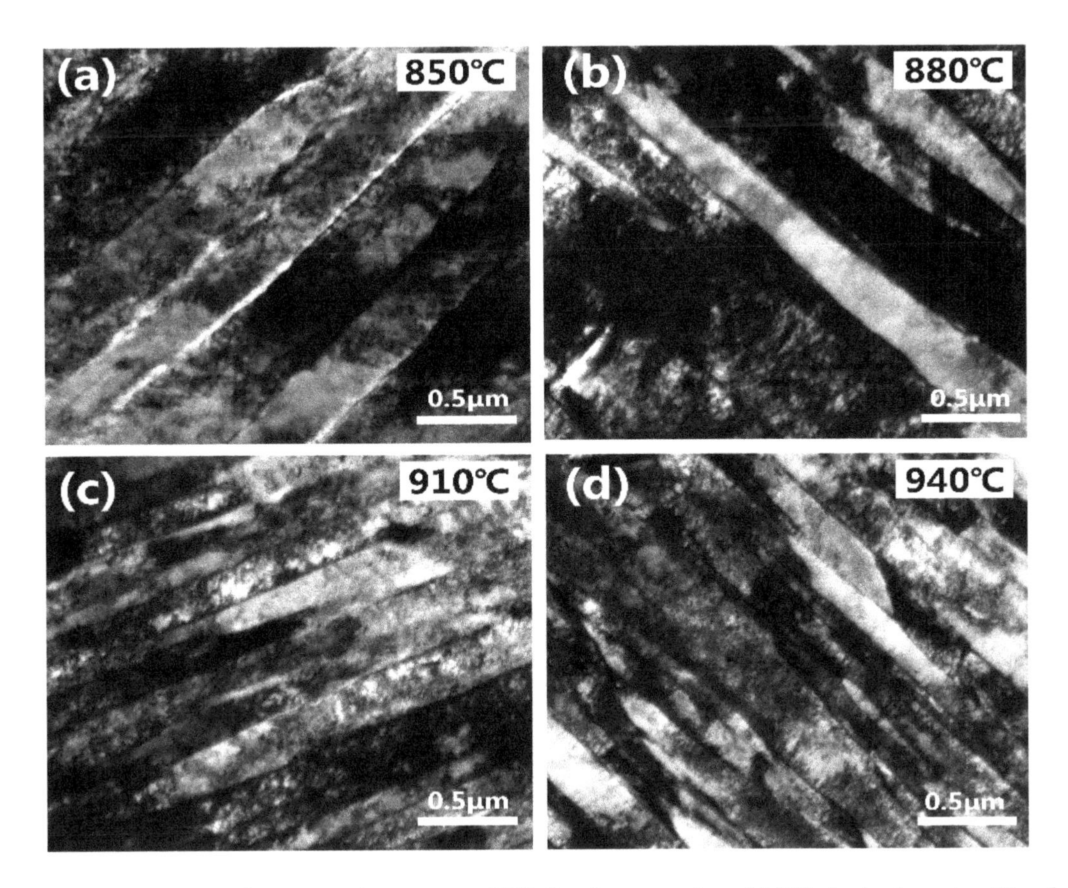

Figure 5. Transmission electron microscopy (TEM) micrographs of MCLA steel austenitized at the different temperatures: (**a**) 850 °C, (**b**) 880 °C, (**c**) 910 °C, and (**d**) 940 °C.

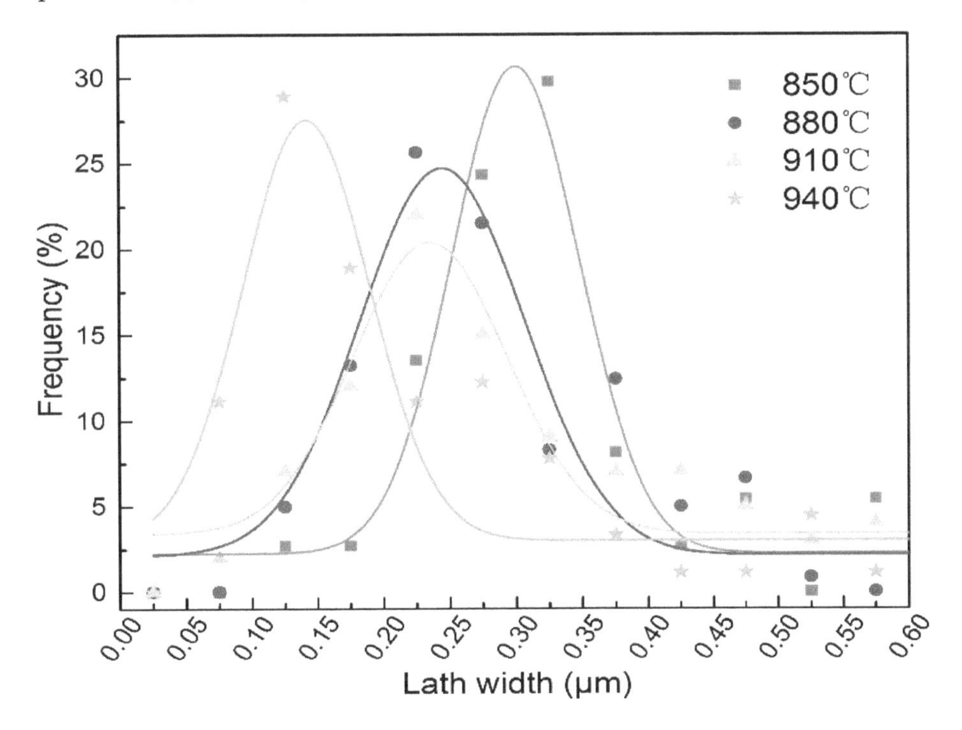

Figure 6. Size distribution of lath width at different austenitizing temperatures.

Figure 7 shows the relationship between d_B, d_L, and PAG size, respectively. The d_B increased linearly and the d_L decreased linearly with the PAG size at different austenitizing temperatures. For example, the d_B changed from 2.28 μm to 2.47 μm as the PAG size increased from 18.3 μm to 29.5 μm, while the d_L decreased from 281 nm to 224 nm. Similar results were reported by Long et al. [13].

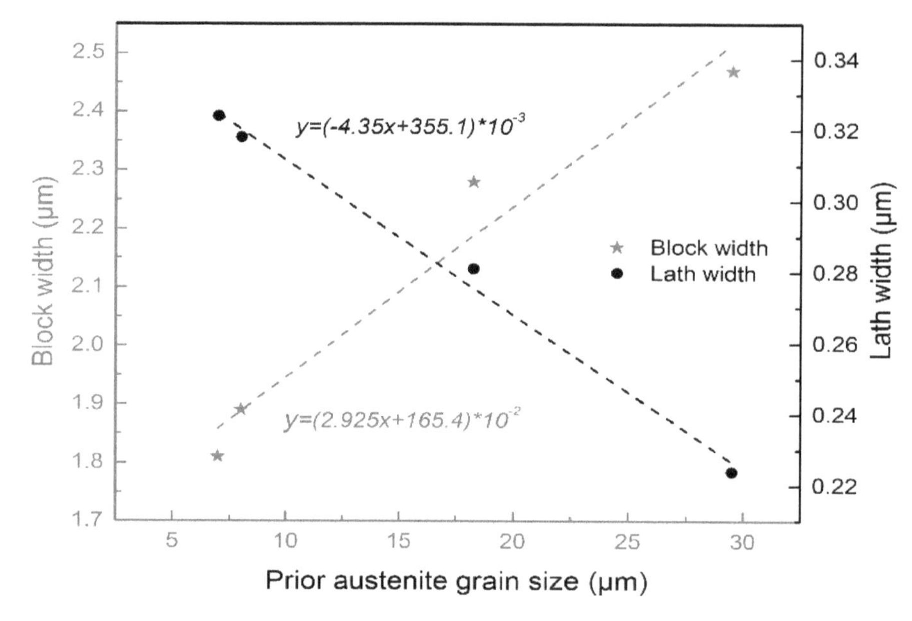

Figure 7. Dependence of block width and lath width on prior austenite grain size in MCLA steel.

3.2. Tensile Properties

Figure 8 shows the relationship between the tensile properties and the austenitizing temperature. When the quenching temperature was increased from 850 °C to 940 °C, the yield strength increased from 1510 MPa to 1591 MPa while the tensile strength increased from 2120 MPa to 2244 MPa. The elongation and section shrinkage were similar for all specimens, with the former being ~8% and the latter being ~35%. The high density of dislocations and boundaries restricted the movement of dislocations, resulting in the poor ductility of all of the samples.

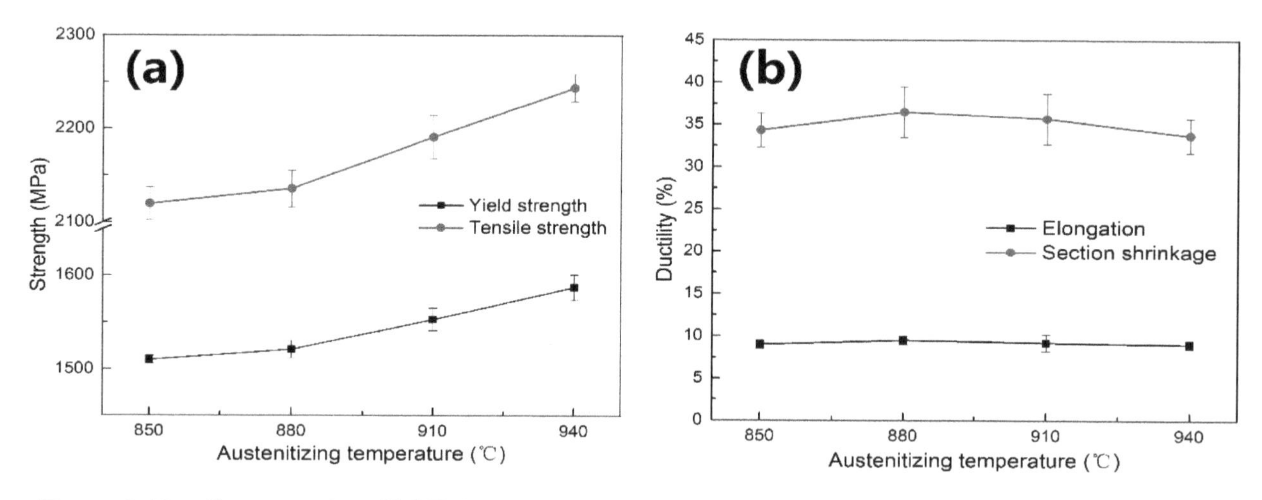

Figure 8. Tensile properties of MCLA martensite steel at different austenitizing temperatures: (**a**) yield strength and tensile strength; (**b**) elongation and section shrinkage.

3.3. Effect of Lath Martensite Microstructures on the Strength

Four strengthening contributions were considered in this work: solid-solution strengthening ($\Delta\sigma_s$), precipitation strengthening ($\Delta\sigma_p$), dislocation strengthening ($\Delta\sigma_d$), and grain boundary strengthening ($\Delta\sigma_g$). The yield strength can be expressed as Equation (1) [22,23]:

$$\sigma_{YS} = \Delta\sigma_0 + \Delta\sigma_s + \Delta\sigma_p + \Delta\sigma_d + \Delta\sigma_g \quad (1)$$

where σ_{YS} is the yield strength (MPa) and $\Delta\sigma_0$ is the intrinsic strength of the matrix (MPa), which was estimated to be 85~88 MPa [24,25].

The precipitation strengthening can be expressed as Equation (2) in light of the Orowan relationship [26,27]:

$$\Delta\sigma_p = \frac{0.538Gbf^{0.5}}{D} ln\left(\frac{D}{2b}\right) \tag{2}$$

where f is the volume fraction of the carbide, D is the mean particle size, and b is the Burgers vector of 0.248 nm. In this study, the carbides dissolved almost completely into the austenite and almost no precipitates remained in the microstructure. Therefore, the value of $\Delta\sigma_p$ can be regarded as 0 MPa while $\Delta\sigma_s$ was similar in all specimens.

$\Delta\sigma_d$ can be expressed as Equation (3) [27]:

$$\Delta\sigma_d = \alpha MGb\rho^{0.5} \tag{3}$$

where M is the Taylor factor, α is a constant of 0.435, G is the shear modulus, and ρ is the dislocation density.

Figure 9 shows the XRD patterns of specimens austenitized at different temperatures. No diffraction peak of retained austenite could be observed in the XRD pattern, meaning that the of retained austenite content was very small. The dislocation densities were calculated according to the XRD results. The calculation method is drawn from references [28,29]. The measured dislocation densities are shown in Figure 9b. It is shown that the dislocation density only changed slightly with the increase of austenitizing temperature, which is consistent with reference [30]. According to Equation (3), the $\Delta\sigma_d$ values were considered to be identical. Therefore, the change of grain boundary strengthening led to an increase of yield strength when the austenitizing temperature was increased. Under such circumstances, the yield strength is used as the value of grain boundary strengthening.

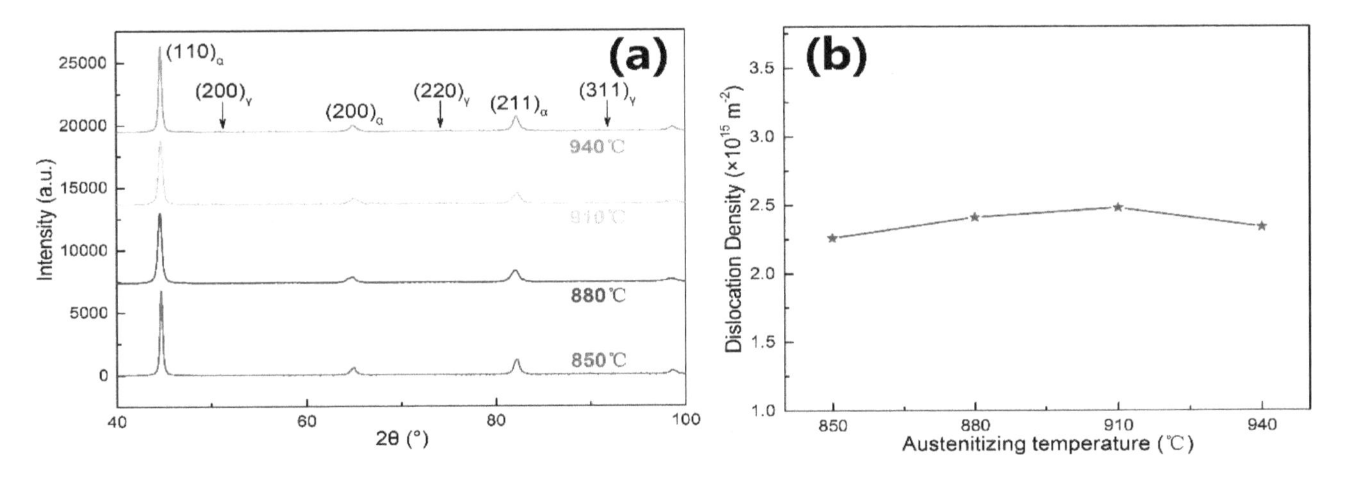

Figure 9. (**a**) X-ray diffraction (XRD) spectra and (**b**) dislocation density of MCLA martensitic steel austenitized at 850–940 °C.

The grain boundary strengthening can be described by the classic Hall–Petch relationship:

$$\sigma_{Ys} = \sigma_0 + k_y d^{-0.5} \tag{4}$$

where k_y is the Hall–Petch slope and d is the EGS. Evidently, the smaller the EGS, the more the boundaries can hinder dislocation motion, and the higher the strength. The relationship between the strength and lath martensite microstructure sizes is shown in Figure 10. The strength decreased linearly with $d_R^{-0.5}$ and $d_B^{-0.5}$, while it increased linearly with $d_L^{-0.5}$. That is, only the relationship between strength and lath width followed the Hall–Petch formula. In other words, the lath width was finally determined as the EGS of strength in the experimental MCLA steel.

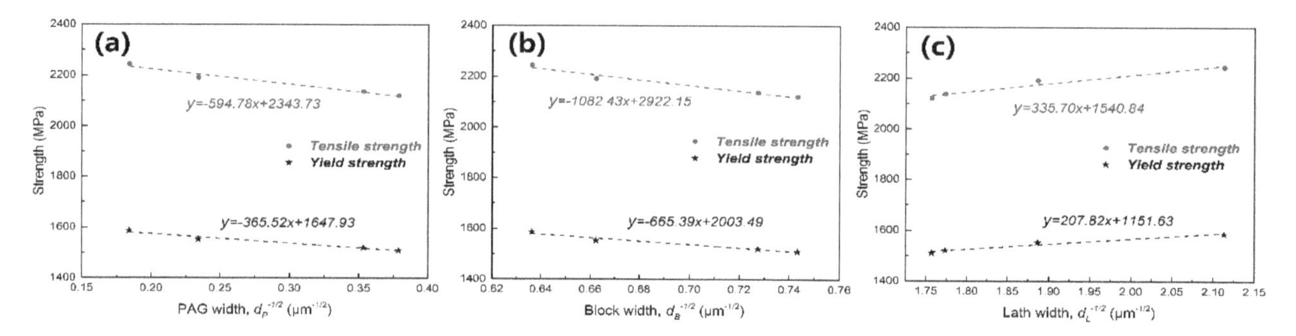

Figure 10. Strength as a function of (**a**) prior austenite grain (PAG) size (d_R), (**b**) block width (d_B), and (**c**) lath width (d_L).

In lath martensite, the PAG boundaries, packet boundaries, and block boundaries are HAGBs, while the lath boundaries are LAGBs. Lath width acting as the EGS means that the LAGBs play a dominant role in hindering dislocation motion. Many previous studies have suggested that only HAGBs can hinder the dislocation motion effectively and cause strengthening [31–33]. However, recent research has come to the opposite conclusion. Du et al. [34] demonstrated that both high- and low-angle grain boundaries act as effective barriers to dislocation movement via uniaxial micro-tensile tests. Chen et al. [35] directly observed that a large number of dislocations accumulate in front of the lath boundaries via in situ TEM experiments, proving that the LAGB is capable of hindering dislocation motion and causing strengthening. These findings act as further proof of our results showing that lath width can act as the EGS in MCLA steel.

3.4. *Effect of Carbon Content on the Effective Grain Size of Strength*

Figure 11 summarizes the relationship between strength and lath martensite microstructure sizes in recent studies [13,14,16,18,30]. As shown in Figure 11a, with increasing of $d_B^{-0.5}$, the strength increased in steels with carbon content below 0.2 wt%, while it decreased linearly in steels with a carbon content above 0.4 wt%. Figure 11b shows the relationship between strength and lath width, revealing the opposite result to that shown in Figure 11a. The relationship between the lath width and strength only followed the Hall–Petch relationship in the medium-carbon steels. Thus, it can be concluded that the EGS of strength seems to be related to the carbon content. In steels with lower carbon content the block size acted as the EGS. While, in steels with higher carbon content, the EGS changed to lath width. That is, the carbon content of steel affects the role of HAGB and LAGB in preventing dislocation movement. When the carbon content is low, HAGBs are the most significant barriers to the dislocation motion, and consequently the hindering effect of LAGB can be ignored. With an increase in carbon content, the hindering effect of LAGB on dislocation movement becomes more important, leading to the gradual transformation of EGS to lath width.

The effect of the Cottrell atmosphere around boundaries may be responsible for this change. Carbon atoms segregate around boundaries and dislocations to reduce distortion energy, leading to the formation of the Cottrell atmosphere [36–38]. During the slipping process, the dislocations are forced to move along with the Cottrell atmospheres, resulting in the so-called "drag effect", which can effectively increase the resistance of the dislocation movement and pin the dislocation. Previous studies have revealed that the drag effect enhances when the carbon content is increased [36,39]. When the carbon content is low (lower than 0.2 wt%), the drag effect of the Cottrell atmosphere becomes weak and the hindering effect of grain boundaries on dislocation motion is mainly dependent on its distorted lattice structure. In this case, the barrier effect of HAGB on dislocation movement is much stronger than that of LAGB. Accordingly, the block width acts as EGS in steels with low carbon content. With the increase of carbon content in steel, the drag effect of the Cottrell atmosphere is greatly enhanced. Accordingly, the difference between the hindering effect of HAGB and LAGB on the dislocation motion, which is caused by the lattice structure, is reduced. When the carbon content is high enough (higher than 0.4 wt%), the LAGB, which holds the absolute advantage in quantity, becomes the dominant factor in hindering dislocation movement. Therefore, the lath width becomes the EGS of strength in steels with high carbon content.

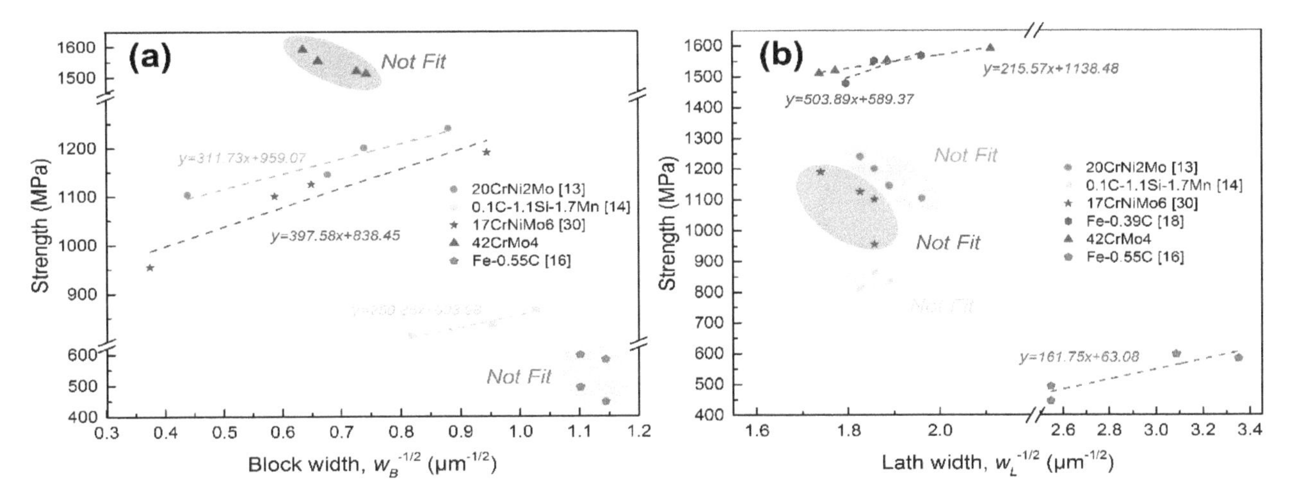

Figure 11. Yield strength as a function of (**a**) block width and (**b**) lath width.

4. Conclusions

We studied the effect of martensite microstructures on the strength of MCLA steel. The results are summarized as follows:

1. When increasing the austenitizing temperature, the PAG size and block size become larger, while the lath width decreases in the experimental MCLA steel. Both the increment of block width and the reduction of the lath width show a linear variation with the increase of the PAG size.
2. When the austenitizing temperature rises, the yield strength and the tensile strength are elevated due to the enhancement of the grain boundary strengthening. The Hall–Petch fitting results reveal that only the relationship between lath width and strength follows the Hall–Petch formula, which indicates that the lath width is the effective grain size of strength in the experimental MCLA steel.
3. Carbon content has a significant effect on the EGS of strength in lath martensite. In terms of low-carbon steels with a carbon content lower than 0.2 wt%, block size acts as the effective grain size, while the EGS tends to be lath width in steels with a high carbon content of over 0.4 wt%. The effect of the Cottrell atmosphere around boundaries is considered to be responsible for this change.

Author Contributions: Conceptualization, C.S., Y.C., and P.F.; methodology, C.S.; validation, C.S., Y.C., and P.F.; formal analysis, C.S.; investigation, C.S., Y.C., and P.F.; resources, P.F. and H.L. (Hongwei Liu); data curation, C.S. and P.F.; writing—original draft preparation, C.S.; writing—review and editing, C.S., Y.C., H.L. (Hanghang Liu), H.L. (Hongwei Liu), and N.D.; visualization, C.S., N.D. and H.L. (Hanghang Liu); project administration, P.F. and H.L. (Hongwei Liu); funding acquisition, H.L. (Hongwei Liu), P.F., and H.L. (Hanghang Liu) All authors have read and agreed to the published version of the manuscript.

References

1. Morris, J.W., Jr.; Kinney, C.; Pytlewski, K.; Adachi, Y. Microstructure and cleavage in lath martensitic steels. *Sci. Technol. Adv. Mater.* **2013**, *144*, 014208.
2. Kinney, C.C.; Pytlewski, K.R.; Khachaturyan, A.G.; Morris, J.W., Jr. The microstructure of lath martensite in quenched 9Ni steel. *Acta Mater.* **2014**, *69*, 372–385. [CrossRef]
3. Sun, C.; Fu, P.X.; Liu, H.W.; Liu, H.H.; Du, N.Y. Effect of tempering temperature on the low temperature impact toughness of 42CrMo4-V steel. *Metals Open Access Metall. J.* **2018**, *8*, 232. [CrossRef]

4. Shibataa, A.; Nagoshi, T.; Sone, M.; Morito, S.; Higo, Y. Evaluation of the block boundary and sub-block boundary strengths of ferrous lath martensite using a micro-bending test. *Mater. Sci. Eng. A* **2010**, *527*, 7538–7544. [CrossRef]
5. Morito, S.; Tanaka, H.; Konishi, R.; Furuhara, T.; Maki, T. The morphology and crystallography of lath martensite in Fe-C alloys. *Acta Mater.* **2003**, *51*, 1789–1799. [CrossRef]
6. Morito, S.; Huang, X.; Furuhara, T.; Maki, T.; Hansen, N. The morphology and crystallography of lath martensite in alloy steels. *Acta Mater.* **2006**, *54*, 5323–5331. [CrossRef]
7. Grange, R.A. Strengthening steel by austenite grain refinement. *Trans. ASM* **1966**, *59*, 26–48.
8. Swarr, T.; Krauss, G. The effect of structure on the deformation of as-quenched and tempered martensite in a Fe-0.2 pct C alloy. *Metall. Trans. A* **1976**, *7A*, 41–48. [CrossRef]
9. Roberts, M.J. Effect of transformation substructure on the strength and toughness of Fe-Mn alloys. *Metall. Trans.* **1970**, *1*, 3287–3294.
10. Wang, C.F.; Wang, M.Q.; Shi, J.; Hui, W.J.; Dong, H. Effect of microstructure refinement on the strength and toughness of low alloy martensitic steel. *J. Mater. Sci. Technol.* **2007**, *23*, 659–664.
11. Morito, S.; Yoshida, H.; Maki, T.; Huang, X. Effect of block size on the strength of lath martensite in low carbon steels. *Mater. Sci. Eng. A* **2006**, *438–440*, 237–240. [CrossRef]
12. Zhang, C.Y.; Wang, Q.F.; Ren, J.X.; Li, R.X.; Wang, M.Z.; Zhang, F.C.; Yan, Z.S. Effect of microstructure on the strength of 25CrMo48V martensitic steel tempered at different temperature and time. *Mater. Des.* **2012**, *36*, 220–226. [CrossRef]
13. Long, S.L.; Liang, Y.L.; Jiang, Y.; Liang, Y.; Yang, M.; Yi, Y.L. Effect of quenching temperature on martensite multi-level microstructures and properties of strength and toughness in 20CrNi2Mo steel. *Mater. Sci. Eng. A* **2016**, *676*, 38–47. [CrossRef]
14. Li, S.C.; Zhu, G.M.; Kang, Y.L. Effect of substructure on mechanical properties and fracture behavior of lath martensite in 0.1C-1.1Si-1.7Mn steel. *J. Alloy. Compd.* **2016**, *675*, 104–115. [CrossRef]
15. Smith, D.W.; Hehemann, R.F. Influence of structural parameters on the yield strength of tempered martensite and lower bainite. *JISI* **1971**, 476–481.
16. Kim, B.; Boucard, E.; Sourmail, T.; Martín, D.S.; Gey, N.; Rivera-Díaz-del-Castillo, P.E.J. The influence of silicon in tempered martensite: Understanding the microstructure–properties relationship in 0.5–0.6 wt.% C steels. *Acta Mater.* **2014**, *68*, 169–178. [CrossRef]
17. Galindo-Nava, E.I.; Rivera-Díaz-del-Castillo, P.E.J. A model for the microstructure behaviour and strength evolution in lath martensite. *Acta Mater.* **2015**, *98*, 81–93. [CrossRef]
18. Wang, J.; Xu, Z.; Lu, X. Effect of the Quenching and Tempering Temperatures on the Microstructure and Mechanical Properties of H13 Steel. *J. Mater. Eng. Perform* **2020**. [CrossRef]
19. Speich, G.R.; Warlimont, H. Yield strength and transformation substructure of low-carbon martensite. *J. Iron Steel Res.* **1968**, *206*, 385–394.
20. Su, T.Y.H. Effect of lath martensite morphology on the mechanical properties steel. *Heat. Treat.* **2009**, *3*, 1–24.
21. Wang, S.X. *Metal Heat Treatment Principles and Process*; Harbin Industrial of Technology Press: Harbin, China, 2009.
22. Chen, J.; Lv, M.Y.; Tang, S.; Liu, Z.Y.; Wang, G.D. Influence of cooling paths on microstructural characteristics and precipitation behaviors in a low carbon V–Ti microalloyed steel. *Mater. Sci. Eng. A* **2014**, *594*, 389–393. [CrossRef]
23. Yen, H.W.; Chen, P.Y.; Huang, C.Y.; Yang, J.R. Interphase precipitation of nanometer-sized carbides in a titanium–molybdenum-bearing low-carbon steel. *Acta Mater.* **2011**, *59*, 6264–6274. [CrossRef]
24. Halfa, H. Recent trends in producing ultrafine grained steels. *J. Miner. Mater. Charact. Eng.* **2014**, *2*, 428–469. [CrossRef]
25. Cheng, X.Y.; Zhang, H.X.; Li, H.; Shen, H.P. Effect of tempering temperature on the microstructure and mechanical properties in mooring chain steel. *Mater. Sci. Eng. A* **2015**, *636*, 164–171. [CrossRef]
26. Peng, H.L.; Hu, L.; Ngai, T.W.; Li, L.J.; Zhang, X.L.; Xie, H.; Gong, W.P. Effects of austenitizing temperature on microstructure and mechanical property of a 4-GPa-grade PM high-speed steel. *Mater. Sci. Eng. A* **2018**, *719*, 21–26. [CrossRef]
27. Yong, Q.L. *Second Phases in Structural Steels*; Metallurgical Industry Press: Beijing, China, 2006.
28. HajyAkbary, F.; Sietsma, J.; B€ottger, A.J.; Santofimia, M.J. An improved X-ray diffraction analysis method to characterize dislocation density in lath martensitic structures. *Mater. Sci. Eng. A* **2015**, *639*, 208–218. [CrossRef]
29. Williamson, G.K.; Smallman, R.E. Dislocation densities in some annealed and cold-worked metals from measurements on the x-ray Debye-Scherrer spectrum. *Philos. Mag.* **1955**, *1*, 34–46. [CrossRef]

30. Wang, C.F. Study of Microstructure Control Unit on Strength and Toughness in Low Alloy Martensite Steel. Ph. D Thesis, Central Iron & Steel Research Institute, Beijing, China, 2008.
31. Shibata, A.; Nagoshi, T.; Sone, M.; Morito, S.; Higo, Y. Micromechanical characterization of deformation behavior in ferrous lath martensite. *J. Alloys. Compd.* **2013**, *577*, S555–S558. [CrossRef]
32. Lim, S.; Shin, C.; Heo, J.; Kim, S.; Jin, H.; Kwon, J.; Guim, H.; Jang, D. Micropillar compression study of the influence of size and internal boundary on the strength of HT9 tempered martensitic steel. *J. Nucl. Mater.* **2018**, *503*, 263–270. [CrossRef]
33. Mine, Y.; Hirashita, K.; Takashima, H.; Matsuda, M.; Takashima, K. Micro-tension behaviour of lath martensite structures of carbon steel. *Mater. Sci. Eng. A* **2013**, *560*, 535–544. [CrossRef]
34. Du, C.; Hoefnagels, J.P.M.; Vaes, R.; Geers, M.G.D. Block and sub-block boundary strengthening in lath martensite. *Scr. Mater.* **2016**, *116*, 117–121. [CrossRef]
35. Chen, S.J.; Yu, Q. The role of low angle grain boundary in deformation of titanium and its size effect. *Scr. Mater.* **2019**, *163*, 148–151. [CrossRef]
36. Hutchinson, B.; Hagström, J.; Karlsson, O.; Lindell, D.; Tornberg, M.; Lindberg, F.; Thuvander, M. Microstructures and hardness of as-quenched martensites (0.1–0.5% C). *Acta Mater.* **2011**, *59*, 5845–5858. [CrossRef]
37. Takahashi, J.; Ishikawa, K.; Kawakami, K.; Fujioka, M.; Kubota, N. Atomic-scale study on segregation behavior at austenite grain boundaries in boron- and molybdenum-added steels. *Acta Mater.* **2017**, *133*, 41–54. [CrossRef]
38. Miyamoto, G.; Goto, A.; Takayama, N.; Furuhara, T. Three-dimensional atom probe analysis of boron segregation at austenite grain boundary in a low carbon steel - Effects of boundary misorientation and quenching temperature. *Scr. Mater.* **2018**, *154*, 168–171. [CrossRef]
39. Waseda, O.; Veiga, R.G.; Morthomas, J.; Chantrenne, P.; Becquart, C.S.; Ribeiro, F.; Jelea, A.; Goldenstein, H.; Perez, M. Formation of carbon Cottrell atmospheres and their effect on the stress field around an edge dislocation. *Scr. Mater.* **2017**, *129*, 16–19. [CrossRef]

9

Effect of Surface Modification of a Titanium Alloy by Copper Ions on the Structure and Properties of the Substrate-Coating Composition

Marina Fedorischeva [1,*], Mark Kalashnikov [1,2], Irina Bozhko [1,2], Olga Perevalova [1] and Victor Sergeev [1,2]

1 Institute of Strength Physics and Materials Science SB RAS, 634055 Tomsk, Russia; kmp1980@mail.ru (M.K.); bozhko_irina@mail.ru (I.B.); perevalova52@mail.ru (O.P.); vs@ispms.tsc.ru (V.S.)
2 Department of Materials Science, National Research Tomsk Polytechnic University, 634050 Tomsk, Russia
* Correspondence: fed_mv@mail.ru or fmw@ispms.tsc.ru

Abstract: To improve the strength properties, adhesion, and the thermal cycling resistance of ceramic coatings, the titanium alloy surface was modified with copper ions under different processing times. It is found that at the maximum processing time, the thickness of the alloyed layer reaches 12 μm. It is shown that the modified layer has a multiphase structure in addition to the main α and β–titanium phases with the intermetallic compounds of the Ti-Cu system. The parameters of the fine structure of the material are investigated by the X-ray diffraction analysis. It has been found that when the surface of the titanium alloy is modified, depletion occurs in the main alloying elements, such as aluminum and vanadium, the crystal lattice parameter increases, the root-mean-square (rms) displacements of the atoms decrease, and the macrostresses of compression arise. A multilevel micro- and nanoporous nanocrystalline structure occurs, which leads to an increase in the adhesion and the thermal cyclic resistance of the ceramic coating based on Si-Al-N.

Keywords: phase composition; structure; surface modification; elemental distribution; ionic treatment; X-ray structural analysis; electron microscopy; multiscale structure

---

1. Introduction

The VT23 alloy is related to the Ti-Al-V-Mo-Cr-Fe system. This is a medium-alloyed (α + β) martensitic-class alloy, which gets the martensitic "α″ structure after hardening from the β-region. This alloy has high ductility, which is an important property for technological applications such as drawing, flanging, and other pressure treatment operations [1].

This property of the titanium alloy allows using it as the basis for deposition of heat-shielding coatings such as Zr-Y-O, Si-Al-N [2–6]. However, in order to improve the thermocyclic resistance and the adhesion properties of the coating, it is necessary to prepare the titanium alloy for coating using different ion-beam technology [7–13].

It is known that across the substrate surface adjacent to the coating, there is a sharp jump in the change in the structural phase state and the physical and mechanical properties of the "heat-protective coating—substrate" system. In this interface region, there appears the maximum localization of the elastic stress. In addition, the state of the substrate surface can significantly affect the formation of the structure and the properties of the coating itself [14,15].

To prepare the surface for deposition of a coating, the surface hardening of metals and alloys is widely used by increasing the density of dislocations in a layer up to 10 microns in depth (a long-range effect [16]) or due to the formation of nanocrystalline intermetallic phases [17]. In the latter case, the element pairs of the "ion beam—processed metal" are selected from those systems in which the formation of intermetallides is possible. Thus, in [17], the ionic synthesis of the intermetallic phases based on the Ni-Ti, Ni-Al, Fe-Al, and TiAl systems was discovered. However, the ionic synthesis in these systems is poorly understood.

One of the effective methods of preparing a substrate for the deposition of coatings is the modification of their surface layer with high-energy beams of metal ions. In this case, treatment with ion beams can change the morphology, phase, and the elemental composition of the surface layer [18].

Therefore, an urgent task is to study the effect of the phase composition, the structural state of the surface layer of a titanium substrate on the structure, and the properties of the coating formed on it, as well as on the thermomechanical characteristics of the whole "coating—substrate" system. This article focuses on the study of the surface modification of the VT23 titanium alloy, which was carried out to improve the thermal cycling resistance and the adhesion strength of the Si-Al-N-based coating.

2. Materials and Methods

Samples were treated under a continuous titanium ion beam with accelerating voltage (1200 V ± 20 V) and ion current (about 15 mA) using the vacuum system KVANT-03MI (Techimplant Ltd., Tomsk, Russia), and the vacuum arc ion source with a copper cathode [19]. The samples were placed in the camera on the object table in front of the ion source for ion bombardment. The temperature of the samples during the ion bombardment was 900 ± 100 K. After the ion beam treatment of the substrate, a Si-Al-N coating was deposited using a magnetron sputtering technique. Parameters of treatment of the titanium alloy by copper ions Technological parameters of processing are shown in Table 1.

Table 1. Parameters of treatment of the titanium alloy by copper ions.

Samples	Bias, V	Treatment Time, min	Fluence, Ion/cm^2
Initial Ti alloy	-	-	-
Treated Ti	−900	3	0.9×10^{18}
Treated Ti	−900	6	1.8×10^{18}
Treated Ti	−900	7.5	2.3×10^{18}

The structural phase state of the ion-modified layers of the samples was investigated by the Transmission Electron Microscope TEM method using a JEOL-2100 device (Jeol Ltd., Tokyo, Japan). Foils for the TEM studies were prepared by the cross-section method using an Ion Slicer EM-09100IS (Jeol Ltd., Tokyo, Japan). To classify the structures, the grain size, and the phase composition, the bright-field images together with the corresponding microdiffraction patterns and the dark-field images were used. The phase composition, the crystalline lattice parameters, the root-mean-square atomic displacement (rms), the macrostresses of the surface layer were determined by X-ray using the DRON-7 device St. (UED-Lab, Petersburg, Russia) [20,21]. X-ray investigation of the modified titanium alloy was carried out under continuous 2θ-scanning with the Bragg–Brentano focusing at Co Kα radiation. The data base JCPDS and PDF-2 (International Centre for Diffraction Data, Campus Blvd, PA, USA) was used for interpretation of the diffractograms.

The symmetric and the asymmetric Bragg–Brentano schemes of X-ray investigation were used. The X-ray diffraction (XRD) registration in an asymmetric mode was carried out at grazing angles of X-ray radiation incidence $\alpha = 3°$. The chemical composition and the element distribution in the titanium surface were determined by the energy dispersive X-ray (EDS) analysis using a microanalyzer INCA-Energy (Oxford Instruments) with the built-in TEM JEOL-2100 and scanning electron microscopy (SEM) LEO EVO-50XVP. The thermal cycling procedure was carried out by heating the samples to 1000 °K in air, followed by holding the sample at this temperature for 1 min and cooling to room temperature.

The VT23 titanium alloy was selected as the material for the study. It contains elements such as: Fe, Cr, Mo, V, Al, as shown in Table 2.

Table 2. The chemical composition of the VT23 alloy (weight %).

Fe	Cr	Mo	V	Ti	Al
0.4–0.8	0.8–1.4	1.5–2.5	4–5	84–89.3	4–6.3

3. Results and Discussion

The VT23 alloy modified with copper ions at different processing times is investigated by X-ray diffraction analysis. Figure 1 shows the diffraction patterns of the titanium alloy modified with copper ions treated for different times (3, 6, and 7.5 min) and the diffraction pattern of the initial titanium alloy VT23. It is seen that the X-ray diffraction patterns of the initial titanium alloy contain the lines of α and β-phases of titanium, which are characteristic of the quenched VT23 alloy. In the composition of the titanium alloy treated with copper ions, in addition to the main phases of α and β-titanium, there are intermetallic phases of the equilibrium state diagram of Ti-Cu [22].

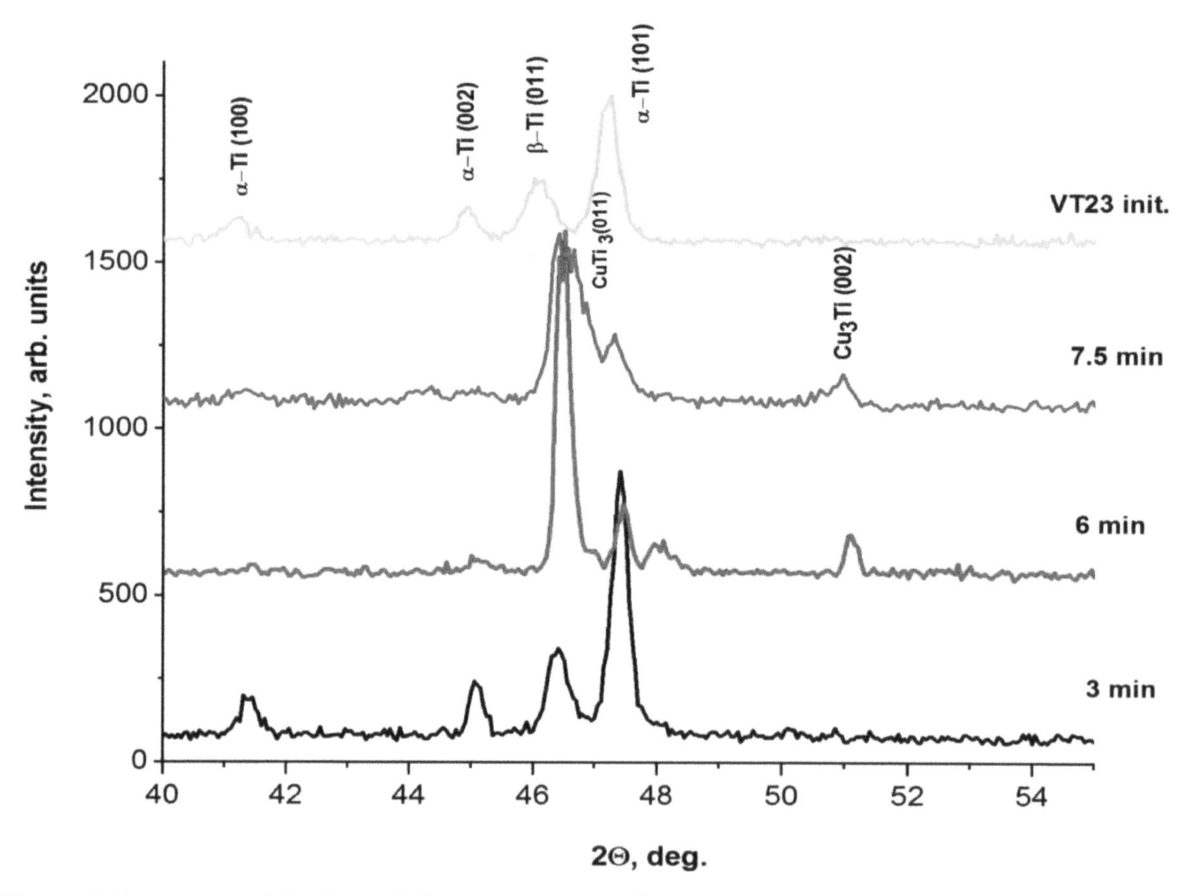

Figure 1. Fragments of the X-ray diffraction patterns of the VT23 alloy in the initial state (1) and upon treatment with copper ions at different processing times (the processing time is indicated on the X-ray diffraction pattern).

It can be seen in the diffraction patterns obtained using the asymmetric X-ray investigation. Figure 1 shows that in the modified layer with a processing time of 7.5 min under the asymmetric X-ray investigation, the intensity of the (111) line of the $CuTi_3$ phase exceeds the intensity of the main peak of the α-phase by several times. This means that the surface layer 1 μm deep mainly consists of the $CuTi_3$ intermetallic compound.

It is interesting to note that at the same processing time, the peak is more extended than all the others. The broadening of the peak can be due to several reasons. First, it is the presence of other phases. At this temperature, in accordance with the phase diagram, the existence of other intermetallic phases, such as Cu_4Ti, Cu_2Ti, is possible, which were subsequently discovered by TEM. Their peaks are close in reflection angles and may overlap. The second reason is a significant grain refinement during more intense bombardment, which we also discovered by the TEM. Concentration inhomogeneity leads to broadening of peaks, since we are dealing with a nonequilibrium process, and the last reason is the creation of higher internal stresses due to defect formation, relaxation and annealing of defects, and accumulation and segregation of impurities. All these reasons can lead to significant peak broadening at maximum processing time.

The crystalline lattice parameter of the α-titanium can be estimated from the average angles of 2θ for the maxima corresponding to this phase (Table 3). The table shows the values of the crystalline lattice parameters of the α-titanium and the c/a ratio for all the investigated alloys.

Table 3. The crystalline lattice parameters, the compressing macrostresses σ //, the rms displacements of atoms $\sqrt{\langle u^2 \rangle}$ for direction <002> of the surface layer at different treatment durations of the titanium alloy with copper ions.

The Structure Parameters	Initial VT23	Treatment Time 3 min	Treatment Time 6 min	Treatment Time 7.5 min
a, nm	0.2893	0.2904	0.2899	0.2918
c, nm	0.4731	0.4705	0.4726	0.4687
c/a	1.63	1.62	1.63	1.61
$\sqrt{<u^2>}$002, nm	0.031	0.023	0.016	0.010
σ//, GPa	-	−1.6	−3.0	-

One can see that the crystalline lattice parameter in the modified titanium is higher than in the VT23 alloy in the initial state. The c/a ratio for this alloy is also higher than for the tabular values of α-titanium (c/a = 1.58). This fact indicates the deformation of the crystal lattice of the titanium alloy during standard heat treatment. The fact that the crystalline lattice parameter of the titanium alloy increases with the ion treatment time suggests that the main dopants, such as aluminum and vanadium, leave the surface layer. At the same time, the rms displacements of the atoms in the modified layer decrease uniformly with increasing ion treatment.

It is shown that the rms displacements of the atoms depend, to a large extent, on the concentration of the alloying elements. As a rule, a decrease in the concentration of the doping atoms in a solid solution leads to a decrease in the rms atomic displacements, which, most likely, is observed in this case with an increase in the duration of the ion treatment. For copper-based solid solutions, it was found that the atomic displacements decrease with a decrease in the concentration of the solid solution [23]. Our assumption has been confirmed by the pattern of the distribution of the elements across the depth of the modified layer.

Figure 2g–i shows that the content of all substitutional elements, such as Al, V, Cr, Mo, first decreases from its initial concentration (table of the composition of the VT23 alloy) in the modified layer, and then gradually reaches its original composition.

The depth of the layer depleted in the alloying elements correlates with the thickness of the modified layer in the first case. With a processing time of 3 min, it is about 150 nm, and then, down to a depth of 540 nm, the content of the alloying elements gradually returns to the previous level (Figure 2g). When the time of the ion treatment is 6 min, the depletion in the alloying elements reaches a depth of about 1 μm and the initial composition, with the thickness of the ion-modified layer being 3.5 μm (Figure 2h). At the maximum processing time, the thickness of the depleted layer is about 1.5 μm and down to 7 μm, with the content of the alloying elements gradually restoring to the initial concentration (Figure 2i). Compressive macrostresses arise in the treated surface, whose magnitudes increase with an increase in the treatment duration of the titanium alloy by an ion beam (Table 3). A similar effect was observed in the work [24], where the authors believe that the stress distribution over the sample depth is such that the alloyed layer is strongly compressed across the surface. The following processes contribute to the formation of a stressed layer (region) upon surface modification: defect formation, relaxation and annealing of defects, accumulation and segregation of impurities, diffusion (thermal and radiation-stimulated), and sputtering.

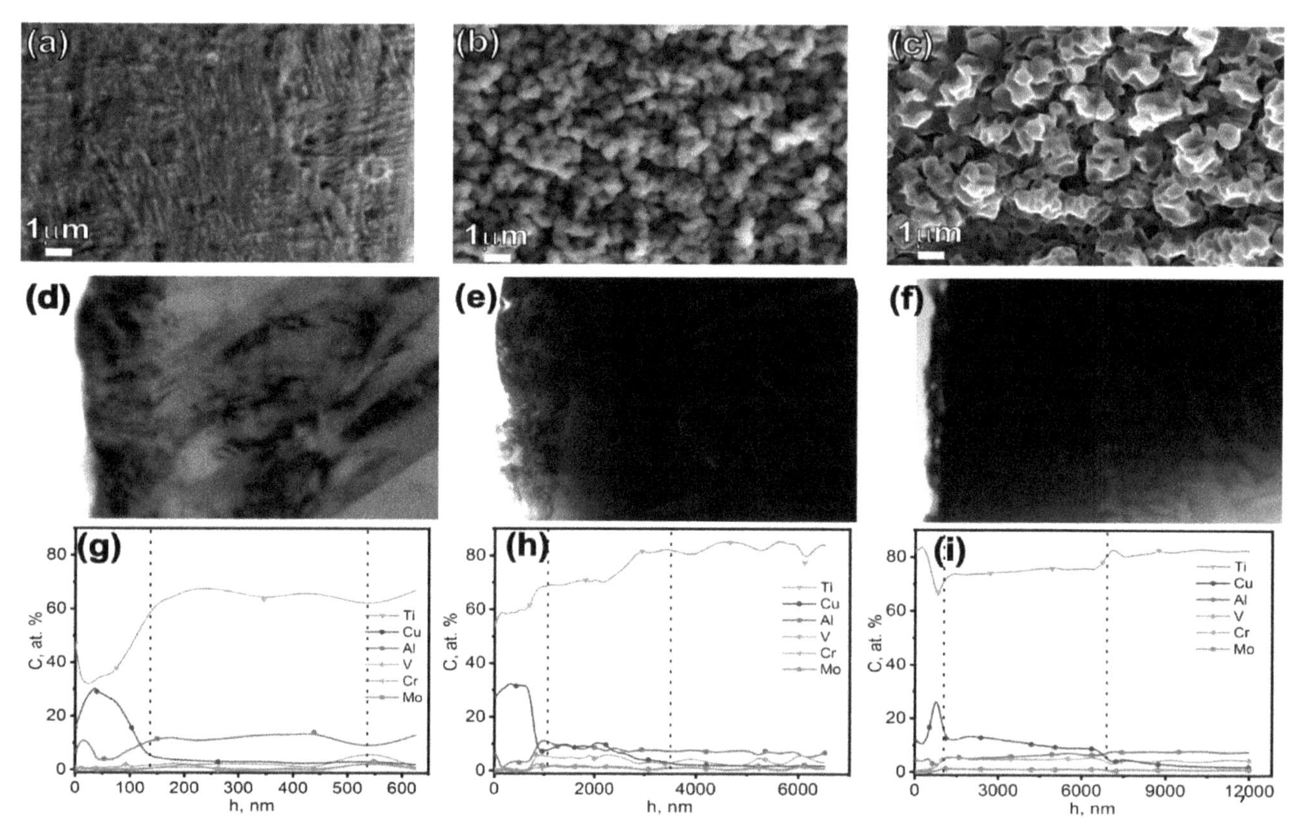

Figure 2. Surface morphology of the treated titanium alloy: 3 min (**a**), 6 min (**b**), 7.5 min (**c**). TEM images of the cross-section of the titanium alloy modified with copper ions with different processing times (**d–f**) and the distribution of the elements in depth: the processing time is 3 min (**g**), 6 min (**h**), 7.5 min (**i**).

Figure 2b,c shows the surface morphology of the titanium alloy treated with copper ions at different processing times. It can be seen that even at a processing time of 3 min, the surface is etched, but the martensite plates are still visible. Upon further processing for 6 min, the structural elements become smaller, the martensitic structure disappears. However, at a processing time of 7.5 min, a structure with a more developed surface takes place; voids are visible between the formations, which should be filled with the material of the deposited coating. In this case, the adhesive properties of the coating are significantly improved, as shown in [25–27].

TEM studies (Figure 3) have confirmed the results of the X-ray phase analysis. It has been established that depending on the processing time, the intermetallic phases of the Cu-Ti equilibrium phase diagram appear in the modified layer. Figure 3a–c shows the cross section of the titanium alloy modified with copper ions for 3 min. One can see that in the upper modified layer, in addition to the main α-Ti phase, there are intermetallic phases of Cu_3Ti, Cu_4Ti_3, and CuTi. The intermetallic phases are identified at a depth of no more than 300 nm (Table 4). The maximum copper content up to 30% is at a depth of 50 nm. Further, in the modified layer, only α-Ti is identified. The depth distribution of the elements shows (Figure 2g) that copper penetrates down to a depth of 600 nm and its amount is not more than 5 at.%. Apparently, it is in solid solution.

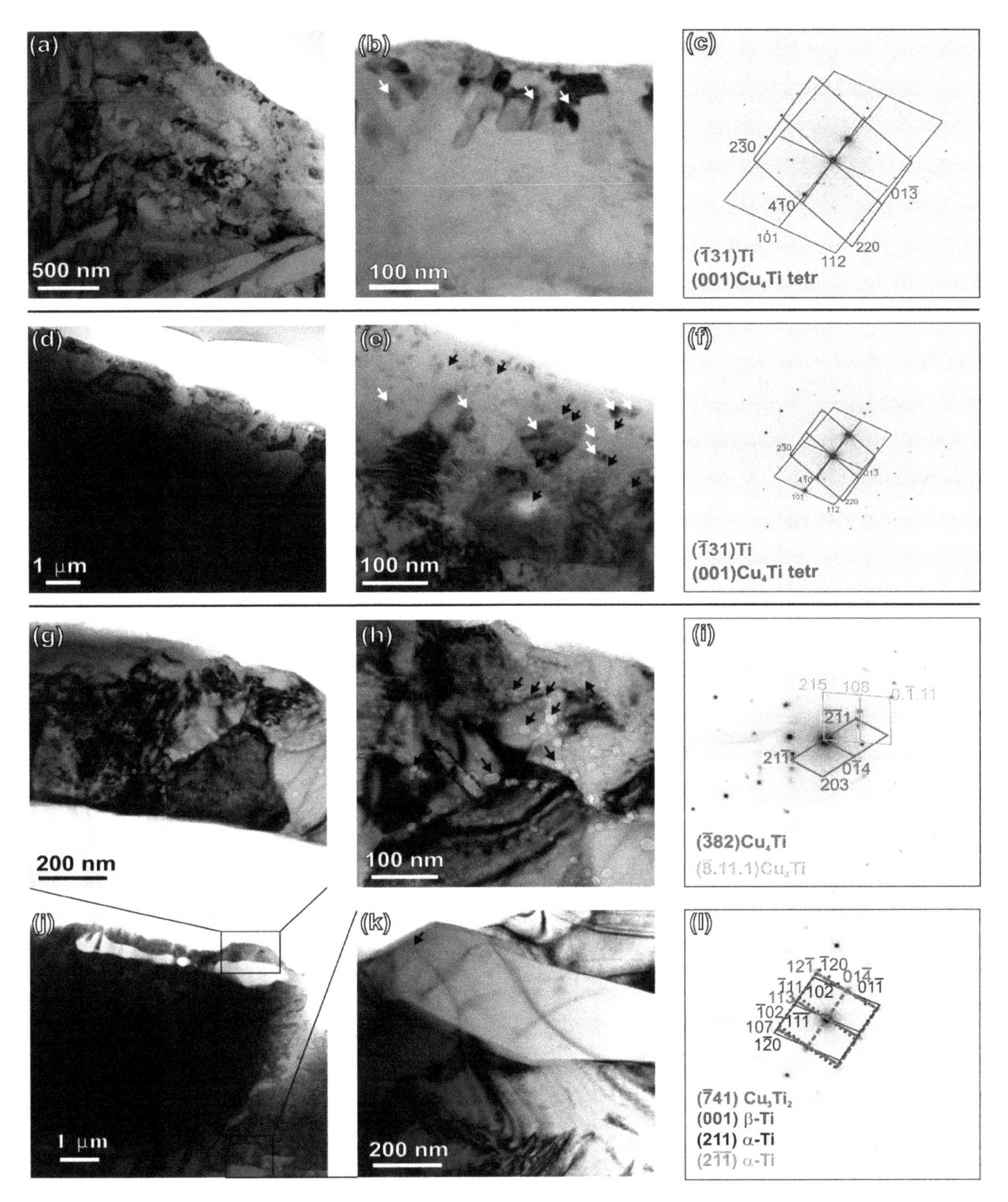

Figure 3. TEM images of the cross-section of the titanium alloy modified with copper ions: during 3 min (**a–c**), bright-field images, microdiffraction, and indication schemes; during 6 min (**d–f**), bright-field images and indication schemes; within 7.5 min (**g–l**). The white arrows show the indicated phases, the black ones—the pores.

Table 4. Phase composition and thickness of Ti modified by Cu ions.

Ion Treatment 3 min		Ion Treatment 6 min		Ion Treatment 7.5 min	
Thickness, nm	Phases	Thickness, nm	Phases	Thickness, nm	Phases
0–150	α-Ti CuTi Cu_4Ti_3 Cu_3Ti	0–250	α-Ti Cu_3Ti Cu_4Ti $CuTi_3$	0–810	α-Ti Cu_2Ti Cu_4Ti
				810–3680	α-Ti CuTi Cu_4Ti
150–500	α-Ti	>250–2000	α-Ti	3680–7500	α-Ti, particles of the Cu_3Ti inside of a-Ti.

For a sample treated with titanium ions for 6 min, the phase formation was found to take place at a depth of 250 nm (Table 4). We can see such phases as Cu_3Ti, Cu_4Ti, $CuTi_3$, and then titanium is observed in the initial state. The maximum amount of copper up to 35 at.% is observed at a distance of 800 nm from the surface, and then it sharply decreases down to 10 at.% and further decreases down to 5 at.% at a distance of 6.5 μm (Figure 2h). In the TEM image, the modified layer is visible up to 3.5 μm from the surface (Figure 3b). In a sample modified during 7.5 min, the pattern is significantly different from the previous ones. The intermetallic phases are observed here almost down to a depth of 4.0 μm. The phases such as Cu_2Ti, Cu_4Ti, CuTi have been identified. Interestingly, the particles of Cu_3Ti within the α-titanium plates are observed in the modified layer at a distance of 6.0 μm from the surface. The maximum amount of copper is observed at a depth of about 1.0 μm, but its concentration at this distance from the surface reaches only 25 at.%, then the copper concentration is 10–12 at.% at a distance of 6.0 μm and down to 12.0 μm the copper concentration remains in the amount of 5 at.% (Figure 2i). It is interesting to note that in this sample, the Cu_3Ti phase is in the α-Ti base phase layer inside the martensite plates at a distance of 7.5 μm from the surface (Table 4).

Figure 3b,e,h shows the pores indicated by black arrows. If in a sample with a short processing time they are only in the upper layer in a small amount, with longer processing times the pores are located everywhere in large quantities. In the sample with a treatment time of 7.5 min, the pore size becomes significantly larger (marked with black arrows). The intermetallic phases present in the modified layer are shown by white arrows.

After a comprehensive study of the titanium alloy modified with copper ions, a Si-Al-N-based coating was applied by magnetron sputtering. For this, the titanium alloy VT 23 is taken in its original state and modified for 7.5 min. Thermal cycling tests were carried out after deposition of the coating. The resistance of the coatings to cracking and peeling under changing temperature was determined from the results of the thermal cycling of the samples according to the following regime: heating the sample up to 1000 °C for 1 min, then cooling for 1 min to 20 °C. For the criterion of the thermal stability of the coatings, the number of the cycles before the destruction of 30% of the coating area of the sample surface was selected [14,28]. After that, the tests were stopped.

The coating, deposited on the initial surface, looks like one containing rectangular hollow cracks. The maximum number of cycles of this coating until catastrophic failure is not more than 500 as shown in Figure 4b. The coating applied to the treated surface contains rounded cracks, which are immediately filled with titanium oxide (in Figure 4d marked with numbers 1 and 2). The amount of oxygen in the cracks (indicated by the number 1) significantly exceeds its value in the coating (number 2). Using the EDS method it was shown that in (1) the oxygen value is about 60 at% and in the coating (2) its amount does not exceed 20 at%. This contributes to the healing of the cracks and prevents their further propagation, which, in turn leads to a high thermal cycling resistance of the coating. It was this coating that withstood 1500 cycles in our investigation (Figure 4d).

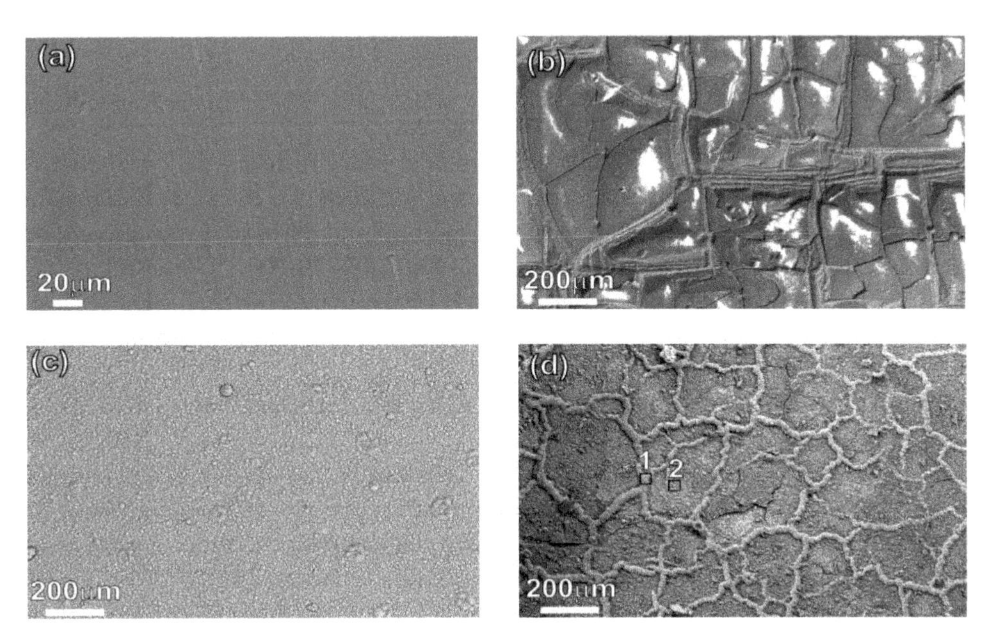

Figure 4. SEM images of coatings on the basis of Si-Al-N: coatings deposited on the initial surface of the titanium alloy: in initial state (**a**), after 500 cycles (**b**). Coating applied to titanium alloy treated with copper ions in initial state (**c**) and after 1500 cycles (**d**).

4. Conclusions

The titanium alloy modified with copper ions at different processing times has been comprehensively studied at various structural levels by the methods of X-ray diffraction analysis and electron and scanning microscopy.

1. It has been established by SEM that the modification of the titanium alloy by copper ions results in the refinement of the structure and the formation of developed micropores.
2. It has been shown by X-ray that the phase composition of the modified surface depends on the processing time. At short times of ionic treatment, the phases rich in titanium ($CuTi_3$) appear. With an increase in the duration of the treatment up to 7.5 min, the reflections of the $CuTi_3$ phase become more intense than those of the titanium substrate, i.e., almost the entire surface consists of an intermetallic compound. The structure parameters of the modified layer change: the crystalline lattice parameters of the titanium alloy increase, the rms displacements of the atom decrease, the compressive macrostresses occur.
3. It has been shown that the depth of the modified layer depends on the processing time. With a minimum processing time of 3 min, the depth of the modified layer is about 150 nm. With a processing time of 6 min, this it is about 3 μm, with a maximum processing time of 7.5 min, it is 7.5 μm, with the copper ions penetrating much deeper (0.6 μm, 7.5 μm, and 12 μm, respectively).
4. It has been found that the intermetallic compounds of the Cu-Ti system are formed not only in the surface layer of the treated surface, but also at a distance of 7.5 μm from the surface located inside the martensite plates, the so-called second-level phases. In the modified surface, the pores are found everywhere for all processing times. As a result, of the above treatment, the multiscale micro- and nanoporous nanocrystalline structure has been produced.
5. The thermocyclic resistance of the Si-Al-N-based coating increases several times if the coating is applied to the surface of the titanium alloy modified with copper ions, which is directly related to an increase in adhesion.

Author Contributions: Conceptualization, V.S. and M.F.; methodology, M.F.; validation, V.S. and O.P.; investigation, M.K., I.B. and M.F.; resources, V.S.; writing—original draft preparation, O.P.; writing—review and editing, M.F., O.P.; funding acquisition, V.S. All authors have read and agreed to the published version of the manuscript.

References

1. Collings, E.W. The physical metallurgy of titanium alloys. In *American Society for Metals*; University of Michigan: Ann Arbor, MI, USA, 1984; Volume 3, pp. 1–261.
2. Huang, X.; Zakurdaev, A.; Wang, E.D. Microstructure and phase transformation of zirconia-based ternary oxides for thermal barrier coating applications. *J. Mater. Sci.* **2008**, *43*, 2631–2641. [CrossRef]
3. Ji, Z.; Haynes, J.A.; Voelkl, E.; Rigsbee, J.M. Phase Formation and Stability in Reactively Sputter Deposited Yttria-Stabilized Zirconia Coatings. *J. Am. Ceram. Soc.* **2001**, *84*, 929–936. [CrossRef]
4. Bartolome, J.F.; Beltran, J.I.; Gutierrez-Gonzalez, C.F.; Pecharroma, C.; Munoz, M.C.; Moya, J.S. Influence of ceramic–metal interface adhesion on crack growth resistance of ZrO2–Nb ceramic matrix composite. *Acta Mat.* **2008**, *56*, 3358–3366. [CrossRef]
5. Enrico, M.; Chiara, S.; Valter, S. Residual Stresses in Alumina/Zirconia Composites: Effect of Cooling Rate and Grain Size. *J. Am. Ceram. Soc.* **2001**, *84*, 2962–2968.
6. Kalashnikov, M.P.; Fedorischeva, M.V.; Sergeev, V.P.; Neyfeld, V.V.; Popova, N.A. Features of surface layer structure of VT23 titanium alloy under bombardment with copper ions. *AIP Conf. Proc.* **2015**, *1683*, 20076.
7. Schmidt, B.; WetzigIon, K. *Beams in Material Processing and Analysis*; Springer: Wien, Austria; Heidelberg, Germany; New York, NY, USA; Dordrecht, The Netherlands; London, UK, 2013; pp. 1–413.
8. Kucheyev, S.O. *Ion.-Beam Processing*; Material Processing Handbook; Joanna, R., Groza, J.F., Shakelford, E.J., Lavernia, M.T., Eds.; Powers, CRC Press: London, UK; New York, NY, USA, 2007; Volume 2, pp. 25–50.
9. Davis, N.A.; Remnev, G.E.; Stinnett, B.W.; Yatsui, K. Intense ion-beam treatment of materials. *MRS Bull.* **1996**, *21*, 58–62. [CrossRef]
10. Chr, A. Straede Ion implantation as an efficient surface treatment. *Nucl. Instrum. Methods Phys. Res.* **1992**, *B68*, 380–388.
11. Zhang, E.; Wang, X.; Chen, M.; Hou, B. Effect of the existing form of Cu element on the mechanical properties bio-corrosion and antibacterial properties of Ti-Cu alloys for biomedical application. *Mater. Sci. Eng.* **2016**, *C69*, 1210–1221. [CrossRef] [PubMed]
12. Walschus, U.; Hoene, A.; Patrzyk, M.; Lucke, S.; Finke, B.; Polak, M.; Lukowski, G.; Bader, R.; Zietz, C.; Podbielski, A.; et al. A Cell-adhesive plasma polymerized allylamine coating reduces the In Vivo Inflammatory Response Induced by Ti6Al4V modified with plasma immersion ion implantation of copper. *J. Funct. Biomater.* **2017**, *8*, 30. [CrossRef] [PubMed]
13. Pierret, C.; Maunoury, L.; Monnet, I.; Bouffard, S.; Benyagoub, A.; Grygiel, C.; Busardo, D.; Muller, D.; Höche, D. Friction and wear properties modification of Ti-6Al-4V alloy surfaces by implantation of multi-charged carbon ions. *Wear* **2014**, *319*, 19–26. [CrossRef]
14. Sergeev, V.P. *Kinetics and Mechanism of the Formation of Nonequilibrium States of Surface Layers under Conditions of Magnetron Sputtering and Ion. Bombardment*; Nanoengineering surface; Lyachko, N.Z., Psahie, S.G., Eds.; Publishing House of the SB RAS: Novosibirsk, Russia, 2008; pp. 227–276.
15. Panin, V.E.; Sergeev, V.P.; Moiseenko, D.D.; Yu, I. Pochivalov, Scientific basis for the formation of heat-protective and wear-resistant the Si-Al-N / Zr-Y-O multilayer coatings. *Fiz. Mezomekh.* **2011**, *14*, 5–14.
16. Didenko, A.N.; Sharkeev, Y.P.; Kozlov, E.V.; Ryabchikov, A.I. *Long-Range Effects in Ion.-Implantable Metallic Materials*; Kolobov, Y.R., Ed.; Scientific and technical literature: Tomsk, Russia, 2004; pp. 4–326.
17. Fedorischeva, M.V.; Kalashnikov, M.P.; Nikonenko, A.V.; Bozhko, I.A. The structure—Phase state and microhardness of the surface layer of the VT1-0 titanium alloys treated by copper ions. *Vacuum* **2018**, *149*, 150–155. [CrossRef]
18. Kurzina, I.A.; Kozlov, E.V.; Sharkeev, Y.P.; Fortuna, S.V.; Koneva, N.A.; Bozhko, I.A.; Kalashnikov, M.P. *Nanocrystalline Intermetallic and Nitride Structures Formed by Ion.-Plasma Exposure*; Koval, N.N., Ed.; T: Publishing House HTJ1: Tomsk, Russia, 2008; pp. 1–317.
19. Sergeev, V.P.; Yanovsky, V.P.; Paraev, Y.N.; Sergeev, O.V.; Kozlov, D.V.; Zhuravlev, S.A. Installation of ion-magnetron sputtering of nanocrystalline coatings (QUANT). *Phys. Mesomech.* **2004**, *7*, 333–336.
20. Gorelik, O.S.; Rastorguev, L.N.; Skakov, Y.A. *X-Ray and Electron-Optical Analysis*; Metallurgiya: Moscow, Russia, 1970; p. 328.
21. Mirkin, L.I. *Spravochnik Po Rentgenostruktunomu Analizu Polikristallov (The Handbook of X-Ray Analysis of Polycrystals)*; Fizmatlit: Moscow, Russia, 1961; p. 864.
22. Lyakishev, N.P. (Ed.) *Diagrams of Binary Metallic Systems*; Mashinostroenie: Moscow, Russia, 1997.

23. Perevalova, O.B.; Konovalova, E.V.; Koneva, N.A.; Kozlov, E.V. *Influence of Atomic Ordering on the Grain Boundary Ensembles in FCC-Solid Solutions*; NTL Publ.: Tomsk, Russia, 2014; p. 247. (In Russia)
24. Domkus, M.; Pranyavichus, L. *Mechanical Stresses in Implanted Solids*; Mokslas: Vilnius, Lithuania, 1990; p. 158.
25. Xu, W.; Lui, E.W.; Pateras, A.; Qian, M.; Brandt, M. In situ tailoring microstructure in additively manufactured Ti-6Al–4V for superior mechanical performance. *Acta Mater.* **2017**, *125*, 390–400. [CrossRef]
26. Guo, G.; Tang, G.; Ma, X.; Sun, M.; Ozur, G.E. Effect of high current pulsed electron beam irradiation on wear and corrosion resistance of Ti6Al4V. *Surf. Coat. Technol.* **2013**, *229*, 140–145. [CrossRef]
27. Shulov, V.A.; Gromov, A.N.; Teryaev, D.A.; Perlovich, Y.A.; Isaenkova, M.G.; Fasenko, V.A. Texture formation in the surface layer of VT6 alloy targets irradiated by intense pulsed electron beams. *Inorg. Mater. Appl. Res.* **2017**, *8*, 387–391. [CrossRef]
28. Perevalova, O.B. Changes in the elastic-stress state and phase composition of TiAlN coatings under thermal cycling. *Inorg. Mater.* **2018**, *9*, 399–409. [CrossRef]

10

Precipitation Stages and Reaction Kinetics of AlMgSi Alloys during the Artificial Aging Process Monitored by Electrical Resistivity Measurement Method

Hong He [1], Long Zhang [1], Shikang Li [1], Xiaodong Wu [1], Hui Zhang [2] and Luoxing Li [1,*]

[1] State Key Laboratory of Advanced Design and Manufacturing for Vehicle Body, College of Mechanical and Vehicle Engineering, Hunan University, Changsha 410082, China; hehong_hnu@hnu.edu.cn (H.H.); long060810304@163.com (L.Z.); kangkangli2009@126.com (S.L.); dongdong830206@163.com (X.W.)

[2] College of Materials Science and Engineering, Hunan University, Changsha 410082, China; zhanghui63hunu@163.com

* Correspondence: luoxing_li@hnu.edu.cn

Abstract: The precipitation process and reaction kinetics during artificial aging, precipitate microstructure, and mechanical properties after aging of AlMgSi alloys were investigated employing in-situ electrical resistivity measurement, Transmission Electron Microscopy (TEM) observation, and tensile test methods. Three aging stages in sequence, namely formation of GP zones, transition from GP zones to β″ phase, transition from β″ to β′ phase, and coarsening of both phases, were clearly distinguished by the variation of the resistivity. It was discussed together with the mechanical properties and precipitate morphology evolution. Fast formation of GP zones and β″ phase leads to an obvious decrease of the resistivity and increase of the mechanical strength. The formation of β″ phase in the second stage, which contributes to the peak aging strength, has much higher reaction kinetics than reactions in the other two stages. All of these stages finished faster with higher reaction kinetics under higher temperatures, due to higher atom diffusion capacity. The results proved that the in-situ electrical resistivity method, as proposed in the current study, is a simple, effective, and convenient technique for real-time monitoring of the precipitation process of AlMgSi alloys. Its further application for industrial production and scientific research is also evaluated.

Keywords: aluminum alloys; aging; precipitation; electrical resistivity; mechanical properties

---

1. Introduction

The light weighting features of the precipitation hardened AlMgSi aluminum alloys trigged tremendous investigation interest in recent years [1–5]. The precipitation behavior during the heat treatment process plays a decisive role in final performance of the AlMgSi alloys. Properties of these alloys, e.g., workability, mechanical strength, electrical conductivity, corrosion resistance etc., are greatly determined by the phase type, morphology, size distribution, and number density of their precipitates [1,3,6–9]. Therefore, a deep understanding of the precipitation behavior during the artificial aging process is crucial.

Significant achievements in the characterization of the precipitation behavior of AlMgSi alloys have been reached since the end of last centenary, with the assistance of advancing technologies, like High Resolution Transmission Electron Microscopy (HRTEM) [5–7,10–13], Atom Probe Tomography (APT) [1,13–17], Differential Scanning Calorimetry (DSC) [18–20], electrical resistivity measurements [21–26], Phase Field Crystal (PFC) modeling [4,24], etc. The precipitation sequence has been established and gradually accepted as: supersaturated solid solution (SSSS)→clusters/GP zones→β″→β′→β [27]. The clusters/GP zones are the early stage precipitates with size of several nanometers, usually formed during natural aging or first minutes of artificial aging process. β″ and β′ phases are formed in the artificial aging process before over ageing. β is the final stable phase. The AlMgSi alloys are usually strengthened by the meta-stable phases. However, debates in this field still remain. For example, detailed crystal structure and composition of the meta-stable phases, precipitation kinetics that is quite related to the natural aging effect, quantitative characterization of the precipitates is under debate [13–16]. The atomic structure and evolution of the early clusters are ambiguous yet [5–7]. Characterization accuracy and effectiveness are still dependent on the methods. Therefore, the investigation on the details of the precipitation process is still in progress.

Recently reported investigations proved that the electrical resistivity measurement method is feasible and convenient for quantitative analysis of the reaction kinetics and precipitate volume fraction in AlMgSi alloys [21,25,26]. When compared to the TEM, APT, and DSC techniques, its merits are: (i) it is sensitive due to the high sensitivity of resistivity to even atomic-scale changes of the microstructure, such as solution atoms, defects, precipitates; (ii) there are no need of complicated sample preparation or sophisticated equipments like TEM system; and, (iii) samples used are much larger than those for HRTEM, APT or DSC, so that the accuracy would be better [26]. The existing resistivity investigations of AlMgSi alloys were mostly conducted after aging and samples were immersed in liquid nitrogen to keep a constant temperature during testing [21–26]. This would cause time interval between the precipitation and the measuring, including the heating and cooling process. Some information about the evolution of the precipitates would be missing, due to the high time and temperature sensitivity of the precipitation in AlMgSi alloys. Transients experienced by the sample also require serious attention, especially when the heating and cooling rates fluctuates.

In the present study, we proposed an in-situ electrical resistivity measuring method, which is much easier and of higher efficiency, for the real-time monitoring of the precipitation behavior of AlMgSi alloys. The sample temperature and electrical resistivity was continuously measured during the whole thermal course from quenching to natural or artificial aging. By this method, details of the precipitation process were analyzed. The relationship between the precipitation-induced resistivity variation during the artificial aging process, mechanical properties after aging, microstructure evolution, and the precipitation behavior was investigated. The industrial and scientific application feasibility of the proposed in-situ electrical resistivity measurement method for describing the precipitation behavior of the AlMgSi alloys was evaluated.

2. Materials and Methods

6063 aluminum profiles with composition of 0.55Mg-0.43Si-0.20Fe-0.001Cu-0.001Mn-0.004Cr-0.01Ti-banance Al (wt %) were used as raw materials. After extrusion, the profiles were immediately quenched from 530 °C to room temperature by the water spray system on the profile extruding machine. Then, they were stored for one month before conducting artificial aging. 150 °C, 175 °C, and 195 °C were selected as low, medium, and high artificial aging temperatures, which can also be considered as representative of the temperatures in industrial production.

Electrical resistance was measured during the whole artificial aging process, on 4 × 35 × 150 mm sheet samples cut from the profiles, by a four-point probe system, as shown in Figure 1. Four pure aluminum wires with 0.6 mm in diameter and 1000 mm in length were used as the probes. The probes were spot welded on the sample surface. For consistency, the welding voltage was kept constant to obtain solder points with the same size. A constant current of 100 mA was applied to the sample by the outside two probes. The resulting potential (in the scale of nano-volt) was measured by the inner two probes and recorded by a counting computer. Then, the sample resistance R under the testing distance was calculated by the relationship of $R = U/I$, where U and I stands of voltage and current respectively. Sample temperature was measured by a thermocouple inserted into a hole with diameter of 1.2 mm drilled on the sample surface. Microstructure and precipitate morphology of samples after aging were investigated through transmission electron microscopy (TEM) and high resolution TEM (HRTEM) observations. Specimens for TEM were first mechanically grinded to about 100 μm in thickness and then further thinned by twin-jet electro-polishing with a solution of 30 vol % nitric acid and 70 vol % methanol at −20 °C, 24 V and 80 mA. TEM observations were performed on a Tecnai F20 TEM system (FEI, Hillsboro, OR, USA) operated at 200 KV. To verify the consistency of the results, at least ten spots in different crystal grains were analyzed for each specimen. Mechanical strength and elongation of samples after aging were measured by uniaxial tensile testing, according to ASMT standards E8M-09. Standard dumbbell-type tensile specimens with gauge length of 60 mm and width of 10 mm were used. Tests were performed on an INSTRON-3382 tensile testing machine (Instron, Norwood, MA, USA) with a stretching speed of 1 mm/min. For each aging situation, three specimens were tested in order to verify the tensile results were in consistence.

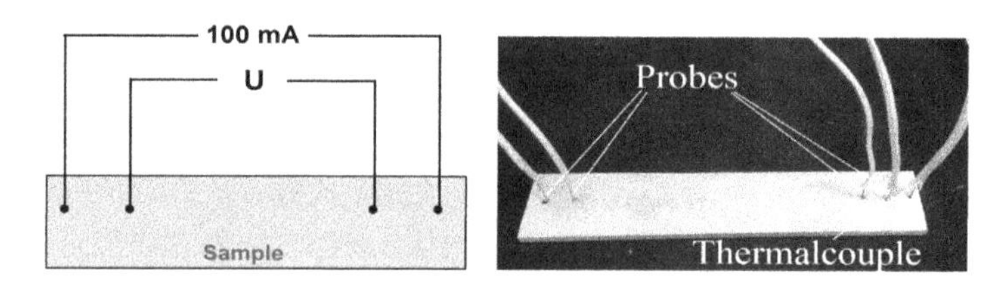

Figure 1. A sketch of the experimental set up for resistance measuring of the sample.

3. Results

3.1. Resistance and Resistivity

The real-time measured electrical resistance and sample temperature during the whole artificial aging process are shown in Figure 2. As shown in Figure 2a, the resistance sharply increased upon heating, and drastically decreased during cooling, a behavior that was caused by the temperature effect on the aluminum alloy matrix. After the aging temperature reached the pre-set value, the resistance gradually decreased as aging time was prolonged. During the aging process, the sample temperature was kept stable, as shown by Figure 2b.

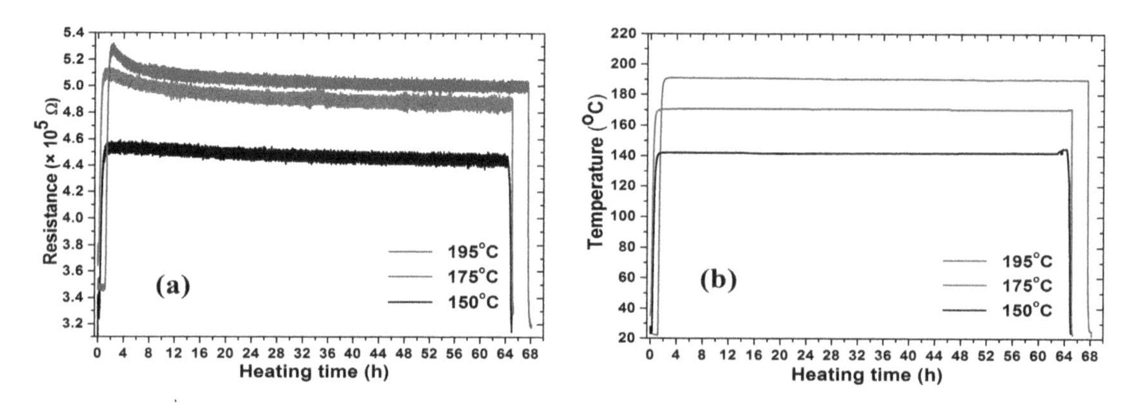

Figure 2. Resistance and sample temperature measurement results: (**a**) variation of the resistance during the whole aging process; and (**b**) sample temperatures during the whole aging process.

The real resistivity of the samples was calculated from the measured resistance and sample size values. The recorded resistance R comprises the system resistance of the instrument R_0, contact resistance between the probes and sample R_c, and the real resistance of the sample R_s, as $R = R_0 + R_c + R_s$. When the temperature reaches the pre-set value, the R_0 and R_c can be considered to be constant. At a fixed temperature, the sample resistance R_s is proportional to the testing distance: $R_s = \rho_s L/A$, where ρ_s is the resistivity, L is the testing distance (namely the distance between the inner probes), A is the cross-sectional area of the sample. By varying the distance at the end of the aging process, series of total resistances R were obtained which should be proportional to the testing distances. Therefore, the fixed R_0 and R_c value can be extracted from the intercept of linear fitting of the resistance values measured on different distances, as shown in Figure 3. By subtracting the R_0 and R_c from R, the R_s was obtained. Then, the resistivity value was calculated using the expression $\rho_s = R_s \times A/L$.

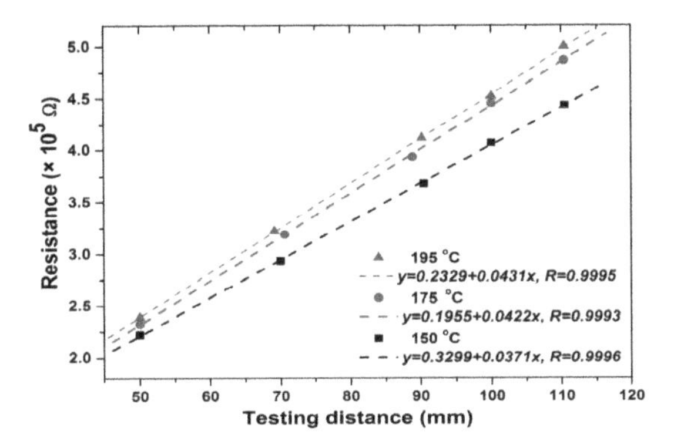

Figure 3. Resistance value of the samples R at the end of aging process measured on different distances and their linear fittings.

The variation of the resistivity with the aging time (in this case, the aging time is set as zero when the sample temperature reached the pre-set value) under different aging temperatures is shown in Figure 4. The electrical resistivity values are around several dozens of nΩ·m, which is in the same order of magnitude of those reported in the literature [24,25]. The resistivity of the samples continuously decreases with aging time, as in the case of the resistance value. By plotting the resistivity versus the logarithmic scale of the aging time, the curve was divided into two or three parts. In each part, the resistivity decreases linearly with the logarithmic scale of aging time, which has also been found in other studies [23,26]. In this way, the whole precipitation process was separated into two or three stages. It reflects that two or three different precipitation reaction types occurred. It can also be seen that these stages end at a different aging time. When the aging temperature is 195 °C and 175 °C, the first stage ends within 0.5 h and 1 h, respectively, while for 150 °C, it is prolonged to about 5 h. When the aging temperatures are 195 °C and 175 °C, the end of the second stage is after 4 h and 11 h, respectively. The end of the third stage, which seems to occur after a long time, is not clearly observed in the present study since the experimental aging time was limited to 60 h. Nevertheless, it can be easily concluded that, as expected, the sequence of reactions is faster when the temperature is higher.

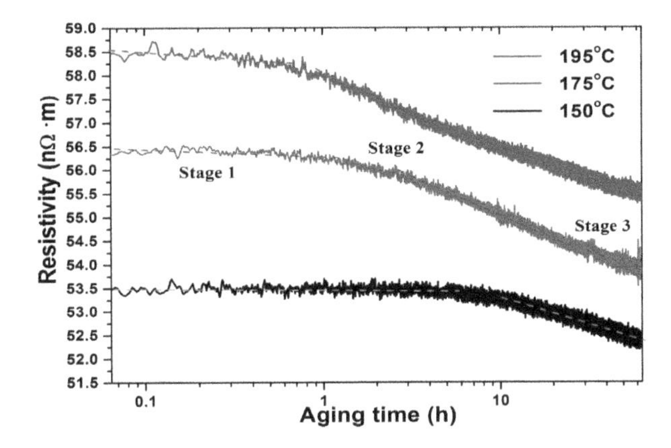

Figure 4. Variation of the resistivity of the samples with aging time under different temperature.

The derivative of the resistivity to the aging time, which would reflect the precipitation reaction rate, is shown in Figure 5. The figure shows that the reaction rate continuously decreases. When the aging temperature is 195 °C or 175 °C, the initial fast decrease in resistivity during the first stage, which comes at completion within several hours, is followed by a slower decrease in the second stage.

The inflection point in the curves distinguishes the transition between the first and the second stages (4–6 h for 195 and 12–14 h for 175 °C, respectively). Such an obvious inflection point is not observed in the curve for 150 °C. The higher the aging temperature, the faster the initial decrease in resistivity. These phenomena should all be related to the precipitation type and reaction kinetics.

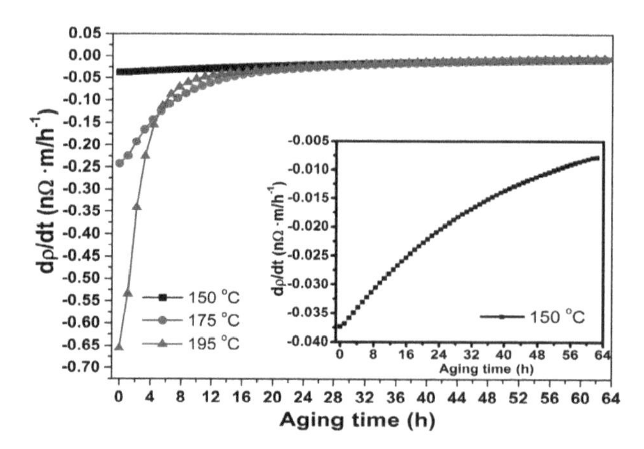

Figure 5. Differentiation of resistivity to aging time under different aging temperature. The inset figure is a magnified picture of the curve for samples aged at 150 °C.

3.2. Mechanical Properties

The relationship between the mechanical properties of the samples and aging temperature and time is shown in Figure 6. The tensile and yield strength firstly increase sharply with the aging time, and then reach a plateau at a maximum value, which is called peak aging strength. A slight decrease of the strength is observed for samples aged at 195 °C when aging time exceeds 4 h, which is called over aging phenomenon [28]. The time required for reaching the peak aging status, at the time when the strength reaches the maximum [28], is shorter when the aging temperature is higher (4 h for 195 °C, 11 h for 175 °C, and 36 h for 150 °C, designated as the peak aging time). The elongation values of the samples show continuous decrease in all the cases. It is also observed that the peak strength is higher when the aging temperature is lower, for example, the maximum tensile strength for samples aged at 150 °C reaches 231 MPa, while it is 209 MPa for those aged at 195 °C. These phenomena should all be related to the precipitation rate, phase type, number density, and size of the precipitates in the matrix [14,16,17], which have obvious different strengthening effects. This would be discussed together with the evolution of the microstructure of the precipitates.

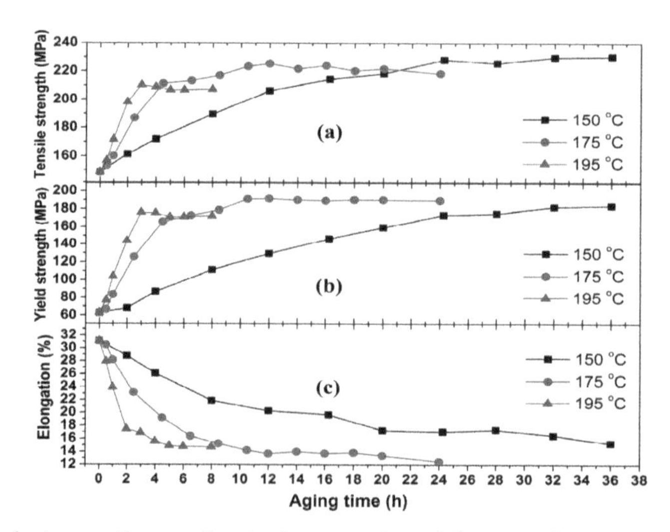

Figure 6. Relationship between the mechanical properties of the samples, aging temperature and aging time (**a**) tensile strength; (**b**) yield strength; and, (**c**) elongation.

3.3. Microstructure of Precipitates

The morphology of the precipitates in the samples after different stages of artificial aging were investigated by TEM. Representative bright field TEM and HRTEM images taken along the <100>$_{Al}$ zone axis are shown in Figure 7. Precipitates were all found to grow along <100> direction of Al matrix. For naturally aged samples without artificial aging, no evidence of large precipitates was found, as shown in Figure 7a. For artificially aged samples under different aging conditions, three kinds of precipitates were observed: spherical GP zones, needle-like β'' phase, and rod-like β' phase, as shown in Figure 7b–f. In the sample not artificially aged, very small clusters is believed to exist in the matrix since the raw materials have been stored for one month, during which the materials must have undergone natural aging [12,21–24]. But, their total morphology is hard to be observed by ordinary TEM images in samples with such low alloy composition, and also due to their small size (around 1–3 m) and highly coherence with the matrix [1,2,10,11]. They are identified by their white contrast in the HRTEM images, as shown in Figure 7f. GP zones were observed to have spherical morphology and size of 3–5 nm in short time artificially aged samples. They can be seen also only in the HRTEM images. But, it was much easier since they have a little larger size and number density. Examples were shown by Figure 7b,g for the sample which has been aged at 175 °C for 1 h. The β'' phase and β' phase were distinguished by their needle-like and rod-like morphology, respectively, and different contrast surroundings in the matrix, as described in the literature [3,11]. This was further verified by their monoclinic and hexagonal crystal structure, respectively [11], identified by HRTEM images, as shown in Figure 7h, taken from the sample aged at 175 °C for 24 h.

The phase type, size and number density evolution of different precipitates with artificial aging time at different temperatures were clearly reflected by the TEM images. After very short time of artificial aging at each selected temperature, also described as stage 1 in Figure 4, the precipitation process starts with the formation of GP zones and the beginning of transformation from GP zones to β'' phase. A typical image is presented in Figure 7b for the sample aged at 175 °C for 1 h. A large amount of GP zones and a little trace of β'' phase with length of about 20 nm were observed. As aging proceeds, the GP zones totally transformed to β'' phase. When the aging time prolongs to the peak aging status, also in the time range described as stage 2 in Figure 4, the β'' phase gradually coarsened. The length of the β'' phase particles in peak aged samples increased to 40–120nm, as shown in Figure 7c, for the sample aged at 195 °C for 3 h. The total number density of the precipitates also increased with aging time. It was verified by comparing TEM images for samples aged at the same temperature but for different aging times, the number density of β'' phase is obviously increased as the aging time increases. After the peak aging time at 175 °C and 195 °C, described as stage 3 in Figure 4, the length of some precipitates abnormally reached several hundreds of nanometers, and the morphology of precipitates became inhomogeneous, as shown in Figure 7d, for the sample aged at 175 °C for 24 h. β' phase also starts to form. However, it is interesting to find that when aging temperature is 150 °C, the length of β'' phase still stays at the level of 15–30 nm, even when the aging time reaches 36 h, as shown in Figure 7e. The small β'' phase is homogenously distributed, and no β' phase is found. This indicates that aging at lower temperature, 150 °C for example, has much lower coarsening and β' phase formation kinetics.

It can be summarized from the TEM results: (i) GP zones form quickly at the early stage of artificial aging process; (ii) for peak-aged samples, the β'' phase is the main precipitate; (iii) as the aging processes, the number density of precipitates increases obviously; (iv) raising the artificial temperature promotes growth of the precipitates and formation of β' phase; and, (v) the transformation from β'' phase to β' phase is harder to occur at low temperature (150 °C) than at higher temperature (175 °C or 195 °C).

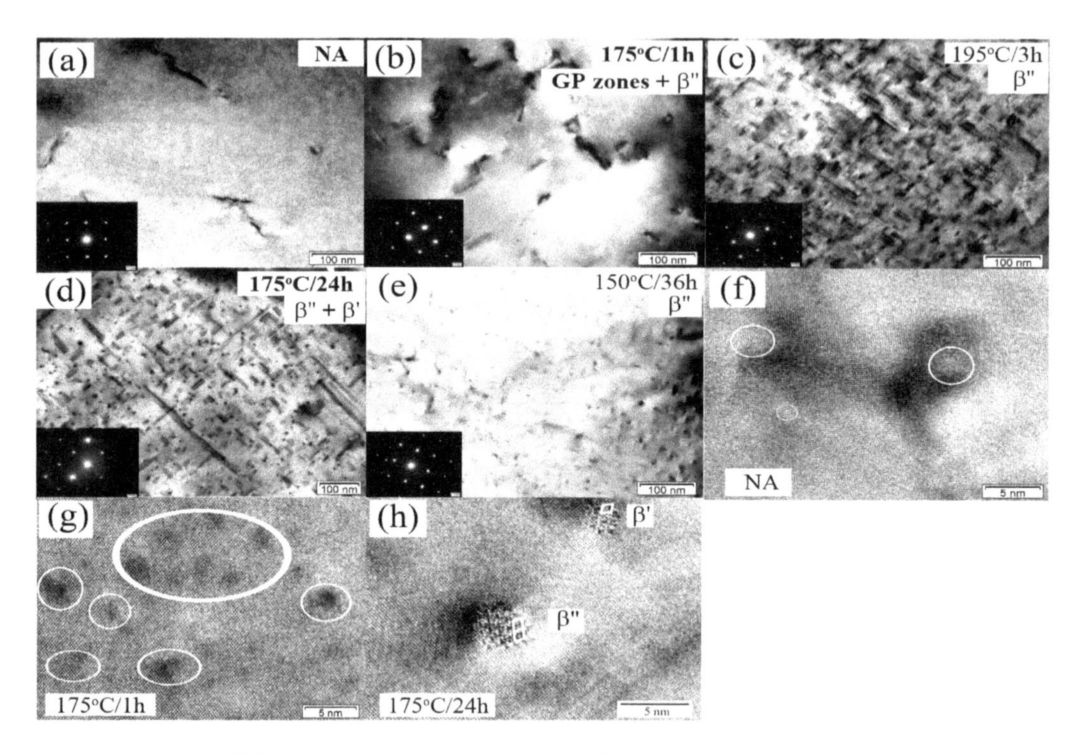

Figure 7. TEM images of the microstructures of selected samples aged at different temperatures and aging times: (**a**) naturally aged (NA) sample without artificial aging; (**b**) 175 °C for 1 h; (**c**) 195 °C for 3 h; (**d**) 175 °C for 24 h; (**e**) 150 °C for 36 h; and, (**f–h**) HRTEM images taken from (**a**,**b**,**d**), respectively.

4. Discussion

All of the experimental results in present study confirmed the same precipitation sequence of AlMgSi alloys, as reported in the literature [27], which was either indirectly reflected in the resistivity measurement results and mechanical properties, or directly observed during the TEM analysis. This sequence is obviously related to the precipitate phase type, particle size, and reaction kinetics, since different precipitates contribute to the electrical resistivity and strengthening effect in a markedly different manner [21,25]. Therefore, the evolution of these precipitates during the whole artificial aging process, and the consistency of the experimental results will be discussed.

4.1. Precipitation and Phase Transformation Process

The precipitation and phase transformation process was well revealed by the resistivity variation curves, mechanical strength, and microstructure evolution. The whole artificial aging process is clearly divided into three stages, as reflected by Figure 4. Precipitate type and morphology, size, density etc. evolve in sequence, with typical characters as shown in Figure 7.

Formation and transition of the precipitates accounts for the variation of electrical resistivity. At the quenched and naturally aged state, the solute atoms exist in the matrix or in the form of clusters (consisted of several atoms), which have different electronic surroundings to the Al atoms, act as serious scattering centers for electron conducting. But, when precipitation occurs, the solute atoms were accumulated to form much larger precipitates. In this way, their electron scattering effect was weakened. Larger precipitates also affect the electron conduction, but it is not so obvious as clusters or solute atoms. As the precipitates grow up, their nearby distance becomes larger. In this way, it provides more convenient pathway for electron conduction and further decreases the resistivity [21–25]. Therefore, the sample has the highest electrical resistivity in the original state. As artificial aging begins, GP zones form quickly from solute atoms or clusters (as shown in Figure 7b), which leads to obvious decline of the resistivity. Therefore, the first stage in Figure 4 is characterized by the formation of GP zones. As aging further proceeds, GP zones, which are also considered as the pre-phase of β″ [1,8,13,27], transforms to the needle-like β″ phase. Since the size of β″ phase (at least 20 nm in length) is much larger than that of GP zones (2–5 nm in diameter), the total number of the precipitates decreases. This process further reduces the electrical resistivity. This process continues to the peak aging status, since the main precipitate is β″ phase for peak aging samples, as shown in Figure 7c. Thus, the second stage in Figure 4 is assigned to transition from GP zones to β″ phase. As the aging time is prolonged to the third stage, the rod-like β′ phase starts to precipitate by consuming the β″ phase, and both of the β″ and β′ phase start to coarsen, as shown by Figure 7d. The distance between neighboring precipitates is further increased, which would facilitate the passage of conducting electrons [23,25]. In this way, the coarsening of the precipitates also contributes to decline of the resistivity. Therefore, the resistivity continuously decreased during the whole artificial aging process, either by formation or phase transformation and coarsening of the precipitates. But, it is worth mentioning that the formation of β″ phase contribute to most decrease of the resistivity, while other transitions have not so obvious influence.

The evolution of the mechanical properties of samples after artificial aging is also related to the formation and phase transition of the precipitates. In the first two stages, the base material is mainly strengthened by small GP zone and/or β″ phase. But, in last stage, the precipitates greatly coarsened. It weakens the dislocation-particle interaction, and thus reduces the effectiveness of their strengthening effect [29,30]. At the early aging status, the precipitate sizes remain very small. The movement of dislocations when encountering precipitate obstacles is mainly in the shearing mode, which has excellent strengthening effect. The larger the amount of fine β″ phase the sample contains, the higher its strength. It shown consistently by the mechanical strength curves and TEM images, as the amount of fine β″ phase increased (before reaching peak-aging), the strength gradually increases. This accounts for the fact that the samples aged at 150 °C, i.e., in a condition that results in large amount of fine β″ phase, exhibit the highest peak strength. If the precipitates size (namely the aspect ratio) reaches and exceeds the critical particle radius, the dislocation movement transforms from the shearing to the bypassing mode, which has not so strong strengthening effect. Although the total number of precipitates increases, coarsening of the precipitates makes the dislocation by-pass easier. Therefore, for samples at the plateau or over-aged status, their strength increases not so obviously, or even decreases.

The evolution of the resistivity, strength, precipitate type and size shows an extremely good consistence. The fast increase in the strength-time curves corresponds to stage 1 and 2 in resistivity variation curves, during which GP zones form and then transform to β″ phase. The plateau in strength curves implies the aging process is in stage 3, in which transition of β″ phase to β′ phase and coarsening of both phases take place.

4.2. Precipitation Kinetics

The precipitation and phase transformation in AlMgSi alloys are atomic diffusion-controlled reaction. The atom diffusion is largely assisted by the spare vacancies, either reserved from quenching or intentionally designed [1,6,12–15,17]. Thus it is obviously driven by vacancy diffusion [1,6,12–15,17]. Activity of the vacancies greatly influences their diffusion and precipitation reaction kinetics, which is also related to temperature and the particular reaction. The resistivity curves in the present study reflect the precipitation kinetics well. Since the stages in the electrical resistivity and mechanical strength curves correspond to precipitation types, their variation trends stand for different reaction kinetics. The reaction kinetics is discussed in two situations: (i) for different reaction types at the same aging temperature; and, (ii) for individual reaction type at different aging temperatures.

The measured resistivity can be expressed by the following equation: $\rho = \rho^* + \sum_i \rho_i C_i + \sum_j P_j f_j$, where the ρ^* is the resistivity of the Al matrix at the measuring temperature; ρ_i and C_i are the specific resistivity and concentration of the ith solution atoms; P and f_j is the scattering power and concentration of the jth precipitate with an average size in the microstructure of pure host metal [26]. The relationship between C_i and f_j can be expressed by $C_i = C_{oi} - N_i f_j$, where the C_{oi} and N_i are the initial concentration of the ith solute atom and its equivalent number in jth precipitate. The precipitation process during artificial aging is considered to follow the Johnson-Mehl, Avrami, Kolmogorov (JMAK) kinetics model [11–13]: $f_r = 1 - \exp(-kt^n)$, where f_r is the relative fraction of the precipitate; k is temperature dependent rate constant; t is the aging time; and, n a numerical exponent for JMAK relationship. For the solution treated and T4 temper AlMgSi alloys, n is always assigned to be 1 [31–34]. In the present study, the original materials were quenched and then stored in room temperature, which equaled to the T4 temper, therefore, n is also considered to be 1. The constant k represents the reaction kinetics, which follows the Arrhenius model: $k = k_o \exp(-Q/RT)$, where k_o, Q, R and T are the pre-exponential constant, apparent activation energy related to the precipitation process, the universal gas constant and temperature, respectively [25,26,31,32]. Therefore, by re-arranging the above relations, the measured resistivity of the sample is found to be in linear relationship to the logarithmic scale of the aging time for an individual reaction, and the linear fitting slope containing the origination of k reflects the reaction kinetics.

As shown in Figure 4, the linear decreasing relationship of the resistivity to aging time represents three types of reactions. The slope of each linear fitting thus reflects the reaction kinetics. It can be seen that the slope for the second stage, which is attributed to formation of β″ phase, is the largest slope for all the curves. This implies that the transformation from GP zones to β″ phase has higher kinetics than other reactions. In the early state, formation of GP zones is determined by the response of clusters that contain large amount of solution atoms. Distance between the small clusters is small, and diffusion of these atoms is rather active, with the assistance of quenched in vacancies. However, it can be either dissolved into the matrix or acts as the nucleation site for GP zones growth [9,18]. In other words, the competition between the dissolution and nucleation processes slows down the formation of GP zones. Once there exists GP zones with size large enough, formation of β″ phase occurs fast. Other studies by DSC have also reported such phenomenon [18]. Therefore, for the first stage, even though the concentration of the clusters at this moment is pretty high, kinetics for formation of GP zones is lower than formation of β″ phase in the second stage. As the reaction comes to the third stage, the solution atoms in the matrix may have been all diffused into the precipitates, phase transition from β″ to β′ and size growing of the precipitates all depends on the atom diffusion rate without the assistance of vacancies. Therefore, this speed of this process is lowered down when compared to the formation of β″ phase. It is also observed in Figure 5. The differential of the resistivity to aging time becomes to be much lower for the last stage. Therefore, the formation of β″ phase have the highest kinetics that is facilitated by both vacancies and existing nucleation sites.

For the same kind of stages aged at different temperatures, the higher the temperature is, the higher kinetics it has. This is seen by the higher slope of the linear fitting of resistivity in Figure 4 and sharper variation rate of the derivation of the resistivity in Figure 5. It is much easier to understand that the atoms or vacancies have higher diffusion capacity under elevated temperatures. Therefore, for each individual reaction stage, it finishes in much shorter time at higher temperature. As a example, it can be seen in Figure 4, the formation of GP zones under 195 °C only takes about 0.5 h, but it requires 5 h when the temperature is 150 °C.

4.3. *Accuracy and Application of the In-Situ Resistivity Measuring Method*

The accuracy of the in-situ resistivity measurement method is evaluated by its data error level. As shown in Figure 3, when testing distance is 110 nm, the R_0 and R_c comprises 4.66%, 4.41%, and 7.44% of R for 195 °C, 175 °C, and 150 °C, respectively. The proportion R_c in R should be some lower than these values. The testing error is mainly caused by the fluctuation of the welded contact surface between the probe and the sample (R_c). It can be further controlled to be even lower by increasing the testing distance. The linear fitting error stays at extremely low level below 0.1% (with fitting level higher than 99.9%). Therefore, the final testing data are confirmed to be highly trustable. The results also showed better accuracy and stability than those reported in the literature [21–26], since the data were in-situ and continuously recorded.

Further applications of the in-situ resistivity measuring method can be both industrial or scientific. On the one hand, it can be employed as a monitoring technique for aluminum profiles manufacturing enterprises. Insufficient aging in partial products that are located in some blind corners in the big industrial aging furnace is often encountered during practical experience. It may be caused by inhomogeneous temperature field, uneven heating, fluctuation of power, or other unknown reasons. It is troublesome because it is unpredictable before ending the aging process and after testing. The usual approach for compensating this problem is by conducting supplementary aging or re-aging, which is time and energy consuming. The method presented in this study would be an effective solution. By calibration of the resistivity of aged samples for any given alloy composition, their target mechanical properties can be marked. Therefore, using the in-situ resistivity measuring method, the aging status, and properties of the products can be monitored in real time. This way, the insufficient aging problem is eliminated. On the other hand, the measured resistivity data may also be useful for further quantitative analysis of the precipitation behavior or modeling of mechanical strength. For example, the relative volume fraction of the precipitates can be simply calculated by its initial, real-time, and final resistivity values (ρ_0, ρ_t, ρ_f) through the expression: $f_r = (\rho_0 - \rho_t)/(\rho_0 - \rho_f)$ [29,31]. It is similar to those using the data obtained by DSC and isothermal calorimetry experiments [19,20]. Moreover, it is suggested that this method would be also useful for analysis of the cluster and GP zones evolution, since the variation of resistivity is highly related to them [24]. Together with other characterization techniques, APT, for example, quantitative investigations would be feasible.

Based on the above discussion, advantages of the in-situ resistivity measurement method are: (i) simplicity, because the sample is easy to prepared; (ii) convenience, no professional skills or sophisticated equipment is required for conducting the measurements, with trustable results in low testing error level; (iii) effectiveness, variation of the obtained resistivity data well reveals the trend of mechanical and microstructure evolution under different aging conditions; and, (iv) high efficiency, for a specific aging situation, only one sample is sufficient for different aging times, while in the reported method, a serial of samples is desirable [21,25].

5. Conclusions

In summary, the present work illustrates a detailed study of the precipitation process of AlMgSi alloys during artificial aging by proposing a simple, convenient, and effective real-time resistivity characterization method. The precipitation processes of AlMgSi alloys during artificial aging, namely formation of GP zones, transition from GP zones to β'' phase, transformation from β'' phase to β' phase, and coarsening of both phases, were clearly separated into three stages based on the variation of the resistivity. Feasibility of monitoring the phase transformation and reaction kinetics of the precipitates by this technique has been proved. It has been revealed that the formation of β'' phase has the highest reaction kinetics. Formation of fine needle-like β'' phase before coarsening and transition to larger β' phase contributes most to the mechanical strength. The proposed method is also recommended for both industrial and scientific applications like real-time monitoring technique for checking the aging completion degree in industrial production, characterization of the cluster and GP zones, and quantitative analysis or modeling of the precipitation process.

Acknowledgments: This work was supported by the National Key Research and Development Program of China [grant number: 2016YFB0101700], the National Natural Science Foundation of China [grant numbers: U1664252, 51475156].

Author Contributions: H.H. and L.L. conceived and designed the experiments; H.H., L.Z. and S.L. performed the experiments; H.H. and S.L. analyzed the data; X.W. contributed to discussion of the data and paper writing; H.Z. and L.L. contributed reagents/materials/analysis tools; H.H. and L.L. wrote the paper together.

References

1. Zandbergen, M.W.; Xu, Q.; Cerezo, A.; Smith, G.D.W. Study of precipitation in Al-Mg-Si alloys by Atom Probe Tomography, I. Microstructural changes as a function of ageing temperature. *Acta Mater.* **2015**, *101*, 136–148. [CrossRef]

2. Chen, J.H.; Costan, E.; Van Huis, M.A.; Xu, Q.; Zandbergen, H.W. Atomic pillar-based nanoprecipitates strengthen AlMgSi alloys. *Science* **2006**, *312*, 416–419. [CrossRef] [PubMed]
3. Pogatscher, S.; Antrekowitsch, H.; Leitner, H.; Ebner, T.; Uggowitzer, P.J. Mechanisms controlling the artificial aging of Al-Mg-Si Alloys. *Acta Mater.* **2011**, *59*, 3352–3363. [CrossRef]
4. Esmaeili, S.; Lloyd, D.J. Modeling of precipitation hardening in pre-aged AlMgSi(Cu) alloys. *Acta Mater.* **2005**, *53*, 5257–5271. [CrossRef]
5. Saito, T.; Ehlers, F.J.H.; Lefebvre, W.; Hernandez-Maldonado, D.; Bjørge, R.; Marioara, C.D.; Andersen, S.J.; Mørtsell, E.A.; Holmestad, R. Cu atoms suppress misfit dislocations at the β″ Al interface in Al-Mg-Si alloys. *Scr. Mater.* **2016**, *110*, 6–9. [CrossRef]
6. Werinos, M.; Antrekowitsch, H.; Kozeschnik, E.; Ebner, T.; Moszner, F.; Löffler, J.F.; Uggowitzer, P.J.; Pogatscher, S. Ultrafast artificial aging of Al-Mg-Si alloys. *Scr. Mater.* **2016**, *112*, 148–151. [CrossRef]
7. Li, K.; Idrissi, H.; Sha, G.; Song, M.; Lu, J.B.; Shi, H.; Wang, W.L.; Ringer, S.P.; Du, Y.; Schryvers, D. Quantitative measurement for the microstructural parameters of nano-precipitates in Al-Mg-Si-Cu alloys. *Mater. Charact.* **2016**, *118*, 352–362. [CrossRef]
8. Buchanan, K.; Colas, K.; Ribis, J.; Lopez, A.; Garnier, J. Analysis of the metastable precipitates in peak-hardness aged Al-Mg-Si(-Cu) alloys with differing Si contents. *Acta Mater.* **2017**, *132*, 209–221. [CrossRef]
9. Marioara, C.D.; Andersen, S.J.; Jansen, J.; Zandbergen, H.W. The influence of temperature and storage time at RT on nucleation of the β″ phase in a 6082 Al-Mg-Si alloy. *Acta Mater.* **2003**, *51*, 789–796. [CrossRef]
10. Wenner, S.; Marioara, C.D.; Ramasse, Q.M.; Kepaptsoglou, D.M.; Hagec, F.S.; Holmestada, R. Atomic-resolution electron energy loss studies of precipitates in an Al-Mg-Si-Cu-Ag alloy. *Scr. Mater.* **2014**, *74*, 92–95. [CrossRef]
11. Li, K.; Béché, A.; Song, M.; Sha, G.; Lu, X.; Zhang, K.; Du, Y.; Ringer, S.P.; Schryvers, D. Atomistic structure of Cu-containing β″ precipitates in an Al-Mg-Si-Cu alloy. *Scr. Mater.* **2014**, *75*, 86–89. [CrossRef]
12. Valiev, R.Z.; Murashkina, M.Y.; Sabirov, I. A nanostructural design to produce high-strength Al alloys with enhanced electrical conductivity. *Scr. Mater.* **2014**, *76*, 13–16. [CrossRef]
13. Liu, C.H.; Lai, Y.X.; Chen, J.H.; Tao, G.H.; Liu, L.M.; Ma, P.P.; Wu, C.L. Natural-aging-induced reversal of the precipitation pathways in an Al-Mg-Si alloy. *Scr. Mater.* **2016**, *115*, 150–154. [CrossRef]
14. Pogatscher, S.; Kozeschnik, E.; Antrekowitsch, H.; Werinos, M.; Gerstl, S.S.A.; Löffler, J.F.; Uggowitzer, P.J. Process-controlled suppression of natural aging in an Al-Mg-Si alloy. *Scr. Mater.* **2014**, *89*, 53–56. [CrossRef]
15. Aruga, Y.; Kozuka, M.; Takaki, Y.; Sato, T. Effects of natural aging after pre-aging on clustering and bake-hardening behavior in an Al-Mg-Si alloy. *Scr. Mater.* **2016**, *116*, 82–86. [CrossRef]
16. Li, H.; Liu, W. Nanoprecipitates and Their Strengthening Behavior in Al-Mg-Si Alloy during the Aging Process. *Metall. Mater. Trans. A* **2017**, *48*, 1990–1998. [CrossRef]
17. Guo, M.X.; Sha, G.; Cao, L.Y.; Liu, W.Q.; Zhang, J.S.; Zhuang, L.Z. Enhanced bake-hardening response of an Al-Mg-Si-Cu alloy with Zn addition. *Mater. Chem. Phys.* **2015**, *162*, 15–19. [CrossRef]
18. Esmaeili, S.; Lloyd, D.J. Characterization of the evolution of the volume fraction of precipitates in aged AlMgSiCu alloys using DSC technique. *Mater. Charact.* **2005**, *55*, 307–319. [CrossRef]
19. Aouabdia, Y.; Boubertakh, A.; Hamamda, S. Precipitation kinetics of the hardening phase in two 6061 aluminium alloys. *Mater. Lett.* **2010**, *64*, 353–356. [CrossRef]
20. Giersberg, L.; Milkereit, B.; Schick, C.; Kessler, O. In Situ Isothermal Calorimetric Measurement of Precipitation Behaviour in Al-Mg-Si Alloys. *Mater. Sci. Forum* **2014**, *794*, 939–944. [CrossRef]
21. Esmaeili, S.; Lloyd, D.J.; Poole, W.J. Effect of natural aging on the resistivity evolution during artificial aging of the aluminum alloy AA6111. *Mater. Lett.* **2005**, *59*, 575–577. [CrossRef]
22. Raeisinia, B.; Poole, W.J.; Lloyd, D.J. Examination of precipitation in the aluminum alloy AA6111 using electrical resistivity measurements. *Mater. Sci. Eng. A* **2006**, *420*, 245–249. [CrossRef]
23. Seyedrezai, H.; Grebennikov, D.; Mascher, P.; Zurob, H.S. Study of the early stages of clustering in Al-Mg-Si alloys using the electrical resistivity measurements. *Mater. Sci. Eng. A* **2009**, *525*, 186–191. [CrossRef]
24. Fallah, V.; Langelier, B.; Ofori-Opoku, N.; Raeisinia, B.; Provatas, N.; Esmaeili, S. Cluster evolution mechanisms during aging in Al-Mg-Si alloys. *Acta Mater.* **2016**, *103*, 290–300. [CrossRef]
25. Esmaeili, S.; Poole, W.J.; Lloyd, D.J. Electrical Resistivity Studies on the Precipitation Behaviour of AA6111. *Mater. Sci. Forum* **2000**, *331–337*, 995–1000. [CrossRef]

26. Esmaeili, S.; Vaumousse, D.; Zandbergen, M.W.; Poole, W.J.; Cerezo, A.; Lloyd, D.J. A study on the early-stage decomposition in the Al-Mg-Si-Cu alloy AA6111 by electrical resistivity and three-dimensional atom probe. *Philos. Mag.* **2007**, *87*, 3797–3816. [CrossRef]
27. Edwards, G.A.; Stiller, K.; Dunlop, G.L.; Couper, M.J. The precipitation sequence in Al-Mg-Si alloys. *Acta Mater.* **1998**, *46*, 3893–3904. [CrossRef]
28. Siddiqui, A.R.; Abdullah, H.A.; Al-Belushi, K.R. Influence of aging parameters on the mechanical properties of 6063 aluminium alloy. *J. Mater. Process. Technol.* **2000**, *102*, 234–240. [CrossRef]
29. Liu, G.; Zhang, G.J.; Ding, X.D.; Sun, J.; Chen, K.H. Modeling the strengthening response to aging process of heat-treatable aluminum alloys containing plate/disc- or rod/needle-shaped precipitates. *Mater. Sci. Eng. A* **2003**, *344*, 113–124. [CrossRef]
30. Myhr, O.R.; Grong, Ø.; Andersen, S.J. Modelling of the age hardening behaviour of Al-Mg-Si alloys. *Acta Mater.* **2001**, *49*, 65–75. [CrossRef]
31. Luo, A.; Lloyd, D.J.; Gupta, A.; Youdelis, W.V. Precipitation and dissolution kinetics in Al-Li-Cu-Mg alloy 8090. *Acta Mater.* **1993**, *41*, 769–776. [CrossRef]
32. Esmaeili, S.; Lloyd, D.J.; Poole, W.J. Modeling of precipitation hardening for the naturally aged Al-Mg-Si-Cu alloy AA6111. *Acta Mater.* **2003**, *51*, 3467–3481. [CrossRef]
33. Cluff, D.R.A.; Esmaeili, S. Prediction of the effect of artificial aging heat treatment on the yield strength of an open-cell aluminum foam. *J. Mater. Sci.* **2008**, *43*, 1121–1127. [CrossRef]
34. Raeisinia, B. A Study of Precipitation in the Aluminum Alloy AA6111. Master's Thesis, Tehran University, Tehran, Iran, 2000.

11

Hot Deformation Behavior and Microstructure Characterization of an Al-Cu-Li-Mg-Ag Alloy

Lingfei Cao [1,2], Bin Liao [3], Xiaodong Wu [1,2,*], Chaoyang Li [1], Guangjie Huang [1] and Nanpu Cheng [4,*]

1 International Joint Laboratory for Light Alloys (Ministry of Education), College of Materials Science and Engineering, Chongqing University, Chongqing 400044, China; caolingfei@cqu.edu.cn (L.C.); licy@cqu.edu.cn (C.L.); gjhuang@cqu.edu.cn (G.H.)
2 Shenyang National Laboratory for Materials Science, Chongqing University, Chongqing 400044, China
3 Alnan Aluminium Co., Ltd., Nanning Guangxi 530031, China; bin.liao@foxmail.com
4 School of Materials and Energy, Southwest University, Chongqing 400715, China
* Correspondence: xiaodongwu@cqu.edu.cn (X.W.); cheng_np@swu.edu.cn (N.C.)

Abstract: The flow behavior of an Al-Cu-Li-Mg-Ag alloy was studied by thermal simulation tests at deformation temperatures between 350 °C and 470 °C and strain rates between 0.01–10 s^{-1}. The microstructures of the deformed materials were characterized by electron backscattered diffraction. Constitutive equations were developed after considering compensation for strains. The processing maps were established and the optimum processing window was identified. The experimental data and predicted values of flow stresses were in a good agreement with each other. The influence of deformation temperature and strain rates on the microstructure were discussed. The relationship between the recrystallization mechanism and the Zener–Hollomon parameter was investigated as well.

Keywords: Al-Cu-Li-Mg-Ag alloy; constitutive equation; EBSD; recrystallization

---

1. Introduction

Al-Cu-Li alloys have great application potentials in the aerospace field due to their excellent properties, such as low density, high strength, and stiffness [1–3]. The addition of 1 wt% Li in aluminum can approximately reduce its density for 3% and enhance its elastic modulus for 6% [4], which makes Al-Li alloys suitable materials for lower wings and fuselages of airplanes. During the production, hot working on these alloys is generally unavoidable, in which the hot working temperature and strain rate are key parameters for quality control. The process design largely depends on a good understanding of the deformation mechanism. Therefore, it is necessary to investigate systematically the deformation behaviors and microstructure evolution at high temperatures for Al-Cu-Li alloys.

The thermal simulation technique has been an effective method to investigate the behaviors of materials under thermo-mechanical loading. Constitutive models can be developed subsequently to describe the relationship among deformation temperature, strain rate, and flow stress [5]. The flow behaviors of aluminum alloys have been widely simulated by Arrhenius type equations [6–10], which is helpful for process design in a specific deformation condition.

Thermal softening, which was a result of dynamic recovery (DRV) and dynamic recrystallization (DRX), and strain hardening have a close relationship with deformation conditions. An improper choice of deformation parameters may induce surface cracks, inhomogeneous distribution of grains, and flow localization. Therefore, processing maps have been established on the basis of power dissipation to avoid undesirable deformation conditions [11]. Yin [12] developed a processing map for an Al-Cu-Li alloy, where the suitable processing windows were set up at 450–500 °C and 0.01–0.1 s^{-1}. Shen [13] investigated the recrystallization of alloy 2397 at 450–550 °C with strain rates of 0.001–0.1 s^{-1}. It was reported that the dynamical recrystallization (DRX) was controlled by a combination of deformation mechanisms, i.e., grain boundary bulging (BLG) in the original grains and the transformation of low-angle grain boundaries (LAGBs) into high-angle grain boundaries (HAGBs) in the recrystallized grains. Nayan [14] established a wider processing map for alloy 2195 and found DRX was the main softening mechanism in the alloy. Two suitable processing windows are located at temperatures between 400–450 °C with strain rates of 10^{-2}–$10^{-1.5}$ s^{-1} and $10^{-0.5}$–10 s^{-1}. Yu et al. studied the recrystallization in 2A97 alloy and found dynamic precipitation of T_1 phase during hot deformation, and the main recrystallization mechanism was discontinuous dynamic recrystallization (DDRX) [7]. Ou et al.'s [15] study on alloy 2060 indicated that the dominant softening mechanism was DRV and the best hot working window for this alloy was a temperature of 380–500 °C and strain rates of 0.01–3 s^{-1}.

It can be summarized that a different softening mechanism was proposed to explain the hot deformation behavior in Al-Cu-Li alloys, and various process maps were developed for suitable hot working windows. However, systematic investigations on the flow behavior, microstructure evolution, and deformation mechanism of 2060-type alloys are still needed, particularly on the deformation at typical production temperatures, strain rates and large strains. The related recrystallization mechanism is seldom reported as well. Therefore, in this work the hot compression tests on an Al-Cu-Li-Mg-Ag alloy were carried out to acquire the stress–strain data in near practical production conditions. The constitutive models were established and the processing maps were developed to investigate the flow behavior. The microstructure was characterized by using the electron back-scattered diffraction (EBSD) technique, and the relationship between lnZ and the recrystallization mechanism was discussed in detail.

2. Materials and Methods

The nominal composition of the experimental alloy was Al-4.0Cu-0.7Li-0.7Mg-0.4Zn-0.3Mn-0.3Ag-0.11Zr (wt%). The material was semi-continuously cast and homogenized at 410 °C for 4 h followed by 490 °C for 24 h, and then was machined into cylinders with a diameter of 8 mm and a height of 12 mm for hot compression tests.

A Gleeble 3500 thermo-mechanical tester (Dynamic Systems Inc., Austin, TX, USA) was used for the hot compression tests. The test temperatures were between 350 °C and 470 °C and the test strain rate was between 0.01 and 10 s^{-1}. In order to ensure the reliability of data through hot compression tests, graphite was applied between the sample and the die to reduce the interface friction. The specimens were heated up with a heating rate of 2.5 °C/s and held for 3 min at the deformation temperature before the hot compression test to ensure a uniform temperature over the entire specimen, then were compressed to a reduction of 60% in thickness, and finally were quenched into water to reserve the hot deformation microstructures. The microstructures of the test specimens were characterized by a field emission scanning electron microscope (SEM, TESCAN MIRI 3, Brno, Czech Republic) with an electron backscatter diffraction (EBSD) detector (EDAX Inc., Oxford Instruments, Abingdon, UK).

The deformed samples were cut parallel to the compression direction. The examined surface was ground with different sizes of SiC papers and then electro-polished in an electrolyte solution comprised of 10% perchloric acid and 90% alcohol at 20 V for 15 s before observation. After exporting the data in the format cpr from AZtec software (version 3.2, EDAX Inc., Abingdon, UK), EBSD maps were obtained by using HKL Channel 5.

3. Results and Discussion

3.1. *Original Microstructure*

Figure 1 shows the EBSD orientation map of the Al-Cu-Li-Mg-Ag alloy after homogenization. The grain size is about 190 μm. In the map, the high angle grain boundaries (HAGBs) are depicted in black lines whereas the low angle grain boundaries (LAGBs) are depicted in white lines. HAGBs and LAGBs are differentiated by the misorientation angle, HAGBs are defined as boundaries with misorientation angles higher than 15°, while LAGBs are defined as misorientation angles lower than 15°. As shown in Figure 1b, HAGBs occupy about 70% of the boundaries and the average misorientation angle is around 30°, so the majority of grain boundaries in the starting material are HAGBs.

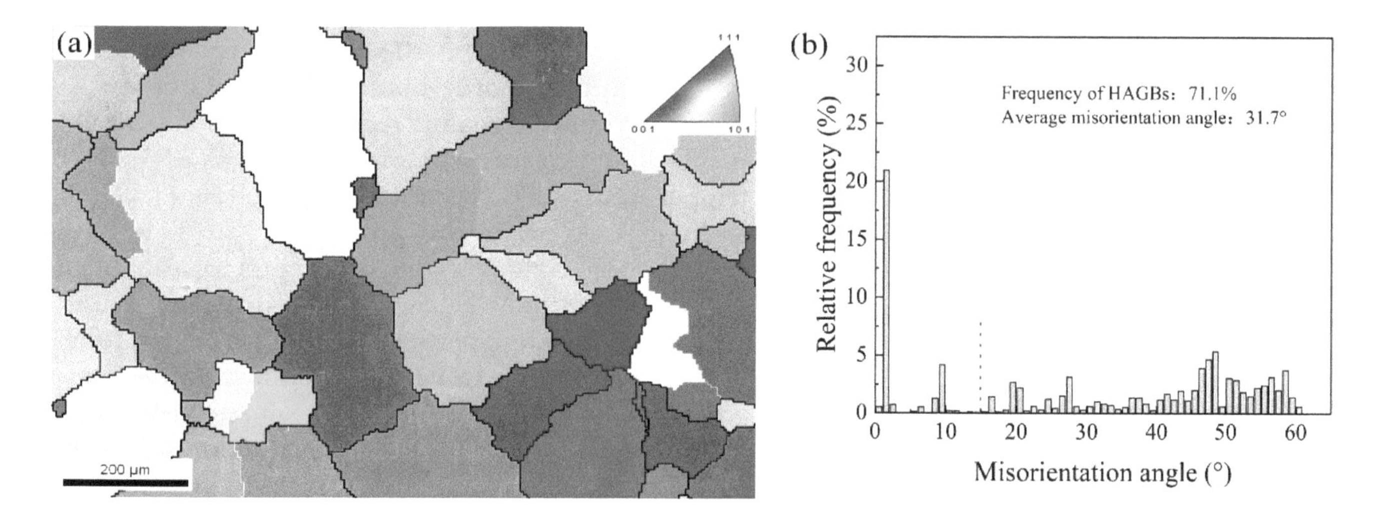

Figure 1. (**a**) Electron back-scattered diffraction (EBSD) map and (**b**) grain boundary misorientation angle of homogenized Al-Cu-Li-Mg-Ag alloy.

3.2. *Flow Behavior*

Figure 2 shows the true stress–true strain curves of the Al-Cu-Li-Mg-Ag alloy during hot compression. In all compression tests, the flow stress decreases progressively with the increase of strain after a peak stress. This is dynamic softening, which is a result of dynamic recovery and recrystallization [16]. At a given strain rate, the flow stress decreases with increasing deformation temperatures, because of a higher softening effect at higher temperatures, and lower resistance to the motion of dislocations. When the sample is compressed at the same temperature, the peak stress decreases with the decreasing strain rates, as a low strain rate extends the deformation time and offers more time for dynamic softening [17]. However, adiabatic heating due to rapid plastic deformation will lead to the flow softening, especially at higher strain rates [18]. Because of this effect, the corrected flow stress shown in Figure 2c,d has been applied to the data used in following sections.

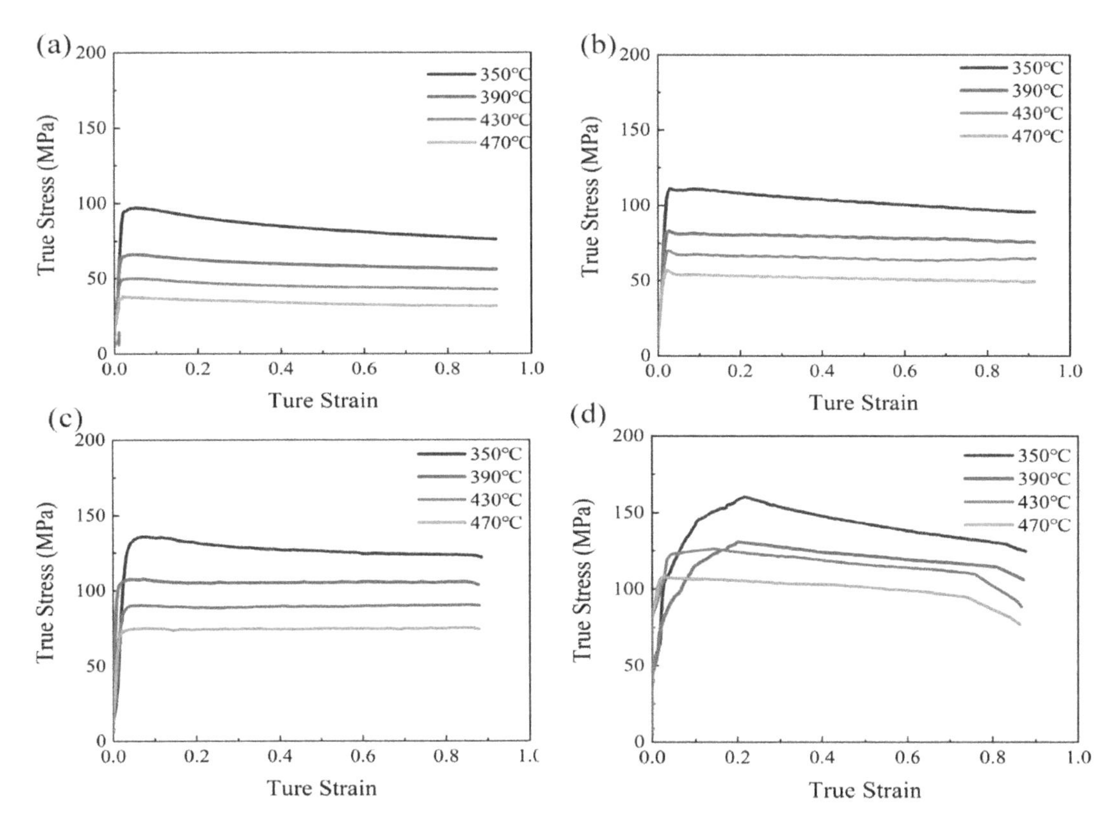

Figure 2. True stress–true strain curves of the Al-Cu-Li-Mg-Ag alloy at various strain rates: (**a**) 0.01 s^{-1}; (**b**) 0.1 s^{-1}; (**c**) 1 s^{-1}; (**d**) 10 s^{-1}.

3.3. Constitutive Equation

The constitutive equation for Al-Cu-Li-Mg-Ag alloy can be developed from the experimental strain–stress curves in Figure 2. In order to understand the influence of deformation temperature and strain rate on the experimental strain–stress curve, the Zener–Hollomon parameter Z is often used [8]:

$$Z = \dot{\varepsilon}\exp(Q/RT) \tag{1}$$

where $\dot{\varepsilon}$ represents the strain rate, Q is the activation energy of hot deformation (kJ/mol), R is the universal gas constant (8.314 kJ mol^{-1}), and T is the deformation temperature. The Z parameter is an important parameter for hot deformation, especially in understanding the microstructure evolution during hot deformation [19].

The relation among stress, strain rate, and temperature during hot deformation can also be expressed by [2]:

$$\dot{\varepsilon} = f(\sigma)\exp(-Q/RT) \tag{2}$$

where f (σ) is a function of stress σ and has different expressions at different stress levels. It is generally expressed using a hyperbolic sine function, as

$$f(\sigma) = A[\sinh(\alpha\sigma)^n] \tag{3}$$

where A, α, and n are material constants and $\alpha = \beta/n_1$. n_1 and β can be calculated by the mean slope of $\ln\ \dot{\varepsilon} - \ln\sigma$ and $\ln\ \dot{\varepsilon} - \sigma$, respectively, so that the value of α can be obtained thereafter.

From Equations (1)–(3), it can be derived that the values of Q and lnZ are

$$Q = R[\frac{\partial \ln\ \dot{\varepsilon}}{\partial \ln[\sinh(\alpha\sigma)]}]_T[\frac{\partial \ln[\sinh(\alpha\sigma)]}{\partial(1/T)}]_{\dot{\varepsilon}} = RnS \tag{4}$$

$$\ln Z = \ln A + n\ln[\sinh(\alpha\sigma)] \tag{5}$$

where n is defined as the relative change of ln $\dot{\varepsilon}$ vs. ln [sinh($\alpha\sigma$)] at constant temperatures and the value of S is defined as the relative change of ln [sinh($\alpha\sigma$)] vs. 1/T at constant strain rates. Therefore, Q and lnZ can be calculated according to Equations (4) and (5) respectively.

The method to determine the material constants listed above is shown hereafter, using the strain at 0.8 as an example. As a result of linear fitting in Figure 3, the average value of n_1 and β are 8.053 and 0.1004 MPa^{-1}, respectively. Therefore, α can be obtained by β/n_1, which is 0.0125 MPa^{-1}. n is 5.886, and Q is 159.9 kJ mol^{-1}. The value of lnZ can then be figured out on the basis of Equation (1) and the results are listed in Table 1.

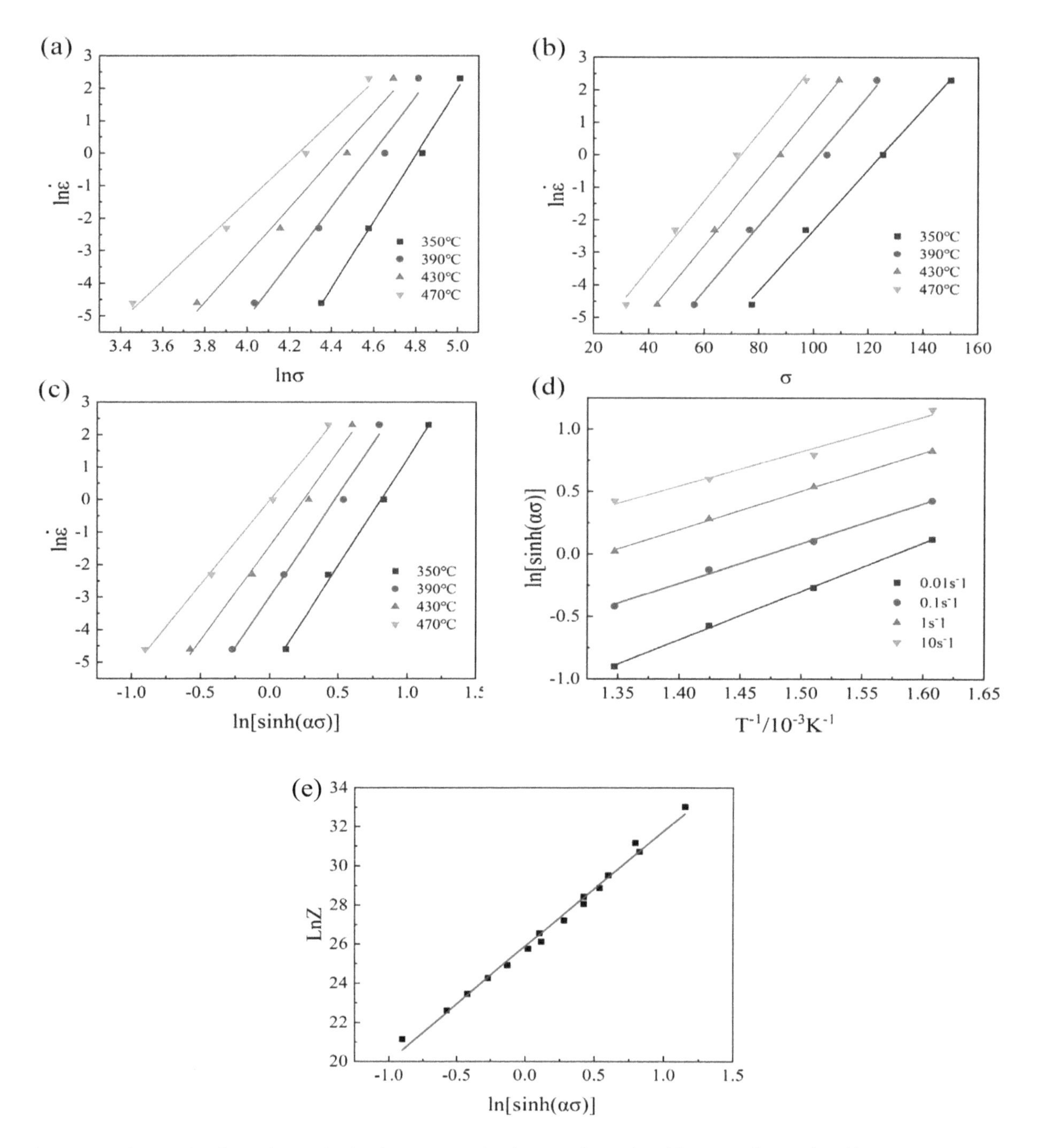

Figure 3. The (**a**) ln $\dot{\varepsilon}$ vs. ln σ, (**b**) ln $\dot{\varepsilon}$ vs. σ, (**c**) ln $\dot{\varepsilon}$ vs. ln[sinh($\alpha\sigma$)]; (**d**) ln [sinh ($\alpha\sigma$)] vs. and (**e**) lnZ vs. ln [sinh ($\alpha\sigma$)] curves at stain of 0.8 for the experimental alloy.

Table 1. lnZ at a strain of 0.8 for different deformation condition.

Strain Rate/s^{-1}	350 °C	390 °C	430 °C	470 °C
0.01	26.116	28.419	30.722	33.024
0.1	24.261	26.563	28.866	31.168
1	22.616	24.919	27.221	29.524
10	21.149	23.452	25.754	28.057

lnA is the value of lnZ at ln[sinh(ασ)] = 0 and can be obtained from Figure 3e. Therefore, the constitutive equation for the Al-Cu-Li-Mg-Ag alloy at the strain of 0.8 is:

$$\dot{\varepsilon} = 1.74 \times 10^{11} [\sinh(0.0125\sigma)]^{5.886} \times \exp(-159.9/RT) \tag{6}$$

For a more general expression, different levels of strains should be considered, as it plays an important role in the flow behavior of aluminum alloy (shown in Figure 2). Based on the method of compensation for strain [20], the constitutive equation predicting flow stresses for the Al-Cu-Li-Mg-Ag alloy can be described as:

$$\sigma = \frac{1}{\alpha}\left\{\left(\frac{Z}{A}\right)^{\frac{1}{n}} + \left[\left(\frac{Z}{A}\right)^{\frac{2}{n}} + 1\right]^{\frac{1}{2}}\right\}$$
$$Z = \dot{\varepsilon}\ \exp\left(\frac{Q}{RT}\right) \tag{7}$$

The material constants (α, Q, n, and lnA) in the above equation at various strains ranging from 0.1 to 0.8 with an interval of 0.1 can be calculated in the similar method. The results are shown in Figure 4. It is clear that these constants are strain dependent and can be correlated by the seventh-order polynomial fitting curves to describe accurately the influence of strains.

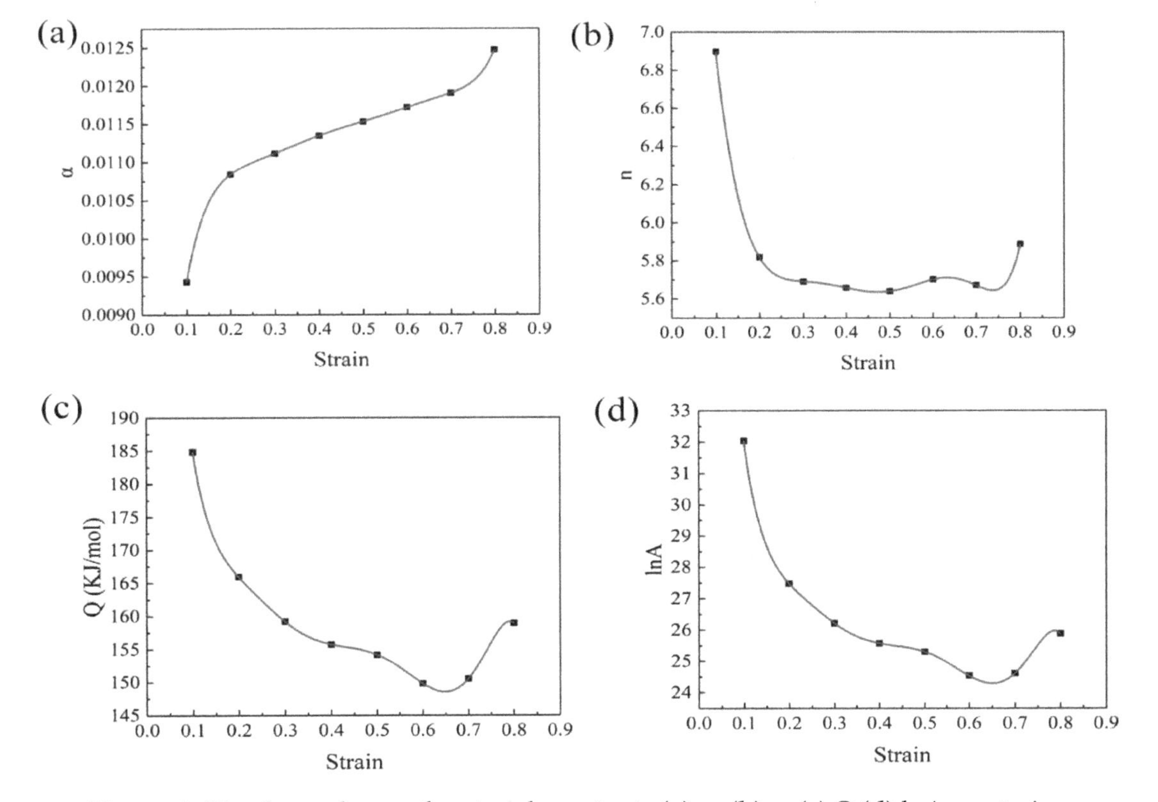

Figure 4. The dependence of material constants (**a**) α; (**b**) n; (**c**) Q;(**d**) lnA on strain.

The predicted values of flow stresses and the experimental data are plotted in Figure 5. The correlation coefficient (R) can be introduced to evaluate the accuracy of the constitutive equations. By fitting the predicted data with the experimental values of flow stresses, the value of R can be figured out as 0.985 with an average relative error (AARE) of 5.4%, which indicates the developed model can predict the high temperature behavior of the experimental Al-Cu-Li-Mg-Ag alloy very well.

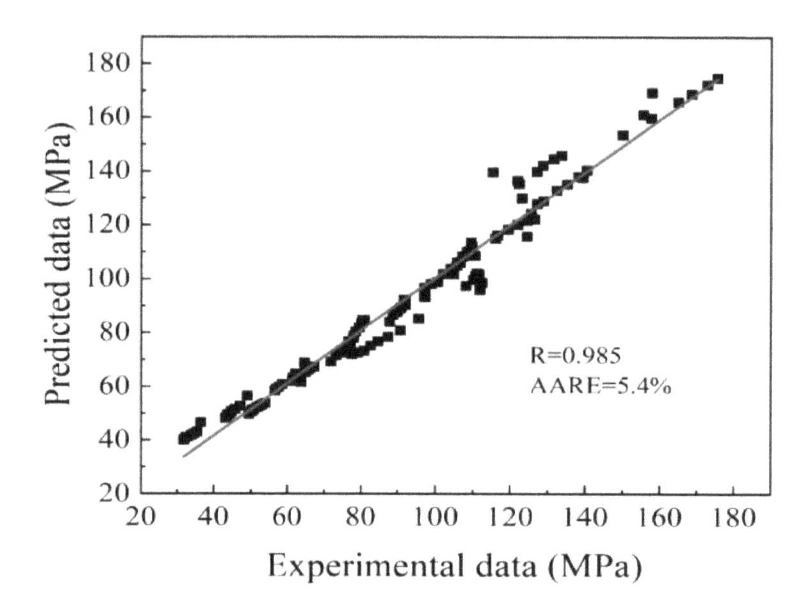

Figure 5. The predicted flow stress vs. the experimental flow stress.

3.4. Hot Processing Map

Hot processing maps can be generated to determine the appropriate process domains for hot working. The values of lg (true stress) versus lg (strain rate) in a compression temperature range between 350 °C and 470 °C are presented in Figure 6a, which are fitted by a cubic spline function. The values of power dissipation (η) are calculated as a function of temperature and strain rate, which are shown as contour lines in Figure 6b and can be used to assess the relationship between power dissipation capacity and microstructural evolution of materials [5]. A dimensionless parameter ξ is also introduced to identify the dependence of flow instability on the temperature and strain rate, and the shaded area in the processing map indicates the unstable domain ($\xi < 0$, Figure 6b).

If a material is deformed within the unstable window, the material is easy to form undesirable microstructures such as adiabatic shear band and intergranular cracking. Therefore, these instability domains ($\xi < 0$) should be avoided when selecting proper parameters for hot working [21,22], while the high power dissipation areas are usually an appropriate processing zone. Based on the processing map, it can be figured out that the optimum process window for this Al-Cu-Li-Mg-Ag alloy locates in the domain at a temperature between 440 °C and 470 °C, and a strain rate between 0.01 s^{-1} and 0.03 s^{-1}. The instable domain is mainly in the window at a temperature range between 350 °C and 430 °C, and a strain rate between 1 s^{-1} and 10 s^{-1}, which is larger than the aluminum alloy 2060 reported, due to greater strains applied in this work [15].

For an accuracy evaluation of processing parameters, the processing map alone is not enough, due to the prediction error of the dynamic materials model [23]. The processing map, composed of power dissipation efficiency and instability domains, connects the microstructure change and the deformation conditions [24,25]. Therefore, microstructure observations are also needed for the purpose of deciding the optimal hot working conditions. Therefore, the microstructures of samples compressed in different domains of the processing map, i.e., the safe domains with high power dissipation (domain A) and low power dissipation (domain B), and unstable domain (domain C) (Figure 6b), will be discussed. The deformation mechanisms will be discussed accordingly.

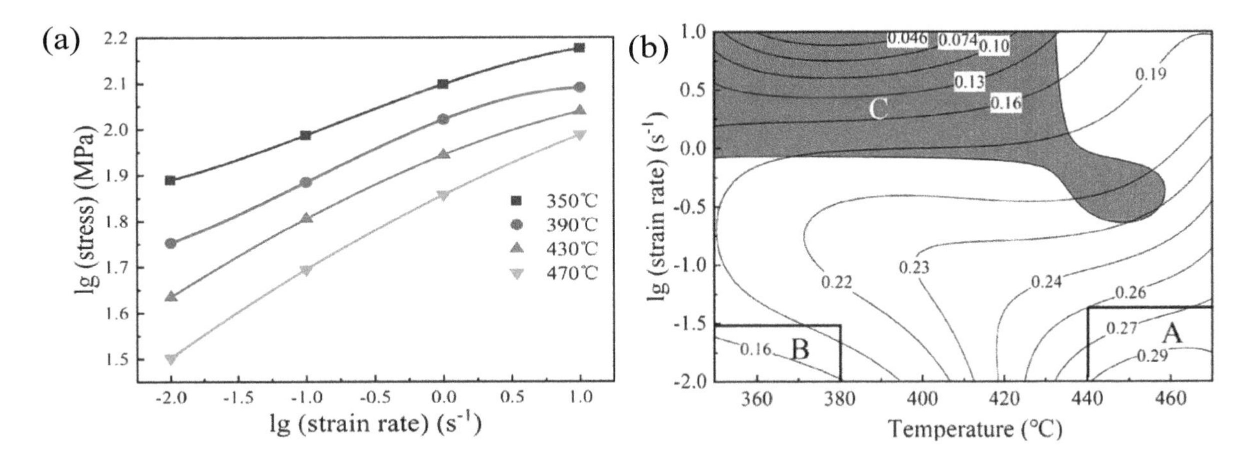

Figure 6. (**a**) The relationship between stress and strain rate, (**b**) processing map at a true strain of 0.8.

3.5. Microstructure Evolution

Figure 7 shows the Zener–Hollomon parameter of the Al-Cu-Li-Mg-Ag alloy deformed at different deformation conditions with a true strain of 0.8. It is obvious that the value of lnZ decreases with the increase of temperature and decrease of strain rate. The highest lnZ is obtained at the area with large strain rates (unstable domain C). The medium values of lnZ (around 25–27) locates at domain B, where both strain rates and deformation temperature are low. The lowest value of lnZ (about 22) locates in domain A, where the deformation temperature is high and the strain rate is low. The influence of deformation temperature and strain rates on microstructure changes will be discussed in the next section.

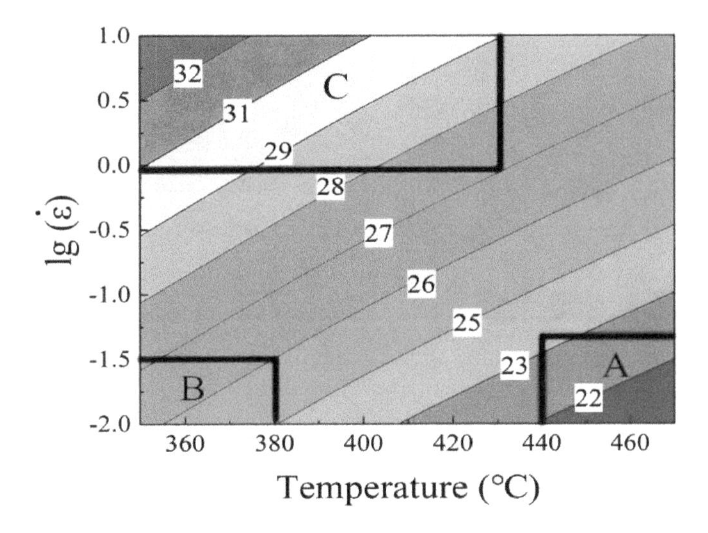

Figure 7. The Zener–Hollomon parameter of the studied alloy at different temperatures and strain rates with a true strain of 0.8. (The values of ln Z are marked on the contour lines; regions A, B, and C correspond the domains shown in Figure 6b).

3.5.1. Effect of Strain Rates

Figure 8 illustrates the effect of strain rate on the microstructure change of the studied Al-Cu-Li-Mg-Ag alloy deformed at 350 °C. The microstructures show typical deformed and recovered structure, i.e., the grains elongate perpendicularly to the compression direction, and many sub-grain boundaries are formed inside the grains. Some recrystallized grains (marked by a black arrow in Figure 8a,b) are observed at lower strain rate (domain A). Adiabatic shear bands, distributed around 45° away from the compression direction as indicated by black rectangles in Figure 8c,d, are observed at higher strain rates (domain C with higher Z value). In this case, severe deformation heat resulted from high strain rates transfer inadequately within limited deformation time, which results in local flow of materials along the plane of the highest shear stress [24,25]. Cracks are expected to generate for the relief of local stress concentrations, thus deformation parameters within domain C are not desirable.

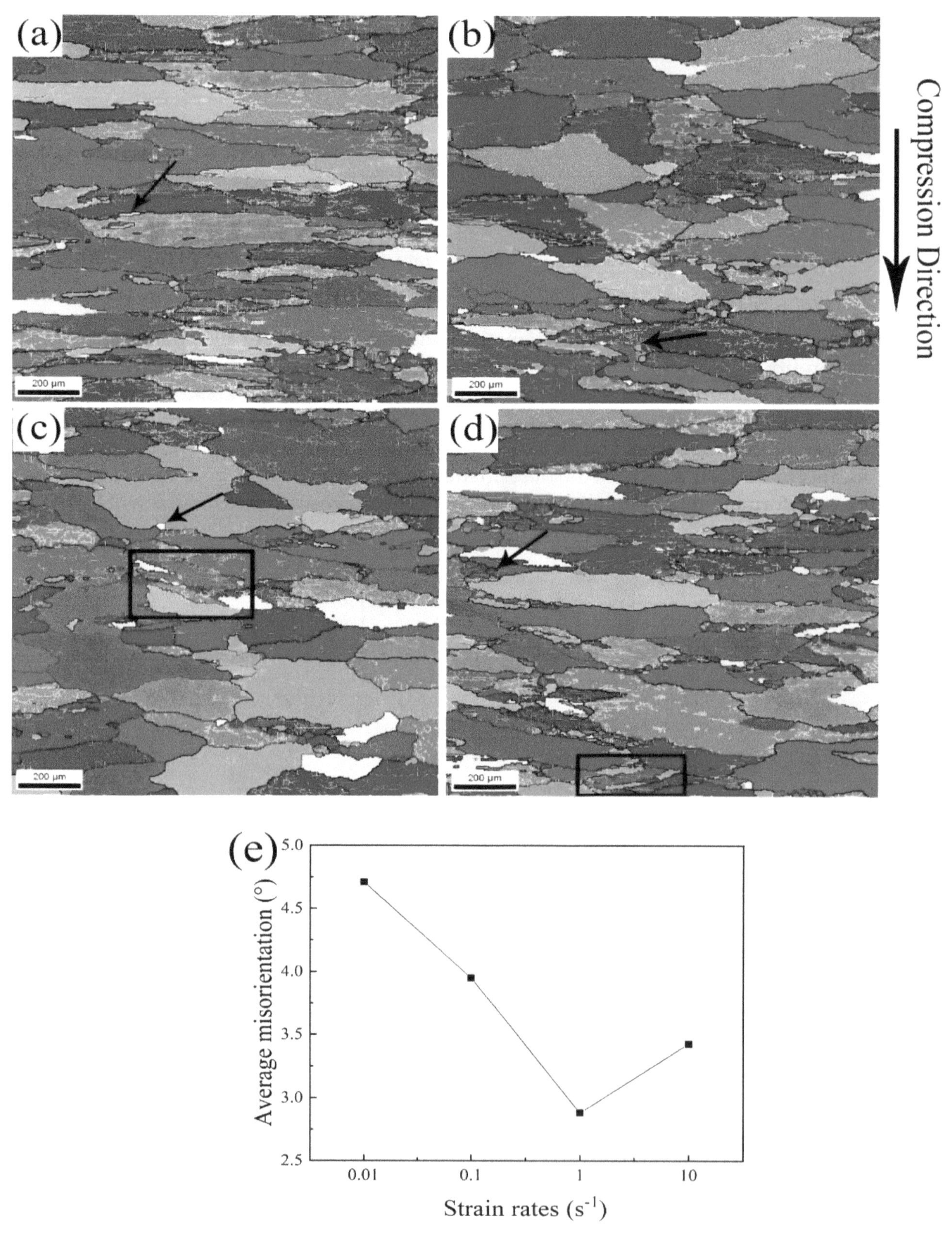

Figure 8. EBSD maps of the Al-Cu-Li-Mg-Ag alloy at 350 °C with various strain rates (**a**) 0.01 s^{-1}; (**b**) 0.1 s^{-1}; (**c**) 1 s^{-1}; (**d**) 10 s^{-1}, and (**e**) average misorientation angle for the four strain rate.

The average misorientation between grains can also be calculated on the basis of EBSD technique. As shown in Figure 8e, the average misorientation decreases firstly with increasing strain rate until strain rates reach 1 s^{-1}, and then increases. Deformation at lower strain rates generally implies that there is adequate time for the dislocation movement, which is evidenced by the massive sub-grain boundaries detected in the grain interior (white lines with misorientation lower than 15° in Figure 8a–c). At higher strain rates (strain rates ≥1 s^{-1}), the adiabatic heating may also be induced, which offers more energy for boundary migration. Thus, the average misorientation values show a non-linear dependence of strain rates.

3.5.2. Effect of Deformation Temperature

Deformation temperature, another factor of Zener–Hollomon parameter Z, has an important influence on the change of microstructures as well. Figure 9 shows the EBSD morphology of the Al-Cu-Li-Mg-Ag alloy deformed at 0.01 s^{-1} and various temperatures. When the sample is deformed at low temperature (350 °C, lnZ = 26.1), dislocation tangling is readily generated, and then dislocation walls develop in the grain interior [5]. So, the microstructure features deformed structures (flat grains) and sub-grains within the grain, as shows in Figure 9a. With increasing temperature, the migration rate of dislocations is enhanced. When the deformation temperature increases to 470 °C (lnZ = 21.1), the improvement of grain boundary mobility promotes the nuclei and growth of recrystallized grains around the original grain boundaries. In addition, dislocation climbs are also enhanced at elevated temperatures, which promote the formation of sub-grain boundaries. Consequently, more recrystallized grains and sub-grains (indicated by white arrows) at the grain boundary and grain interiors (indicated by black arrows) are found than in other lower temperatures (Figure 9a–d). The degree of recrystallization and LAGB% increase with increasing temperature in Figure 9e,f. The fraction of DRX is low regardless of the deformation temperature, which is consistent with the results of Ling [15], which state that the dynamic recovery is the main reason for dynamic softening.

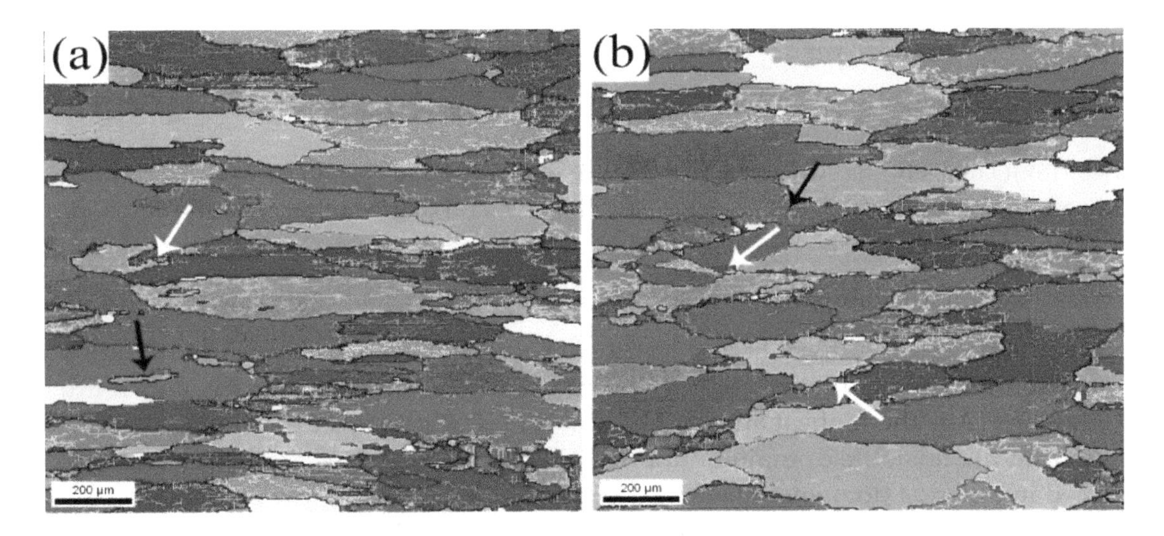

Figure 9. *Cont.*

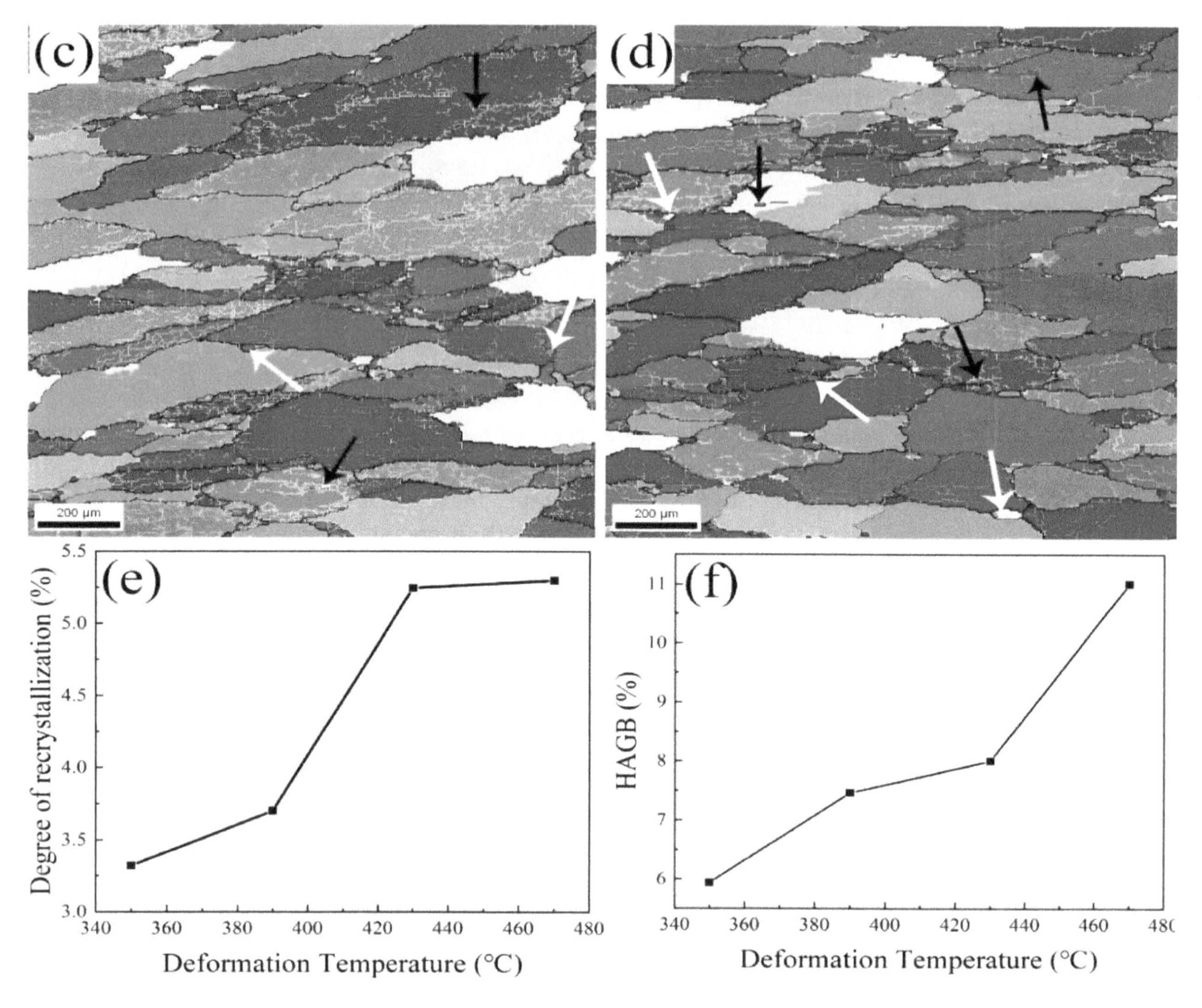

Figure 9. EBSD maps of the Al-Cu-Li-Mg-Ag alloy deformed at a strain rate of $0.01s^{-1}$ and various temperatures (**a**) 350 °C, (**b**) 390 °C, (**c**) 430 °C, (**d**) 470 °C, (**e**) degree of recrystallization and (**f**) the frequency of HAGB.

3.5.3. Recrystallization Mechanism

Based on the processing map (Figure 6b) three representative domains identified by the value of lnZ are chosen to investigate the recrystallization mechanism. Domains A and B are optimum processing regions (A is the highest and B is the lowest value of power dissipation) while domain C is all areas within the unstable domain boundary. Figure 10 illustrates the EBSD morphology of the alloy deformed in different domains: unstable domain C (Figure 10a,b), low power dissipation domain B (Figure 10c), and high power dissipation domain A (Figure 10d). As known to us, the continuous dynamic recrystallization (CDRX) is an important dynamic recrystallization mechanism for high stacking fault energy materials during thermo-mechanical processing. A characteristic of CDRX is the increase of average misorientation angle with the increasing strain [26]. The point-to-origin (cumulative) misorientation is obtained from Figure 10a–d in the grain with similar orientation and shown in Figure 10e. An obvious increase of misorientation gradient is observed inside the grains. Misorientation to the initial grain boundary higher than 15° is observed in all samples, which is strong evidence for CDRX. The fraction of CDRX can be evaluated by the fraction of misorientation angles between 10° and 15° [26–29]. Figure 10f shows the fraction of misorientation angle between 10° and 15° in the four samples, which significant increases compared to the original sample and decreases with decreasing lnZ, suggesting the fraction of CDRX decreases with decreasing lnZ.

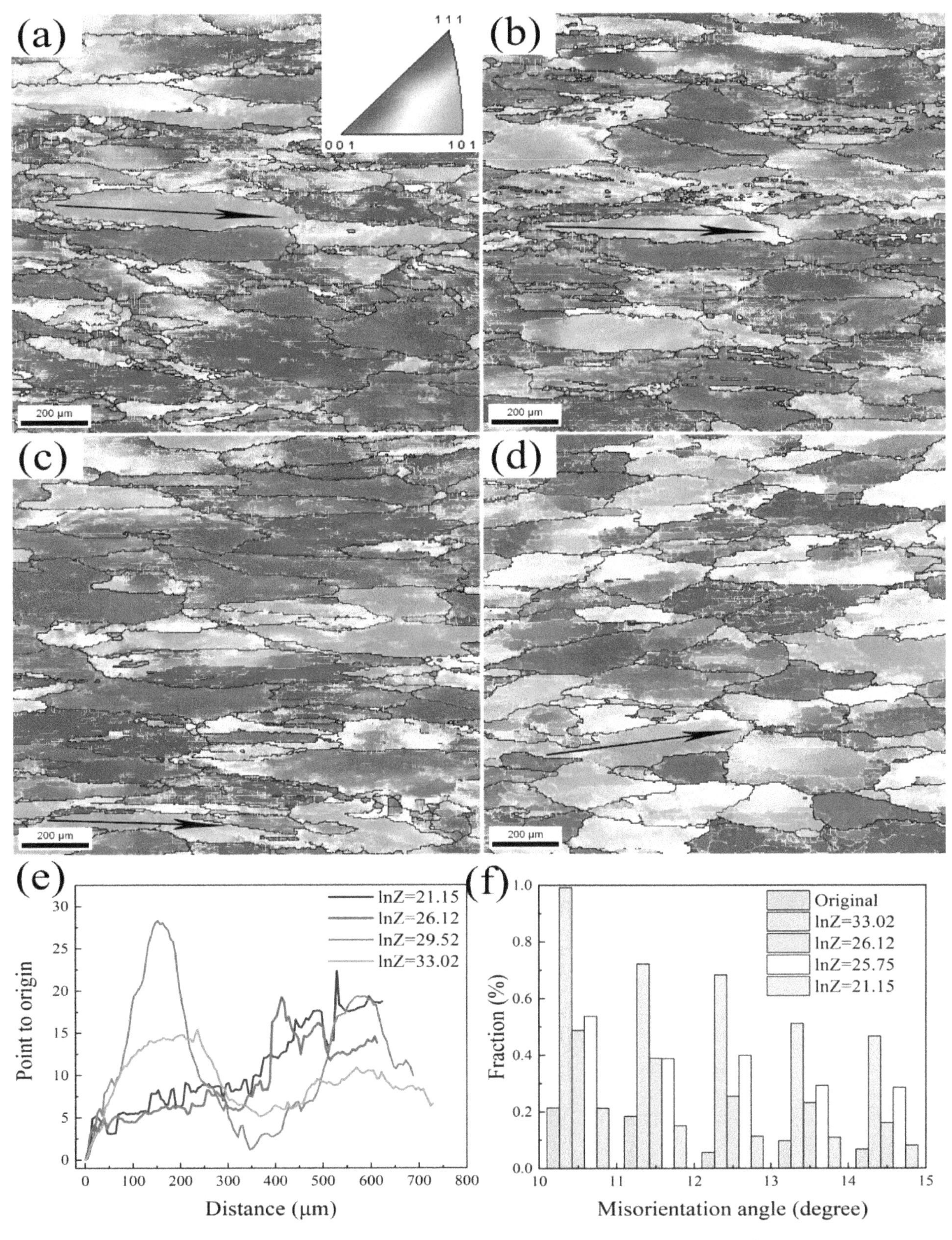

Figure 10. Inverse pole figure maps of the Al-Cu-Li-Mg-Ag alloy under various deformation conditions: (**a**) lnZ = 21.15; (**b**) lnZ = 25.75; (**c**) lnZ = 26.12; (**d**) lnZ = 33.02; (**e**) the changes of misorientation angle along the black lines in (**a**–**d**) and (**f**) the proportion of HAGBs.

Another representative dynamic recrystallization mechanism is discontinuous dynamic recrystallization (DDRX). Grain boundary bulging (BLG), which shows a bow out of grain boundary, is a typical nucleation mechanism of DDRX [30]. The grains with BLG feature, marked by blue arrows, are observed in different deformation conditions, together with the CDRX featured grains indicated by black arrows in Figure 11. The number of grains with BLG features increases with the decreasing of lnZ. Lower values of lnZ usually occur at higher deformation temperatures or lower strain rates, in which case the sub-grain size increases and the dislocation density decreases. Therefore, it can be deduced that the main recrystallization mechanism is gradually transformed from CDRX to DDRX with the decreasing value of lnZ.

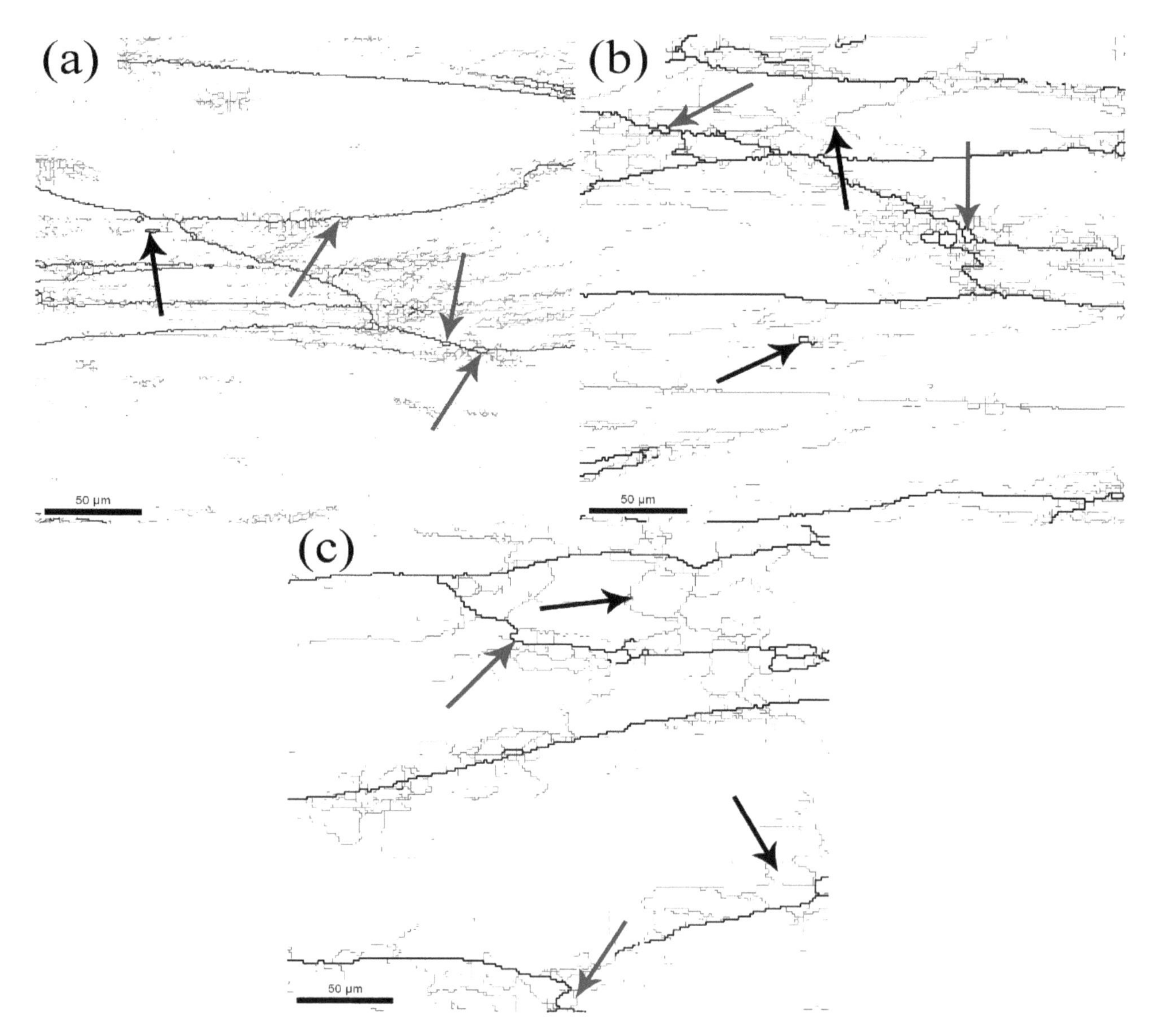

Figure 11. Grain boundary maps of the Al-Cu-Li-Mg-Ag alloy under various lnZ: (**a**) lnZ = 21.15; (**b**) lnZ = 26.12; (**c**) lnZ = 33.02.

Quite a few insoluble constituent particles exist in the Al alloys after homogenization treatments. Most of these insoluble particles are crushed by external force during hot processing and distributed along the processing direction [31]. These particles have a complicated effect on the microstructure evolution during hot processing. On one side, these particles may promote the recrystallization by particle stimulated nucleation. On the other hand, they may inhibit recrystallization by hindering the motion of the grain boundary. The influence of hot deformation parameters on the constituent particles is important and should have significant influence on the microstructure evolution during hot deformation, and should be clarified in the future.

4. Conclusions

In this work, an Al-Cu-Li-Mg-Ag alloy was deformed in a temperature range of 350–470 °C and a strain rate range of 0.01–10 s^{-1}. The flow behavior and microstructure evolution were carefully studied. The conclusions are presented below.

(1) A constitutive equation was developed based on the experimental true stress–strain curves, and for the alloy deformed at a strain of 0.8 it was expressed as

$$\dot{\varepsilon} = 1.74 \times 10^{11}[sinh(0.0125\,\sigma)]^{5.886} \times exp[-159.9/RT]$$

(2) The hot processing map was established and the optimum process window was located in the domain with a temperature between 440 °C and 470 °C, and a strain rate between 0.01 s^{-1} and 0.03 s^{-1}.

(3) Dynamic recovery was the main softening mechanism for the experimental alloy. Dynamic recrystallization was also observed, and CDRX was the main recrystallization mechanism at higher lnZ while DDRX was the main recrystallization mechanism at lower lnZ.

Author Contributions: X.W. and L.C. formulated the experimental design; B.L. and C.L. carried out the Gleeble tests and the microstructure characterization; B.L. analyzed the experimental data and wrote the original draft; X.W. and N.C. discussed the results and made the revisions; L.C. and G.H. applied for the financial support. All authors have read and agreed to the published version of the manuscript.

References

1. Alexopoulos, N.D.; Migklis, E.; Stylianos, A.; Myriounis, D.P. Fatigue behavior of the aeronautical Al–Li (2198) aluminum alloy under constant amplitude loading. *Int. J. Fatigue* **2013**, *56*, 95–105. [CrossRef]
2. Li, B.; Pan, Q.; Yin, Z. Characterization of hot deformation behavior of as-homogenized Al–Cu–Li–Sc–Zr alloy using processing maps. *Mater. Sci. Eng. A* **2014**, *614*, 199–206. [CrossRef]
3. Yu, X.X.; Zhang, Y.R.; Yin, D.F.; Yu, Z.M.; Li, S.F. Characterization of hot deformation behavior of a novel Al–Cu–Li alloy using processing maps. *Acta Metall. Sin.* **2015**, *28*, 817–825. [CrossRef]
4. Rioja, R.J.; Liu, J. The evolution of Al-Li base products for aerospace and space applications. *Metall. Mater Trans A* **2012**, *43*, 3325–3337. [CrossRef]
5. Liao, B.; Cao, L.F.; Wu, X.D.; Zou, Y.; Huang, G.J.; Rometsch, P.A.; Couper, M.J.; Liu, Q. Effect of heat treatment condition on the flow behavior and recrystallization mechanisms of aluminum alloy 7055. *Materials* **2019**, *12*, 311. [CrossRef]
6. Zheng, X.; Luo, P.; Chu, Z.; Xu, J.; Wang, F. Plastic flow behavior and microstructure characteristics of light-weight 2060 Al-Li alloy. *Mater. Sci. Eng. A* **2018**, *736*, 465–471. [CrossRef]
7. Yu, W.; Li, H.; Du, R.; You, W.; Zhao, M.; Wang, Z.A. Characteristic constitution model and microstructure of an Al–3.5Cu–1.5Li alloy subjected to thermal deformation. *Mater. Charact.* **2018**, *145*, 53–64. [CrossRef]
8. Zhao, Q.; Chen, W.; Lin, J.; Huang, S.S.; Xia, X.S. Hot deformation behavior of 7A04 aluminum alloy at elevated temperature: Constitutive modeling and verification. *Int. J. Mater. Form.* **2019**, *13*, 293–302.
9. Chen, L.; Zhao, G.; Yu, J.; Zhang, W. Constitutive analysis of homogenized 7005 aluminum alloy at evaluated temperature for extrusion process. *Mater. Des.* **2015**, *66*, 129–136. [CrossRef]
10. Li, J.; Li, F.; Cai, J.; Wang, R.; Yuan, Z.; Xue, F. Flow behavior modeling of the 7050 aluminum alloy at elevated temperatures considering the compensation of strain. *Mater. Des.* **2012**, *42*, 369–377. [CrossRef]
11. Liu, Y.; Geng, C.; Lin, Q.; Xiao, Y.; Xu, J.; Kang, W. Study on hot deformation behavior and intrinsic workability of 6063 aluminum alloys using 3D processing map. *J. Alloys Compd.* **2017**, *713*, 212–221. [CrossRef]
12. Yin, H.; Li, H.; Su, X.; Huang, D. Processing maps and microstructural evolution of isothermal compressed Al–Cu–Li alloy. *Mater. Sci. Eng. A* **2013**, *586*, 115–122. [CrossRef]
13. Shen, B.; Deng, L.; Wang, X. A new dynamic recrystallisation model of an extruded Al–Cu–Li alloy during high-temperature deformation. *Mater. Sci. Eng. A* **2015**, *625*, 288–295. [CrossRef]
14. Nayan, N.; Murty, S.V.S.N.; Chhangani, S.; Prakash, A.; Prasad, M.J.N.V.; Samajdar, I. Effect of temperature and strain rate on hot deformation behavior and microstructure of Al–Cu–Li alloy. *J. Alloys Compd.* **2017**, *723*, 548–558. [CrossRef]

15. Ou, L.; Zheng, Z.; Nie, Y.; Jian, H. Hot deformation behavior of 2060 alloy. *J. Alloys Compd.* **2015**, *648*, 681–689. [CrossRef]
16. Zhang, C.S.; Wang, C.X.; Guo, R.; Zhao, G.Q.; Chen, L.; Sun, W.C.; Wang, X.B. Investigation of dynamic recrystallization and modelling of microstructure evolution of an Al–Mg–Si aluminum alloy during high-temperature deformation. *J. Alloys Compd.* **2019**, *773*, 59–70. [CrossRef]
17. Zhao, J.; Deng, Y.; Tang, J.; Zhang, J. Influence of strain rate on hot deformation behavior and recrystallization behavior under isothermal compression of Al–Zn–Mg–Cu alloy. *J. Alloys Compd.* **2019**, *809*, 151788. [CrossRef]
18. Liu, L.; Wu, Y.; Gong, H.; Dong, F.; Ahmad, A.S. Modified kinetic model for describing continuous dynamic recrystallization behavior of Al 2219 alloy during hot deformation process. *J. Alloys Compd.* **2020**, *817*, 153301. [CrossRef]
19. Liu, S.H.; Pan, Q.L.; Li, H.; Huang, Z.Q.; Li, K.; He, X.; Li, X.Y. Characterization of hot deformation behavior and constitutive modelling of Al–Mg–Si–Mn–Cr alloy. *J. Mater. Sci.* **2019**, *54*, 4366–4383. [CrossRef]
20. Xiang, Y.; Xiao, S.; Tang, Z.; Zhou, Y. The flow behavior of homogenizated Al-Mg-Si-La aluminum alloy during hot deformation. *Mater. Res. Express* **2019**, *6*, 066563. [CrossRef]
21. Wang, Y.; Zhao, G.; Xu, X.; Chen, X.; Zhang, C. Constitutive modeling, processing map establishment and microstructure analysis of spray deposited Al-Cu-Li alloy 2195. *J. Alloys Compd.* **2019**, *779*, 735–751. [CrossRef]
22. Hu, H.E.; Wang, X.Y.; Deng, L. High temperature deformation behavior and optimal hot processing parameters of Al–Si eutectic alloy. *Mater. Sci. Eng. A* **2013**, *576*, 45–51. [CrossRef]
23. Jenab, A.; Karimi Taheri, A. Experimental investigation of the hot deformation behavior of AA7075: Development and comparison of flow localization parameter and dynamic material model processing maps. *Int. J. Mech. Sci.* **2014**, *78*, 97–105. [CrossRef]
24. Zhang, F.; Sun, J.L.; Shen, J.; Yan, X.D.; Chen, J. Flow behavior and processing maps of 2099 alloy. *Mater. Sci. Eng. A* **2014**, *613*, 141–147. [CrossRef]
25. Kai, X.; Chen, C.; Sun, X.; Wang, C.; Zhao, Y. Hot deformation behavior and optimization of processing parameters of a typical high-strength Al–Mg–Si alloy. *Mater. Des.* **2016**, *90*, 1151–1158. [CrossRef]
26. Huang, K.; Logé, R.E. A review of dynamic recrystallization phenomena in metallic materials. *Mater. Des.* **2016**, *111*, 548–574. [CrossRef]
27. Chamanfar, A.; Alamoudi, T.M.; Nanninga, E.N.; Misiolek, Z.W. Analysis of flow stress and microstructure during hot compression of 6099 aluminum alloy (AA6099). *Mater. Sci. Eng. A* **2019**, *743*, 684–696. [CrossRef]
28. Mandal, S.; Bhaduri, A.K.; Sarma, V.S. A study on microstructural evolution and dynamic recrystallization during isothermal deformation of a Ti-modified austenitic stainless steel. *Metall. Mater. Trans. A* **2010**, *42*, 1062–1072. [CrossRef]
29. Liao, B.; Wu, X.D.; Cao, L.F.; Huang, G.J.; Wang, Z.A.; Liu, Q. The microstructural evolution of aluminum alloy 7055 manufactured by hot thermo-mechanical process. *J. Alloys Compd.* **2019**, *796*, 103–110. [CrossRef]
30. Shimizu, I. Theories and applicability of grain size piezometers: The role of dynamic recrystallization mechanisms. *J. Struct. Geol.* **2008**, *30*, 899–917. [CrossRef]
31. Mcqueen, H.J.; Spigarelli, S.; Kastner, M.E.; Evangelista, E. *Hot Deformation and Processing of Aluminum Alloys*; CRC Press: Boca Raton, FL, USA, 2011.

12

Microstructure Evolution in Super Duplex Stainless Steels Containing σ-Phase Investigated at Low-Temperature using In Situ SEM/EBSD Tensile Testing

Christian Oen Paulsen [1,2,*,†], Runar Larsen Broks [1,†], Morten Karlsen [1,3,†], Jarle Hjelen [1,4,†] and Ida Westermann [1,2,†]

1 Department of Materials Science and Engineering, Norwegian University of Science and Technology (NTNU), NO-7491 Trondheim, Norway; runarbro@stud.ntnu.no (R.L.B.); mortenka@statoil.com (M.K.); jarle.hjelen@ntnu.no (J.H.); ida.westermann@ntnu.no (I.W.)
2 Centre for Advanced Structural Analysis (CASA), Norwegian University of Science and Technology (NTNU), NO-7491 Trondheim, Norway
3 Equinor ASA, NO-7053 Trondheim, Norway
4 Department of Geoscience and Petroleum, Norwegian University of Science and Technology (NTNU), NO-7491 Trondheim, Norway
* Correspondence: christian.o.paulsen@ntnu.no
† These authors contributed equally to this work.

Abstract: An in situ scanning electron microscope (SEM) study was conducted on a super duplex stainless steel (SDSS) containing 0%, 5% and 10% σ-phase. The material was heat treated at 850 °C for 12 min and 15 min, respectively, to achieve the different amounts of σ-phase. The specimens were investigated at room temperature and at −40 °C. The microstructure evolution during the deformation process was recorded using electron backscatter diffraction (EBSD) at different strain levels. Both σ-phase and χ-phase were observed along the grain boundaries in the microstructure in all heat treated specimens. Cracks started to form after 3–4% strain and were always oriented perpendicular to the tensile direction. After the cracks formed, they were initially arrested by the matrix. At later stages of the deformation process, cracks in larger σ-phase constituents started to coalesce. When the tensile test was conducted at −40 °C, the ductility increased for the specimen without σ-phase, but with σ-phase present, the ductility was slightly reduced. With larger amounts of σ-phase present, however, an increase in tensile strength was also observed. With χ-phase present along the grain boundaries, a reduction of tensile strength was observed. This reduction seems to be related to χ-phase precipitating at the grain boundaries, creating imperfections, but not contributing towards the increase in strength. Compared to the effect of σ-phase, the low temperature is not as influential on the materials performance.

Keywords: in situ tensile testing; super duplex stainless steel; SDSS; low-temperature; σ-phase; SEM; EBSD; microstructure analysis

1. Introduction

Duplex stainless steels (DSS) consist of two phases: austenite and ferrite. The two phases, in combination with the alloying elements, result in a steel with superior mechanical properties and corrosion resistance compared to steels with similar cost. DSS was first developed by the oil and gas industry for use in the North Sea. Here, it is typically used in process pipe systems and fittings exposed to corrosive environments at elevated temperature (up to 150 °C in H_2S atmosphere) [1].

DSS typically contains 22% Cr, 5% Ni and 0.18% N, to achieve the desired phase composition and corrosion properties. If better corrosion properties are required, super duplex stainless steel (SDSS) can be used instead. This alloy contains a higher amount of Cr, Ni and N typically 25%, 7% and 0.3%, respectively. In order to achieve the desired phase composition, it is annealed at 1050 °C, and left there until a 50-50 phase balance between ferrite and austenite is obtained. During cooling after the heat treatment, precipitation of intermetallic phases (σ, χ, π and R) may occur. These intermetallics have been found to considerably reduce the mechanical properties and corrosion resistance of the material [2–13]. The most common of these phases is the σ-phase. Even small amounts (<0.5%) of σ-phase will significantly reduce the fracture toughness [2,14]. This reduction, combined with the short time it takes for the phase to form and the deteriorating effect on corrosion properties, is what makes σ-phase a dangerous and strongly unwanted intermetallic.

The χ precipitates on ferrite/ferrite grain boundaries and occurs before the σ-phase [13,15]. In addition, the χ-phase is a metastable phase, consumed during σ-phase precipitation [13]. σ-phase typically forms on austenite/ferrite boundaries, but can also form on ferrite/ferrite boundaries. The σ-phase forms in the temperature range 675 °C–900 °C. After 10 min at 850 °C, small amounts of σ-phase will start to precipitate [6,16,17]. σ-phase has a body centered tetragonal crystal structure with the lattice parameters a = 8.8 Å and c = 4.55 Å, while the χ-phase has a body centered cubic crystal structure with lattice parameter a = 8.8 Å [18]. The lattice parameters of both are significantly larger than the 2.87 Å and 3.65 Å for ferrites and austenite, respectively [19,20]. The chemical composition of σ-phase includes, in addition to Fe, approximately 30–60% Cr and 4–10% Mo. The χ-phase differs from σ-phase with a higher Mo content and a lower content of Cr [15]. As a result, since χ-phase has a higher atomic weight, it is possible to distinguish it from σ-phase in a scanning electron microscope (SEM) using Z-contrast. In such an image, the χ-phase will appear brighter. Since Cr and Mo are stabilizing elements for ferrite, the σ and χ-phase will form at the expense of ferrite. Following the eutectoid reaction $\alpha \rightarrow \sigma + \gamma$ or $\alpha \rightarrow \chi + \gamma$, an increase in the austenite phase will also occur [15,17]. The surrounding area will be depleted of Cr and Mo, which are important elements for corrosion protection and, as a consequence, leaving the material exposed. This is especially troublesome in SDSS since these are mostly selected to operate in areas requiring a corrosion resistance superior to DSS.

A study by Børvik et al. [2] looked into the low-temperature effect on σ-phase in DSS. It was found that the temperature had a minor effect on the tensile ductility, while increasing amounts of σ-phase in the structure considerably reduced the ductility. Another study by Kim et al. [21] investigated the low-temperature mechanical behavior of SDSS containing σ-phase. Here, the material was tested in a universal tensile test machine, equipped with a sub-zero chamber. After the specimens were tested, the microstructure was investigated, comparing the amount of σ-phase present with microcrack length. Microcracks were found to have propagated through the entire σ-phase, relating the crack length to the size of σ-phase inclusions. As in Børvik et al. [2], the influence of temperature was observed to be minor. In addition, in the tensile tests performed at -50 °C, no strain-induced martensite was produced.

In the present study, in situ SEM tensile tests have been conducted on an SDSS with 0%, 5% and 10% σ-phase present in the microstructure. The tensile tests were carried out at both room temperature and sub-zero temperature (-40 °C). The microstructure was monitored with secondary electron imaging and electron backscatter diffraction (EBSD). Images were acquired at different loading steps. From these results, it is possible to observe the microstructure evolution and study the effects of χ-phase and σ-phase on the microstructure during the deformation process.

2. Materials and Methods

2.1. Material and Heat Treatments

The investigated material in this study was a 2507 SDSS, with the chemical composition listed in Table 1. This pipe was manufactured by welding a rolled plate along the length of the pipe. The microstructure of the steel investigated here contained more ferrite than austenite, 56.3% and

43.7%, respectively. The grain size in the two phases is also different, with ferrite having larger grains compared with austenite grains. These results have been summarized in Table 2. In addition, the grains have different morphology in different directions. Figure 1 gives a phase map of the pipe in three different directions. Here, LD, RD, and TD are abbreviations for longitudinal direction, radial direction, and transverse directions, respectively. The meaning of these are illustrated in Figure 2a.

Table 1. Chemical composition of 2507 SDSS.

Element	C	Si	Mn	P	S	Cr	Mo	Ni	Cu	W	N
wt%	0.018	0.42	0.52	0.017	0.001	25.55	3.46	8.28	0.72	0.52	0.25

Table 2. Microstructure statistics summarized. The data was collected from EBSD scans.

Average	Ferrite	Austenite	Overall
Composition	56.3%	43.7%	100%
Grain size	9 μm	6.5 μm	7.9 μm

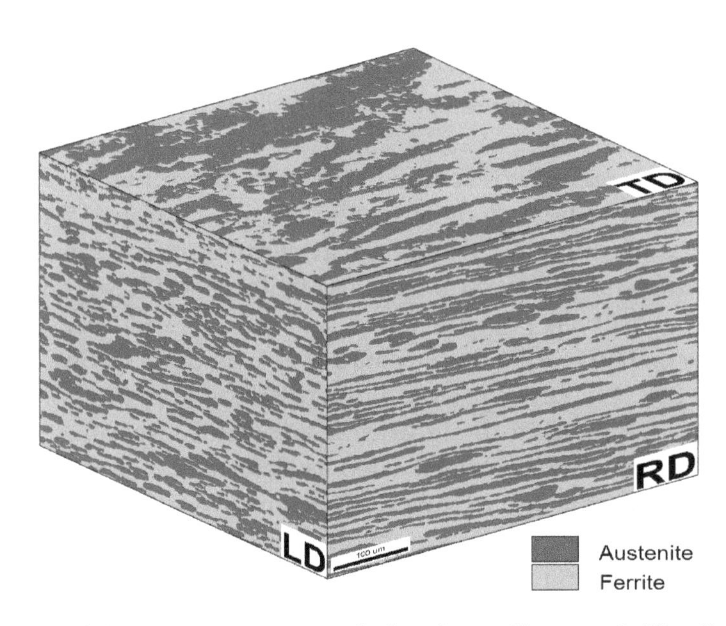

Figure 1. An illustration of the microstructure with the phases illustrated. The dimensions of the cube are 500 μm × 500 μm × 500 μm, green representing ferrite and red representing austenite. In the bottom right corner of each side, the plane normal is given. LD, RD, and TD are illustrated in Figure 2a.

Specimens being used for EBSD analysis need a completely smooth surface, where the deformation layer at the surface has to be removed. For SDSS this was done by grinding and polishing down to 1 μm, followed by electropolishing. The settings used are summarized in Table 3. Specimens were spark eroded from a 10 mm thick pipe to the dimensions in Figure 2b. All specimens were parallel to the length of the pipe, as illustrated in Figure 2a. The observed plane in the specimen during an in situ experiment has TD as plane normal.

Table 3. Parameters used during the electropolishing.

Electrolyte	Struers A2
Voltage [V]	20
Time [s]	15
Temperature [°C]	22

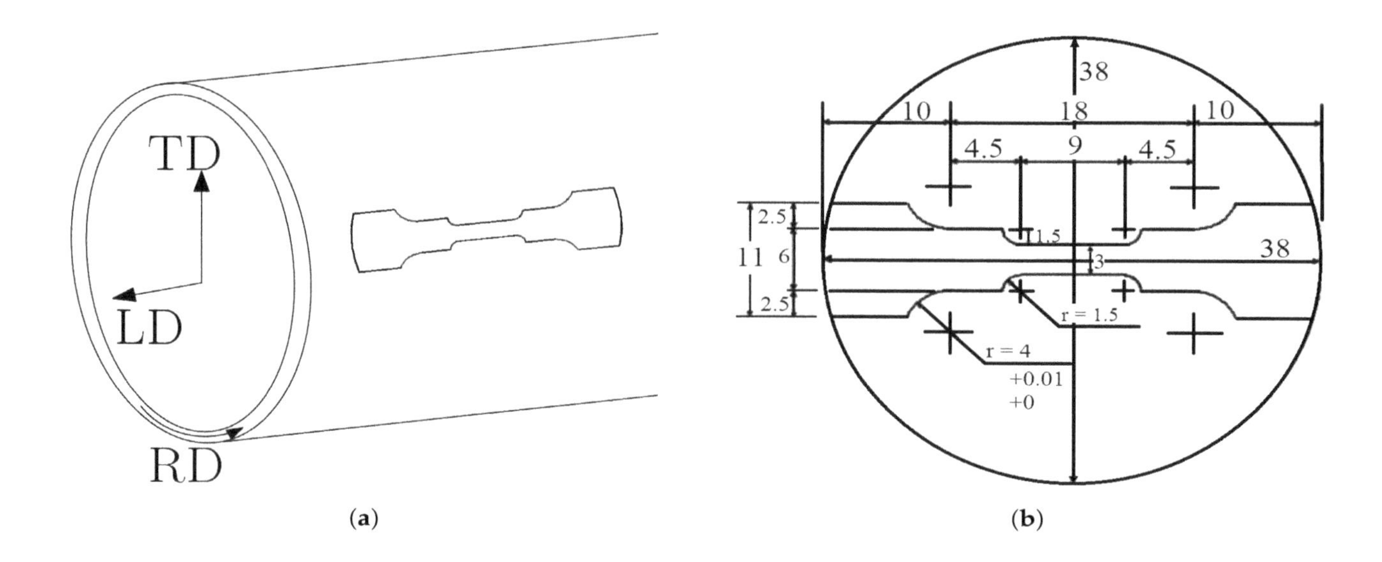

Figure 2. (**a**) an illustration on how the specimens were taken from the SDSS pipe and gives the definition of LD, RD and TD. The specimen dimensions is magnified compared to the pipe for illustration purpose; (**b**) specimen geometry, with all measurements in mm. The specimen had an original thickness of 2 mm before grinding and polishing.

The material was heat treated to achieve different amounts of σ-phase in the structure. Specimens were placed in a pre-heated oven at 850 °C for 12 min, 15 min, 20 min and 25 min. Cooling was performed by quenching in a water bath at room temperature. The heat treatment and the resulting amount of σ-phase achieved are summarized in Table 4. Phase maps from EBSD scans were used to quantify amounts of σ-phase present. These will not be exact measurements since they were only taken from the surface. The results from Elstad [16] was used to determine the heat treatment procedures used in this work. However, σ-phase precipitation was not constant with the same heat treatment being performed. Resulting in significant variation in σ-phase content during the heat treatment. All specimens in this work are heat treated as described in Table 4, but only specimens with amounts roughly in the region indicated in the third column were used for the in situ tests. However, the deviation was less than 1% for the 5% specimen, measured by EBSD. For the specimens with larger amounts of σ-phase present, the deviation was 1–2%.

Table 4. Heat treatment and resulting amount of σ-phase in the specimens tested.

Temperature [°C]	Time [min]	Amount σ-Phase [%]
-	-	0
850	12	5
850	15	10
850	20	15
850	25	20

2.2. Materials Characterization

During this experiment, the microstructure was monitored using secondary electron imaging and EBSD. Images were acquired at different loading steps. At each step, the same area (350 µm × 350 µm) was recorded with EBSD, using a step size of 1 µm. From these results, it is possible to observe the microstructure evolution and study the effects of σ-phase and χ-phase on the microstructure during the deformation process. The microscope used was a Field Emission SEM Zeiss Ultra 55 Limited Edition (Jena, Germany) with a NORDIF UF-1100 EBSD detector (Trondheim, Norway), with the microscope settings given in Table 5. Z-contrast imaging mode was used in order to distinguish σ-phase from χ-phase during the experiments.

Table 5. SEM parameters used during EBSD acquisition.

Acceleration Voltage [kV]	**20**
Working distance [mm]	24.6–25.4
Tilt angle [°]	70
Aperture size [μm]	300
Probe current [nA]	65–70

2.3. Tensile Testing

The specimen was deformed using a spindle-driven in situ tensile device. This device was placed inside the vacuum chamber of the SEM to monitor the microstructure. In situ tensile tests were carried out at both room temperature and at −40 °C for specimens containing 0%, 5% and 10% σ-phase. Tensile tests were also performed on specimens containing 15% and 20% of σ-phase, however, no in situ investigation or low-temperature testing was carried out on these specimens, due to their purely brittle behavior. The in situ tensile tests were carried out with a constant ramp speed of 1 μm/s. This corresponds to a strain rate of 1.11×10^{-4} s^{-1}. For further reading and previous use of the in situ device, the reader is referred to [22,23]. When performing the sub-zero experiments, a cold finger was attached to the specimen as shown in Figure 3. This cold finger is made from 99.99% Cu. It goes from the specimen, through the microscope door, into a dewar filled with liquid N. The blue and white wire seen in Figure 3 is a thermocouple. It was placed between the screw-head and specimen throughout the experiment. The temperature was measured to be in the interval −35 °C and −45 °C for all specimens. However, the fluctuations in temperature are assumed to be due to the variable thermal resistance between the thermocouple and specimen. This variation is a result of the thermocouple shifting position during straining. The temperature is assumed constant and reported as −40 °C in this paper.

Figure 3. Cold finger attached to the specimen with a thermocouple placed between the screw-head and specimen.

3. Results

3.1. Tensile Properties and Fracture Surfaces

The tensile test curves for the specimens tested in this work are shown in Figure 4. As seen from these curves, the specimens with more than 10% σ-phase present exhibit a purely brittle behavior at room temperature and do not deform plastically before fracture. For that reason, these specimens are not suitable for in situ and low-temperature investigations. Hence, only specimens containing roughly 5% and 10% σ-phase are further investigated. The specimens containing 0% σ-phase are included as a reference.

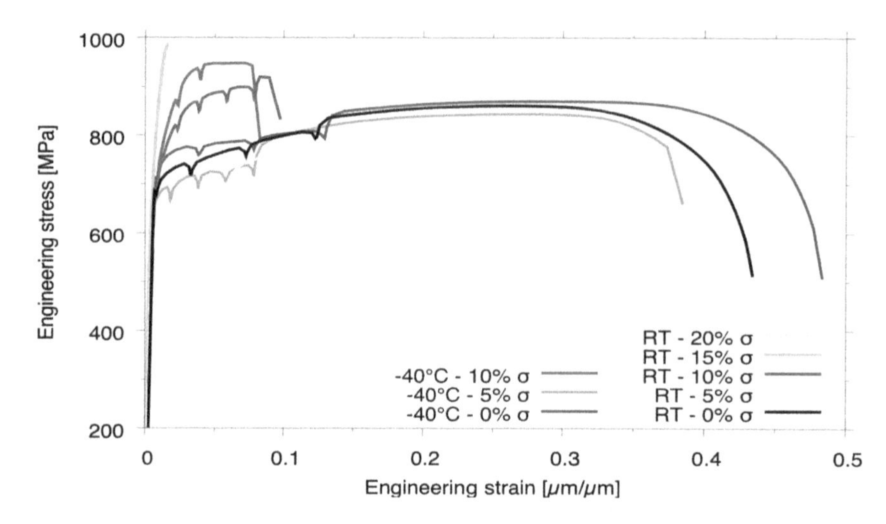

Figure 4. Tensile test curves obtained during the in situ tensile tests. The drops in the curves are when the test is paused for acquiring EBSD data.

The stress–strain curves in Figure 4 show that small amounts of σ-phase greatly affect the tensile properties of the material. Another observation is the short time at the critical temperature it takes before the material is completely brittle (cf. Table 4). Specimens containing 15% and 20% σ-phase only deforms elastically before fracture. A general remark is that the yield strength increases at low temperature and the strain at fracture decrease with an increasing amount of intermetallic phases. Conversely, for the material not heat treated, there is an increase of fracture strain. In addition, the drops in the curves are from when the tensile test is paused for EBSD acquisition. A curious observation from the tensile test curve is that the tests containing 5% σ-phase have a lower yield strength and ultimate tensile strength (UTS). In Figure 5, the microstructure of one of these specimens can be seen. Along the grain boundaries, the χ-phase has precipitated as a thin continuous layer of approximately 200 nm thickness. This image was acquired during the test at room temperature, after 4% strain. In the center of the image is a σ-phase island, with two cracks marked with white circles. The χ-phase also contains numerous small cracks, seen in the black circles in Figure 5, which seem to contribute towards a reduction in strength. When the amount of σ-phase increases, it also adds towards increased tensile strength.

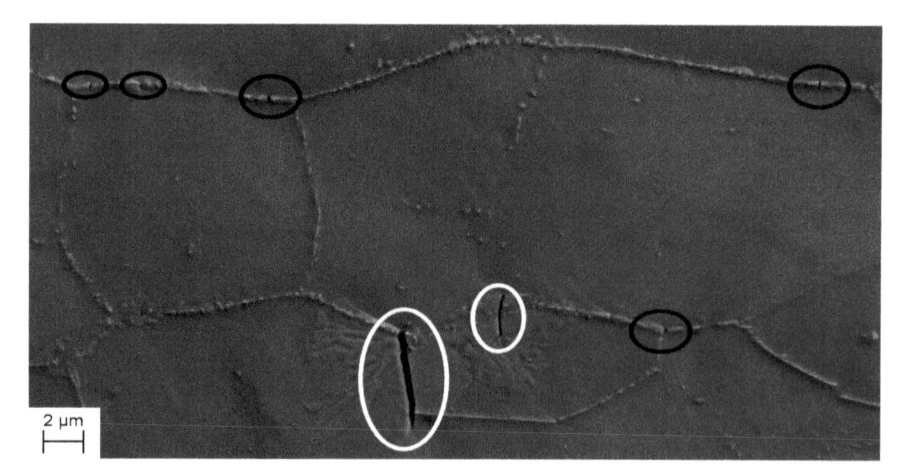

Figure 5. A close up micrograph from the specimen containing 5% σ-phase, after 4% strain, tested at room temperature. Along the grain boundaries, the χ-phase can be found, and, in the center, a larger island of σ-Phase is seen. The white circles show cracks in the σ-phase and the black circles show the cracks in the χ-phase.

In Figure 6, the fracture surfaces of the specimens tested at room temperature with 0%, 5% and 10% σ-phase are presented. To the left is an overview of the total surface area and to the right is a close-up image showing the fracture surface at a higher magnification. The reference sample exhibits classic ductile fracture features, with a large reduction of area and the typical cup and cone dimpled structure at the surface. This is also expected when compared to the tensile test curve (Figure 4). In the specimen with 5% σ-phase present, Figure 6b, some reduction in area is observed—however, not as great as in the test with 0% σ-phase present. In addition, here the fracture surface appears to be mixed between a ductile dimpled structure and a brittle faceted structure. Conversely, the specimen containing 10% σ-phase, Figure 6c, has all the characteristics of a brittle fracture. There is little to no reduction in area and completely faceted fracture surface, despite having a 10% fracture strain.

Figure 6. Fracture surfaces for the tensile test specimens with (**a**) 0%, (**b**) 5% and (**c**) 10% σ-phase, tested at room temperature. To the left is the total fracture area and to the right is a close-up of the fracture surface.

3.2. Microstructure Evolution

During the tensile tests, specimens containing different amounts of σ-phase were recorded using secondary electron imaging and EBSD to observe the microstructure throughout the deformation process. EBSD scans were obtained at the same area at approximately 0%, 2% and 6% strain of all tested specimens. Each of the tensile test curves in Figure 4 showed a drop when the test was paused for the acquisition of EBSD scans and secondary electron imaging. An observation is that the specimen with 0% σ-phase, tested at −40 °C, has a greater fracture strain than the specimen tested at room temperature.

In all specimens containing σ-phase, cracks were observed throughout the microstructure. These were observed to form after 3–4% strain in all the specimens, initiating in the σ-phase. Typical size of the cracks is seen in Figure 7. In Figure 8, two micrographs acquired at 6% and 10% strain show several micro-sized cracks in the σ-phase. During further straining, these cracks widen and appears to propagate deeper into the specimen. The ferrite and austenite grain boundaries act as a barrier for the cracks to propagate further. However, the larger constituents of σ-phase in the matrix contain large cracks, which eventually will propagate through the matrix. This is seen in the center of both frames in Figure 8. The microcracks in Figure 8a grow, and, in Figure 8b (black circle), they have coalesced, forming one large crack. A close-up of this crack is shown in Figure 9. This is a phase map superimposed on to an image quality (IQ) map from EBSD, acquired with a step size of 50 nm. From this map, it can be seen that the crack propagates along grain boundaries when it is moving through the matrix. When the cracks start to coalesce, the material is close to fracturing, as the volume fraction of cracks is increasing fast. The micrograph in Figure 8b was acquired after 10% strain. The specimen fractured after being strained less than 1% further. It is possible to see how the cracks in the white circles widen from Figure 8a to Figure 8b. Presumably, they are propagating through the thickness of the material.

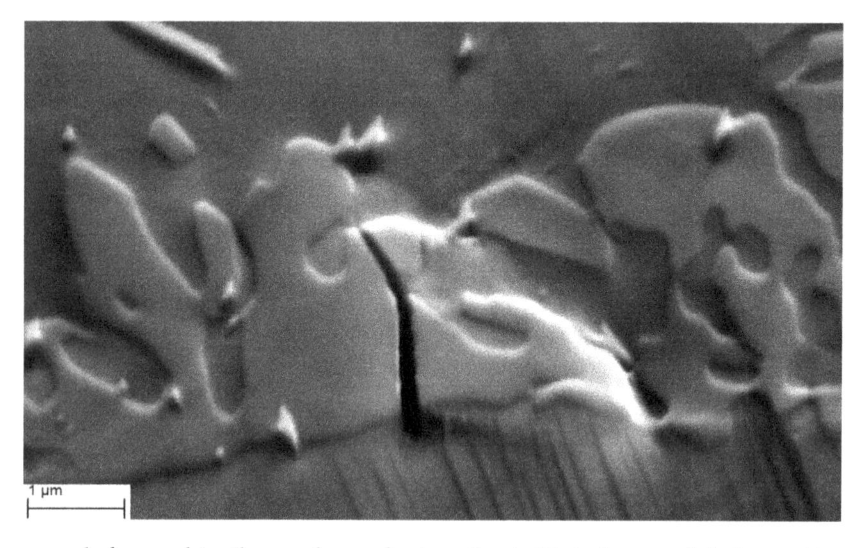

Figure 7. Micro-crack formed in the σ-phase during the initial stages of deformation. This frame is acquired after 6% strain, during the low temperature test with 5% σ-phase.

In Figure 10, the grain orientation spread (GOS) in the different tests are shown. All curves are obtained after 6% global strain. The GOS gives a quantitative description of the crystallographic orientation gradients in individual grains [24,25]. It is found by calculating the average orientation deviation of all points in a grain from the average grain orientation. A higher spread would indicate that those grains are accommodating a larger deformation compared to a lower spread. However, as seen from the graphs, there is, in general, a low spread, with peaks for all tests around 1–2°. One notable deviation is the curves from the experiment at −40 °C with 0% σ-phase present in Figure 10b. These grains seem to accommodate more deformation, with a larger GOS distribution compared to other curves in Figure 10. Another observation is that the ferrite and austenite phases have nearly identical curves in the low-temperature test in Figure 10b, while the phases are behaving differently at room temperature (Figure 10a). During the room temperature tests, all curves for ferrite grains have a taller peak compared to austenite grains. In addition, the specimen with 5% σ has a higher GOS peak-value compared to the specimen containing 10% σ-phase when tested at −40 °C.

At room temperature, the austenite for both tests is fairly similar, while the ferrite is accommodating more deformation in the specimen with 5% σ-phase.

(**a**) 6% strain (**b**) 10% strain

Figure 8. Micrographs of cracks formed in the σ-phase, taken from the test carried out at room temperature with 10% σ-phase present. Some cracks are restricted by the matrix while some propagate and coalesce. In the green circle, a heavily deformed austenite grain with cross-slip is seen. The white circles show microcracks restrained by the matrix and the large crack in the black circle was formed when many smaller cracks coalesce. A close-up of this crack is shown in Figure 9.

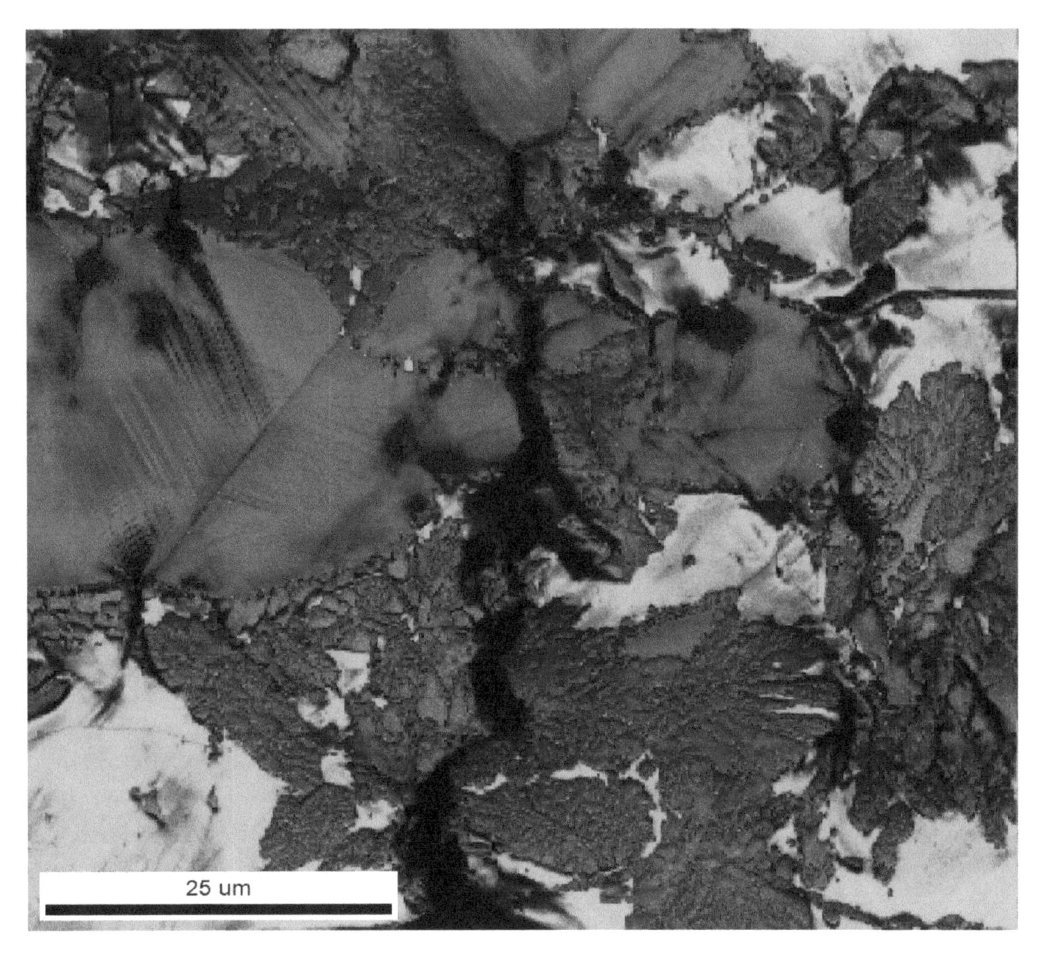

Figure 9. A close-up of the crack shown in the black circle in Figure 8b. This is a phase map with an IQ map overlay, acquired by EBSD. The red is austenite, green is ferrite and mustard yellow is σ-phase. The EBSD-scan of this area was acquired with a step size of 50 nm.

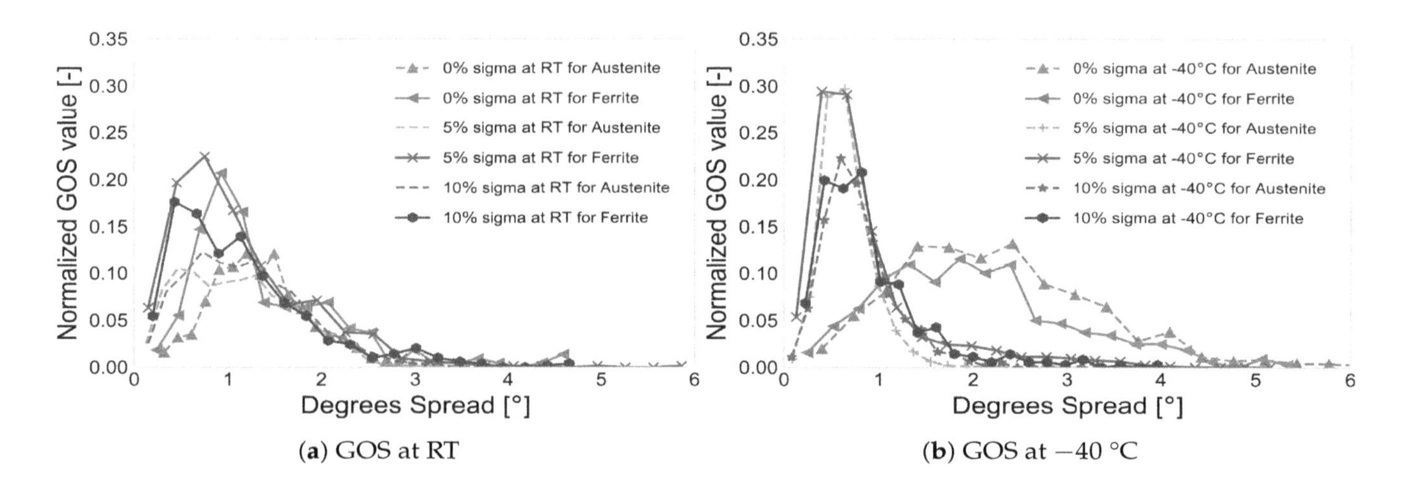

Figure 10. The grain orientation spread curves for the different specimens. (**a**) are the specimens tested at room temperature and (**b**) are the specimens tested at −40 °C. All curves were taken after 6% strain. The solid lines are ferrite and the dashed lines are austenite.

4. Discussion

During this work, different specimens of super duplex stainless steel, containing varying amounts of σ-phase have been investigated, during an in situ SEM tensile test. Each specimen was taken from a pipe segment and heat treated to get different amounts of σ-phase present. In general, it took roughly 10 min for the intermetallic phases to start forming at 850 °C. During the next 10 min, approximately 15% of σ-phase had precipitated, and the material had changed to an utterly brittle behavior, as seen in Figure 4. It proved hard to meet our targets of 5% and 10% σ-phase, sometimes achieving 0% after 13 min and other times 15% after 15 min at 850 °C. However, when the σ-phase starts to precipitate, it forms fast. Since duplex steels are being heat treated, typically at 1050 °C, to achieve its final microstructure and often goes through other heat treatment, e.g., welding, a thorough control of the cooling rate is crucial. In addition, no ductile-to-brittle transition was observed in this work. This was also the case in the work of Børvik et al. [2] and Kim et al. [21]. In these works, DSS and SDSS, respectively, were tested at −50 °C and no transition was observed. This means that, if the material has a ductile-to-brittle transition temperature, it is lower than −50 °C.

As seen from Figure 4, additions of σ-phase significantly reduce the ductility. This phenomenon is also well documented by others in previous studies [2,3,5,7,9]. However, in this study, the microstructure has been closely monitored during the tensile test to elucidate how it is accommodating the σ-phase in relation to deformation. The GOS in grains from the austenite and ferrite (shown in Figure 10) suggests that the presence of σ-phase and low temperature (−40 °C) is influencing the deformation behavior of the matrix. A consequence of presence of σ-phase is a lower fraction of ferrite. This altered phase balance, in combination with much harder particles containing numerous cracks explains this difference in behavior between specimens with and without σ-phase present. However, the primary concern is the brittle nature of σ-phase. Cracks were observed in the σ-phase at 3–4% strain in all specimens, and all cracks were oriented perpendicular to the tensile direction. During the initial stages, the surrounding matrix restricts the growth of the crack. As the material is strained further, the cracks continues to widen. Eventually, the cracks start to propagate and coalesce. In specimens with higher amounts of σ-phase, the propagation occurs earlier, following the shorter distance to the nearest σ-phase inclusion. In addition, the σ-phase particles are larger and the cracks, therefore, grow to a larger size.

The influence of temperature seems to make the σ-phase somewhat more brittle, resulting in a higher UTS and lower ductility. Austenite and ferrite grains seem to behave similarly during the low-temperature tests with σ-phase present when studying Figure 10b. However, during the test at room temperature, the ferrite accommodates more deformation compared to the austenite. This is seen from the curves in Figure 10a. The reason for the ferrite being more active is believed to be due to the fact that ferrite has 48 active slip systems at room temperature. Conversely, austenite has 12 slip systems and they are not dependent on temperature. With more slip systems available, there are more ways for the dislocations to propagate. In addition, the specimens without any σ-phase present have a larger GOS compared to the specimens containing σ-phase. This indicates that the presence of σ-phase in the structure is retarding the deformation of ferrite and austenite. This is also observed through visual inspection of micrographs. There are more slip lines present, at equal strain level, in specimens without σ-phase present.

An observation of a specimen with 0% σ-phase, tested at −40 °C, has a greater fracture strain than the specimen tested at room temperature. It could be expected that the ferrite would have a brittle behavior at this temperature. A reason for this behavior might be due to the fact that SDSS is a highly alloyed material, containing elements improving the low-temperature performance of ferrite. In addition, the presence of austenite will improve low-temperature performance. It has been reported in several studies that austenitic steels have increased ductility at −50 °C in static uniaxial tensile tests [26–28].

Looking at the tensile test curve in Figure 4 for the tests with 5% σ-phase, a lower tensile strength compared to the curve without any σ-phase present is observed. Conversely, a greater amount of σ-phase gives a contribution towards increased strength. An explanation for this can be the relative amount of χ-phase present. As seen from the black circles in Figure 5, the χ-phase precipitates along grain boundaries and is very brittle containing many cracks. These cracks result in the observed reduction of tensile strength. However, the size of the cracks in χ-phase are subcritical and does not contribute towards a large reduction in ductility. The specimen containing 5% σ-phase is still a very ductile material, with a fracture strain of 35%–38%. This is in contrast to previously reported literature. As mentioned in the Introduction, it has been reported that specimens with only 0.5% σ-phase have significantly reduced fracture toughness. However, as discussed in Børvik et al. [2] and Børvik et al. [3], DSS are more sensitive towards σ-phase with respect to fracture toughness than to tensile ductility. In this work, all specimens were tested strain rate of $1.11 \times 10^{-4}\ s^{-1}$. In addition, the tensile tests were paused at certain intervals to acquire images and EBSD scans. In Børvik et al. [2], an increase in flow stress of about 30% was found for DSS when the strain rate was increased from $5 \times 10^{-4}\ s^{-1}$ to $50\ s^{-1}$ based on tensile tests.

No strain-induced martensite was observed in any of the specimens investigated in this work. This indicates a very stable austenitic phase. However, this is not unexpected, since the σ-phase is formed at the expense of ferrite, not austenite. The alloying elements added to stabilize the austenitic phase are still present in the matrix. In the work by Kim et al. [21], there was also no martensite observed.

5. Conclusions

- The cracks in χ-phase contribute towards a lower flow stress but were not of critical size concerning a large reduction in tensile ductility. The specimens with small amounts of χ-phase and σ-phase still retained a ductility of 35%.
- Visible cracks start to form after 3–4% strain, regardless of σ-phase content and they all form perpendicular to the tensile direction.
- During the initial stages of deformation, the cracks are constrained by the ferrite/austenite matrix. However, during the later stages, these cracks start to propagate through the material and coalesce. This occurs moments before fracture.
- The ferrite accommodates more deformation than austenite at room temperature tests; however, during low-temperature tests, both phases have a more equal behavior during deformation.
- At low temperature, with σ-present, the material had slightly higher flow stress and lower ductility. However, the amount of σ-phase present is the most important aspect when it

comes to duplex steels. It alters the phase balance of ferrite and austenite and deteriorates the mechanical properties.

Author Contributions: C.O.P. is the first author and analyzed the data and wrote the paper. The experiments were performed by C.O.P. and R.L.B. I.W., M.K., and J.H. conceived, designed and supervised the experiments. In addition, they contributed to the interpretation of data and editing the paper.

Abbreviations

The following abbreviations are used in this manuscript:

CASA	Centre for Advanced Structural Analysis
DSS	Duplex Stainless Steel
EBSD	Electron Backscatter Diffraction
GOS	Grain Orientation Spread
IQ	Image Quality
LD	Longitudinal Direction
NTNU	Norwegian University of Science and Technology
RD	Radial Direction
RT	Room Temperature
SDSS	Super Duplex Stainless Steel
SEM	Scanning Electron Microscope
TD	Transverse Direction
UTS	Ultimate Tensile Strength

References

1. NORSOK Standard. Materials Selection. 2004. Available online: http://www.standard.no/pagefiles/1174/m-dp-001r1.pdf (accessed on 11 June 2018).
2. Børvik, T.; Lange, H.; Marken, L.A.; Langseth, M.; Hopperstad, O.S.; Aursand, M.; Rørvik, G. Pipe fittings in duplex stainless steel with deviation in quality caused by sigma phase precipitation. *Mater. Sci. Eng. A* **2010**, *527*, 6945–6955, doi:10.1016/j.msea.2010.06.087. [CrossRef]
3. Børvik, T.; Marken, L.A.; Langseth, M.; Rørvik, G.; Hopperstad, O.S. Influence of sigma-phase precipitation on the impact behaviour of duplex stainless steel pipe fittings. *Ships Offshore Struct.* **2016**, *11*, 25–37, doi:10.1080/17445302.2014.954303. [CrossRef]
4. Lee, Y.H.; Kim, K.T.; Lee, Y.D.; Kim, K.Y. Effects of W substitution on ς and χ phase precipitation and toughness in duplex stainless steels. *Mater. Sci. Technol.* **1998**, *14*, 757–764, doi:10.1179/mst.1998.14.8.757. [CrossRef]
5. Kim, S.B.; Paik, K.W.; Kim, Y.G. Effect of Mo substitution by W on high temperature embrittlement characteristics in duplex stainless steels. *Mater. Sci. Eng. A* **1998**, *247*, 67–74, doi:10.1016/S0921-5093(98)00473-0. [CrossRef]
6. Lopez, N.; Cid, M.; Puiggali, M. Influence of σ-phase on mechanical properties and corrosion resistance of duplex stainless steels. *Corros. Sci.* **1999**, *41*, 1615–1631, doi:10.1016/S0010-938X(99)00009-8. [CrossRef]
7. Chen, T.H.; Yang, J.R. Effects of solution treatment and continuous cooling on σ-phase precipitation in a 2205 duplex stainless steel. *Mater. Sci. Eng. A* **2001**, *311*, 28–41, doi:10.1016/S0921-5093(01)00911-X. [CrossRef]
8. Chen, T.H.; Weng, K.L.; Yang, J.R. The effect of high-temperature exposure on the microstructural stability and toughness property in a 2205 duplex stainless steel. *Mater. Sci. Eng. A* **2002**, *338*, 259–270, doi:10.1016/S0921-5093(02)00093-X. [CrossRef]
9. Zucato, I.; Moreira, M.C.; Machado, I.F.; Lebrão, S.M.G. Microstructural characterization and the effect of phase transformations on toughness of the UNS S31803 duplex stainless steel aged treated at 850 °C. *Mater. Res.* **2002**, *5*, 385–389, doi:10.1590/S1516-14392002000300026. [CrossRef]
10. Cvijović, Z.; Radenković, G. Microstructure and pitting corrosion resistance of annealed duplex stainless steel. *Corros. Sci.* **2006**, *48*, 3887–3906, doi:10.1016/j.corsci.2006.04.003. [CrossRef]
11. Michalska, J.; Sozańska, M. Qualitative and quantitative analysis of σ and χ phases in 2205 duplex stainless steel. *Mater. Charact.* **2006** *56*, 355–362, doi:10.1016/j.matchar.2005.11.003. [CrossRef]

12. Souza, C.M.; Abreu, H.F.G.; Tavares, S.S.M.; Rebello, J.M.A. The σ phase formation in annealed UNS S31803 duplex stainless steel: Texture aspects. *Mater. Charact.* **2008**, *59*, 1301–1306, doi:10.1016/j.matchar.2007.11.005. [CrossRef]
13. Pohl, M.; Storz, O.; Glogowski, T. Effect of intermetallic precipitations on the properties of duplex stainless steel. *Mater. Charact.* **2007**, *58*, 65–71, doi:10.1016/j.matchar.2006.03.015. [CrossRef]
14. Calliari, I.; Zanesco, M.; Ramous, E. Influence of isothermal aging on secondary phases precipitation and toughness of a duplex stainless steel SAF 2205. *J. Mater. Sci.* **2006**, *41*, 7643–7649, doi:10.1007/s10853-006-0857-2. [CrossRef]
15. Escriba, D.; Materna-Morris, E.; Plaut, R.; Padilha, A. Chi-phase precipitation in a duplex stainless steel. *Mater. Charact.* **2009**, *60*, 1214–1219, doi:10.1016/J.MATCHAR.2009.04.013. [CrossRef]
16. Elstad, K.R. In Situ Tensile Testing During Continuous EBSD Mapping of Super Duplex Stainless Steel Containing Sigma Phase. 2016. Available online: http://hdl.handle.net/11250/2418016 (accessed on 25 May 2018).
17. Stradomski, Z.; Dyja, D. Sigma Phase Precipitation in Duplex Phase Stainless Steels. 2009. Available online: http://www.ysesm.ing.unibo.it/Abstract/57Dyja.pdf (accessed on 25 May 2018).
18. Padilha, A.F.; Rios, P.R. Decomposition of Austenite in Austenitic Stainless Steels. *ISIJ Int.* **2002**, *42*, 325–327, doi:10.2355/isijinternational.42.325. [CrossRef]
19. Cahn, R.W.; Haasen, P.; Kramer, E.J. *Materials Science and Technology: A Comprehensive Treatment—Volume 1: Structure of Solids*; Wiley-VCH: Weinheim, Germnay, 2005.
20. Cahn, R.W.; Haasen, P.; Kramer, E.J. *Materials Science and Technology: A Comprehensive Treatment—Volume 7: Constitution and Properties of Steel*; Wiley-VCH: Weinheim, Germnay, 2005.
21. Kim, S.K.; Kang, K.Y.; Kim, M.S.; Lee, J.M. Low-temperature mechanical behavior of super duplex stainless steel with sigma precipitation. *Metals* **2015**, *5*, 1732–1745, doi:10.3390/met5031732. [CrossRef]
22. Karlsen, M.; Hjelen, J.; Grong, Ø.; Rørvik, G.; Chiron, R.; Schubert, U.; Nilsen, E. SEM/EBSD based in situ studies of deformation induced phase transformations in supermartensitic stainless steels. *Mater. Sci. Technol.* **2008**, *24*, 64–72, doi:10.1179/174328407X245797. [CrossRef]
23. Karlsen, M.; Grong, Ø.; Søfferud, M.; Hjelen, J.; Rørvik, G.; Chiron, R. Scanning Electron Microscopy/Electron Backscatter Diffraction—Based Observations of Martensite Variant Selection and Slip Plane Activity in Supermartensitic Stainless Steels during Plastic Deformation at Elevated, Ambient, and Subzero Temperatures. *Metall. Mater. Trans. A* **2009**, *40*, 310–320, doi:10.1007/s11661-008-9729-5. [CrossRef]
24. Jorge-Badiola, D.; Iza-Mendia, A.; Gutiérrez, I. Study by EBSD of the development of the substructure in a hot deformed 304 stainless steel. *Mater. Sci. Eng. A* **2005**, *394*, 445–454, doi:10.1016/j.msea.2004.11.049. [CrossRef]
25. Mitsche, S.; Poelt, P.; Sommitsch, C. Recrystallization behaviour of the nickel-based alloy 80 A during hot forming. *J. Microsc.* **2007**, *227*, 267–274, doi:10.1111/j.1365-2818.2007.01810.x. [CrossRef] [PubMed]
26. Byun, T.; Hashimoto, N.; Farrell, K. Temperature dependence of strain hardening and plastic instability behaviors in austenitic stainless steels. *Acta Mater.* **2004**, *52*, 3889–3899, doi:10.1016/J.ACTAMAT.2004.05.003. [CrossRef]
27. Lee, K.J.; Chun, M.S.; Kim, M.H.; Lee, J.M. A new constitutive model of austenitic stainless steel for cryogenic applications. *Comput. Mater. Sci.* **2009**, *46*, 1152–1162, doi:10.1016/J.COMMATSCI.2009.06.003. [CrossRef]
28. Park, W.S.; Yoo, S.W.; Kim, M.H.; Lee, J.M. Strain-rate effects on the mechanical behavior of the AISI 300 series of austenitic stainless steel under cryogenic environments. *Mater. Des.* **2010**, *31*, 3630–3640, doi:10.1016/j.matdes.2010.02.041. [CrossRef]

13

Improvement of the Crack Propagation Resistance in an $\alpha + \beta$ Titanium Alloy with a Trimodal Microstructure

Changsheng Tan [1,*], Yiduo Fan [1], Qiaoyan Sun [2] and Guojun Zhang [1,*]

[1] School of Materials Science and Engineering, Xi'an University of Technology, Xi'an 710048, China; 2190120044@stu.xaut.edu.cn
[2] State Key Laboratory for Mechanical Behavior of Materials, Xi'an Jiaotong University, Xi'an 710049, China; qysun@xjtu.edu.cn
* Correspondence: cstan@xaut.edu.cn (C.T.); zhangguojun@xaut.edu.cn (G.Z.)

Abstract: The roles of microstructure in plastic deformation and crack growth mechanisms of a titanium alloy with a trimodal microstructure have been systematically investigated. The results show that thick intragranular α lath and a small number of equiaxed α phases avoid the nucleation of cracks at the grain boundary, resulting in branching and fluctuation of cracks. Based on electron back-scattered diffraction, the strain partition and plastic deformation ahead of the crack tip were observed and analyzed in detail. Due to the toughening effect of the softer equiaxed α phase at the grain boundary, crack arresting and blunting are prevalent, improving the crack growth resistance and generating a relatively superior fracture toughness performance. These results indicate that a small amount of large globular α phases is beneficial to increase the crack propagation resistance and, thus, a good combination of mechanical property is obtained in the trimodal microstructure.

Keywords: titanium alloy; trimodal microstructures; strain partition; crack propagation

1. Introduction

Due to their high strength, good corrosion and fatigue resistance, titanium alloys have been extensively used for aerospace engineering [1,2]. During applications in the aircraft industry, two typical microstructures of bimodal microstructure and lamellar microstructure are widely used for titanium alloys [2,3]. Generally, the crack growth resistance is significantly influenced by the volume fraction and size of the equiaxed α phase as well as the grain boundary [4–6] in bimodal microstructures and equiaxed microstructures. The propagation of microvoids can be restricted by softer coarse α particles [7]. For titanium alloys, this is significantly strengthened by the fine secondary α phase [3,8]; crack growth is mainly affected by the thickness of the lamellar α phase, grain boundary α (GB α) and the α colony size of the lamellar microstructure [4,9]. High fracture toughness could be achieved in a lamellar microstructure with large α plates and the finest lamellar spacing [10–12]. It has been reported [11,13,14] that α plates are an effective microstructure unit for controlling fracture toughness as they can effectively deflect the crack propagation path. In conclusion, the fracture toughness of titanium alloy is extremely sensitive to microstructural parameters, such as the prior β grain size, α morphology, the width of the grain boundary of α phase and α laths and so on. It has been found that the lamellar shape of α phase promotes high toughness, while an equiaxed α phase results in low toughness; however, the ductility is degraded with the lamellar α phase and improved with the equiaxed α phase [15]. Due to these contradictions, the microstructure that has high fracture toughness may lead to an unsatisfactory decrease in other properties.

Recently, a new type of microstructure, named "trimodal microstructure", has been reported. This microstructure contains a primary globular α phase, a lamellar α phase and a transformed β matrix (β_{trans}: secondary α phase and β phase) [16,17]. Hosseini et al. [18] found that the excellent comprehensive mechanical properties of the trimodal microstructure are achieved when compared with the widmanstätten microstructure and bimodal microstructure. However, there is little investigation of the deformation behavior ahead of the crack tip, crack formation and crack growth in the trimodal microstructure during loading. Especially in the two-phase titanium alloy with high strength and toughness.

Therefore, in this work, research on the plastic deformation ahead of the crack tip as well as on the detailed essential relationship between microstructure and crack propagation behavior of the Ti-6Al-2Sn-2Zr-3Mo-1Cr-2Nb-0.1Si titanium alloy with the trimodal microstructure was carried out systematically. The present results can be applied to predict the microstructural features that are required to obtain the desired mechanical properties.

2. Experimental Program

The initial titanium alloy was supplied by the Northwest Institute for Non-ferrous Metals Research of China. The ingot with diameter of 450–500 mm was obtained after 3 vacuum-consumable electric arc smelting processes. Subsequently, forging was performed more than a dozen times on the ingot and then the bar with a diameter of about 350 mm was obtained after forging. The titanium alloy was finally forged in a two-phase region of 30–50 °C (lower than the phase transition point), and it was then strengthened with a solution and aging treatment. Vacuum smelting can remove defects and obtain component uniformity. The chemical composition (wt.%) of H and O was 0.001% and 0.075%. The content of C was low than 0.005%. In present work, the phase transition temperature (Tβ) was determined by the metallographic method with continuous heating. The α to β transformation temperature was about 945 ± 5 °C. Firstly, these samples were heated at 930 °C for 2 h and then air cooled. Several primary equiaxed α grains (α_p) were retained and some α laths with 0.5 to 1 μm in width were obtained during this process. Then, the low temperature aging at 600 °C for 4 h was performed and cooled by air to room temperature. This process was performed to obtain the secondary α precipitation from the residual β matrix. Consequently, a tri-modal microstructure (TM) was developed consisting of an intermixture of primary equiaxed α grains (α_p), α lath (α_l) and a transformed β matrix (secondary α phase (α_s) in the β matrix). A plate-shaped tensile specimen (width: 6 mm; thickness: 2 mm; gage length: 15 mm) was determined according to the national standard of the People's Republic of China (GB/T 228-2010) and performed on INSTRON 1195 Testing Machine (INSTRON, Boston, MA, USA). At least three individual tests were carried out to increase the accuracy of the tensile property. The compact tensile specimens with a size of 25 mm × 60 mm × 62.5 mm were used in the present work. After the corresponding heat treatment, the specimens with a notch of a "V" shape were machined into the compact tensile specimens. Firstly, a prefabricated crack of 2 mm in length was carried out on MTS 810 machines with sinusoidal waveforms at room temperature (a stress ratio of R = 0.1). The tensile test was carried out on the Instron 5895 testing machine (INSTRON, Boston, MA, USA). Three specimens were carried out to increase the accuracy of K_{IC}. Finally, scanning electron microscopy (SEM, HITACHI SU6600, Tokyo, Japan) were applied to observe the fractographic surface and analyze the crack growth behavior. The microstructures were etched by a corrosion solution of $HF:HNO_3:H_2O$ = 1:2:5. The microstructures ahead of the crack tip were investigated in detail using a field emission gun SEM (Carl Zeiss Microscopy GmbH 73447, Carl Zeiss AG, Jena, Germany) equipped with an electron backscattered diffraction (EBSD) system.

3. Results

3.1. Microstructures before Deformation

Figure 1 displays the SEM microstructural features before deformation of the TM. The TM contains three different α morphologies, namely, the equiaxed primary α phase (α_p), the lamellar α phase (α_l) and the acicular secondary α phase (α_s), as shown in Figure 1a. A few α_p are located at the prior β grain boundary. Figure 1b displays the magnified image of the morphology of α_l and α_s in the β_{trans} matrix, as indicated by the blue box. The grain size, volume fraction and aspect for α_p are about 5.9 μm, 5.0% and 1.5, respectively. The thickness for α_l and α_s is about 554 nm and 133 nm, respectively. The volume fraction for α_l and the β_{trans} matrix is about 32.0% and 62.0%, respectively.

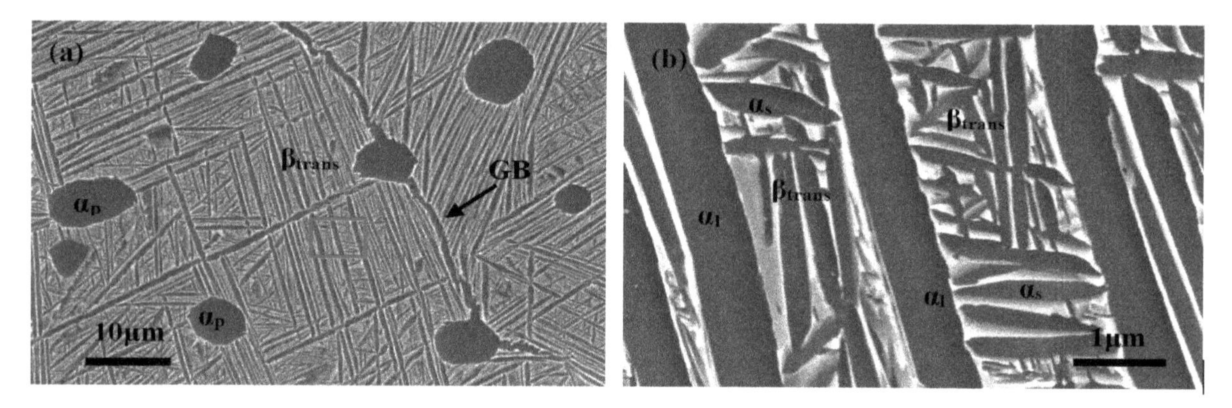

Figure 1. The SEM features of the trimodal microstructure (TM) before deformation: (**a**) three different α morphologies, namely, the primary equiaxed α phase (α_p), the lamellar α phase (α_l) and the secondary α phase(α_s); (**b**) the morphology of α_l and α_s in the β_{trans} matrix.

3.2. Mechanical Properties of the Alloy with a TM

The tensile engineering stress–strain curves at room temperature of the TM are shown in Figure 2a, which indicates that the average value of yield strength, tensile strength and elongation of the TM are about 1067 MPa, 1186 MPa and 12.1%, respectively. Figure 2b displays the force–displacement curve. The fracture toughness of the TM is approximately 62 MPa·m$^{1/2}$ and is significantly higher than the bimodal microstructure and bi-lamellar microstructure [19], which indicates that a small number of the equiaxed primary α phases are beneficial to the improvement of fracture toughness. The mechanical properties and the error of these measurements are listed in Table 1. It can be concluded that a good comprehensive mechanical property is achieved for the TM in the present work, which has been reported in other studies [16,18].

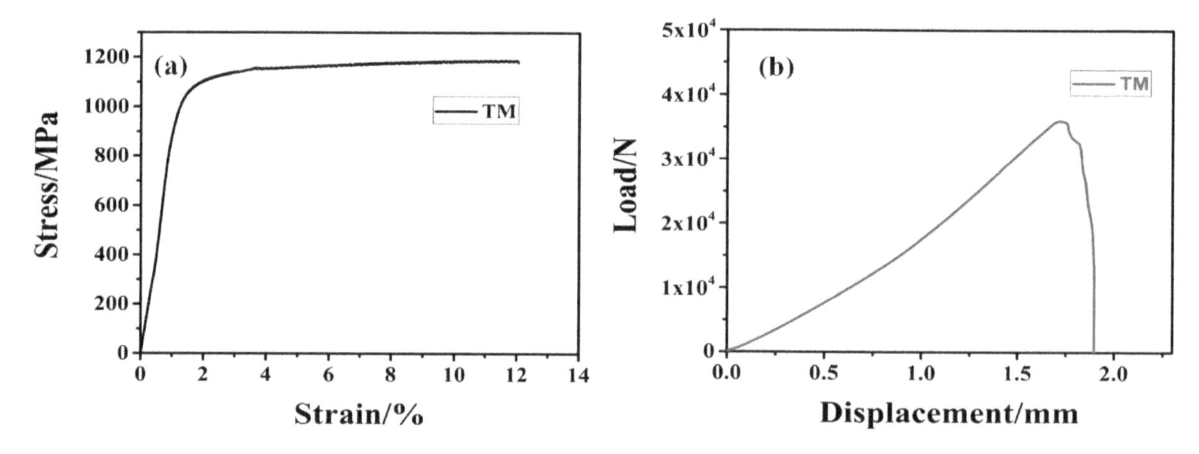

Figure 2. The comparison of mechanical properties of the TM and BLM: (**a**) the engineering stress–strain curves; (**b**) the force–disposition curves.

Table 1. The mechanical properties of the bimodal microstructure and bi-lamellar microstructure.

Microstructures	Yield Strength/MPa	Tensile Strength/MPa	Elongation/%	Fracture Toughness/MPa·m$^{1/2}$
TM	1067 ± 24	1186 ± 4	12.1 ± 1	62 ± 1

3.3. Fractographic Analyses and the Crack Propagation Behavior

The characteristics of fracture surfaces will provide important information to clarify the fracture mechanism during the failure process. To reveal the influence of the microstructure on the fracture mechanisms, the samples with different combinations of α phase were opted for the fracture analysis. Fracture morphology of the TM is shown in Figure 3. As can be seen, the fracture surface could be clearly separated into two apparent zones the: crack source zone and crack growth zone (Figure 3a). To further observe the detailed information regarding the fracture surface, some local fractographs of the TM were obtained, as shown in Figure 3.

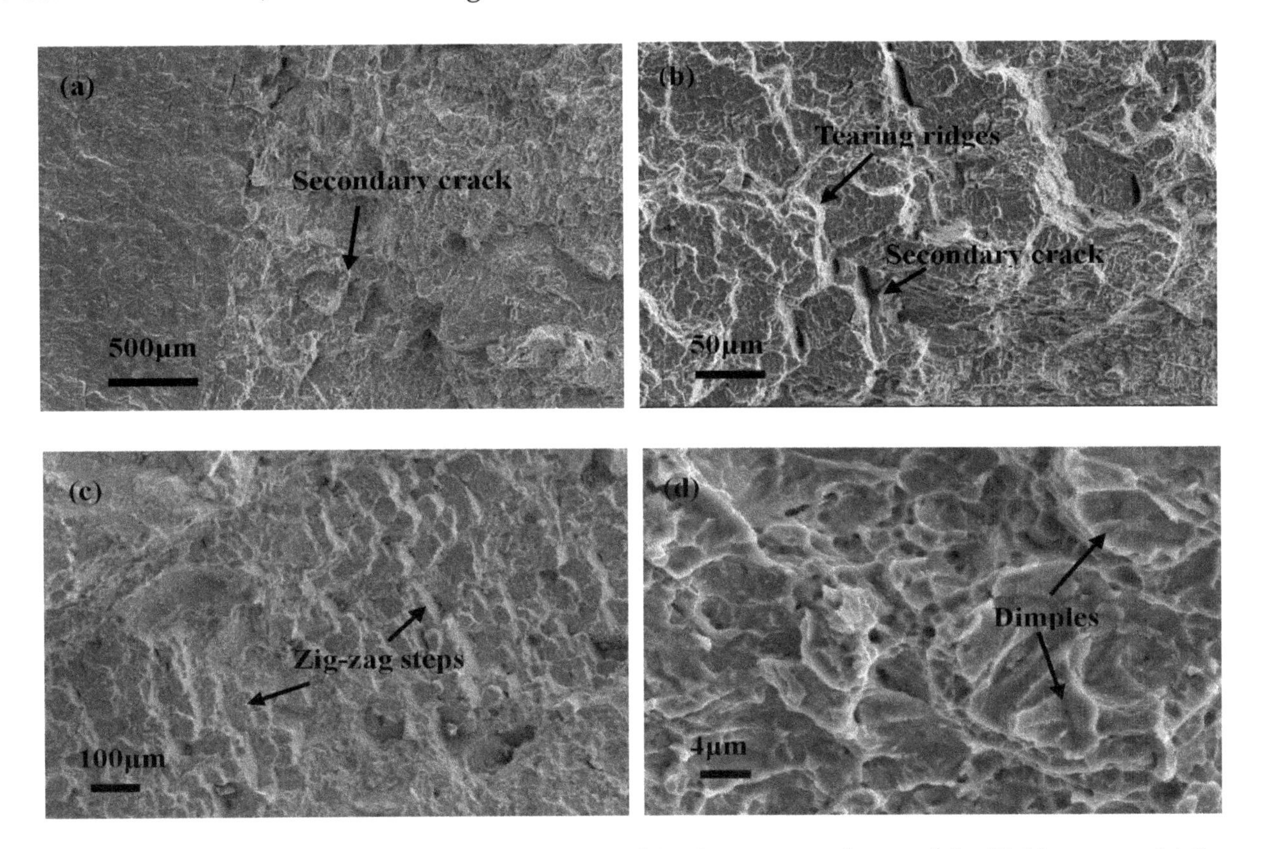

Figure 3. The SEM morphological characteristics of the fracture toughness of the TM fractures: (**a**) the transition region from fatigue crack to tensile crack propagation, (**b**) tear ridges and secondary cracks, (**c**) large fracture steps with a zig-zag fracture pattern, (**d**) big dimples surrounded by ridges.

The fracture surface of the TM is characterized by a high amount of ductile tearing ridges and secondary cracks resulting in transgranular crack propagation of the whole crack propagation region, which indicates a dimple-type fracture (ductile fracture), as shown in Figure 3a,b. The large number of tearing ridges that appear on the fracture surface and the fibrous zones imply a transgranular fracture (Figure 3b), which has previously been reported in titanium alloy [3,20]. The fracture toughness can be improved by the tearing ridges, which indicates that the higher the number of ridges, the higher the achievable K_{IC} of the titanium alloy [21]. Furthermore, a large amplitude of fracture steps with zig-zag fracture patterns are present in the crack propagation region (Figure 3c), which indicates that the crack alters direction and causes crack bifurcation, zigzagging and formation of secondary cracks, that is, much more fracture energy is consumed [22,23]. Thus, it can be concluded that the considerable steps were induced by large crack deviation. Figure 3d exhibits a significant number of inhomogeneous and deep ductile dimples and microvoids, which reveals that the fracture is caused by the typical ductile mechanism of microvoid nucleation, growth, and coalescence [24].

It has been reported that the fracture toughness was mainly affected by the tortuosity of the crack path and plastic deformation ahead of the crack tip [25]. In general, the improvement in tortuosity and plasticity is beneficial to the process of enhancing the fracture toughness of material [26]. The crack front profiles of the fractured specimen of TM are shown in Figure 4. Figure 4a is the main crack propagation path of the TM. The letters b, c, d and e denote the regions that are selected for detailed observation and analysis. The larger deflection and bifurcation of the crack with the local regions are achieved, as shown in Figure 4b, and the fluctuation of the crack can be up to 60 μm. Due to the existence of the equiaxed α phase at the grain boundary, the main crack does not further propagate along the β grain boundary, but changes the direction of propagation away from the GB, even though some microcracks initiate at the equiaxed α interface in front of the crack tip (Figure 4c). This result indicates that the equiaxed α phase at the GB could be the obstacle to the crack growth and lead to branching of the crack of about 90 μm in length. Figure 4d displays the characteristics of the secondary crack initiation near the main crack. It is suggested that although the crack is created at the grain boundary, it connects with the interface crack of intragranular α laths, thus avoiding the propagation along the β grain boundary. It has been reported that cracks tend to grow by passing through rather than cutting the thick α lath [15]. This can be further illustrated by Figure 4e, in which the crack mainly propagates along the long axis of the α laths. Additionally, it deviates from the grain boundary with an angle of 58° when it encounters the equiaxed α phase located at the GB (Figure 4e), which improves the crack growth resistance and is beneficial to improving the fracture toughness of the microstructure. These results indicate that a small amount of equiaxed primary α phases located at the grain boundary is instrumental in the deflection and bifurcation of the main crack propagation.

Figure 5a shows that the interface between the primary α lath and the β_{trans} matrix is the preferable location for crack creation. Cracks mainly form at the α lath interface and propagate along the long axis of the primary α lath, forming cracks related to the orientation of the primary α lath. The width of the primary α lath in the TM is almost equal to the thickness of the grain boundary α, and there are a small number of equiaxed α phases with large sizes at the β grain boundary, as shown in Figure 5b. Figure 5c,d (c and d regions in Figure 5b) represent the plastic deformation behavior of the primary equiaxed α phase at the GB and trimodal β grain boundaries, respectively. The equiaxed α phase at the β grain boundary is relatively soft, and reaches yielding first, producing abundant of slip bands. It has been reported that the multi-slip bands with different orientation are preferably created in the equiaxed primary α phase [27,28]. Cracks easily grew along the slip bands in the equiaxed α phase, because the slip bands provided a low energy channel for crack propagation [29]. As displayed in Figure 5c, there is a certain intersection angle of about 70 degrees between the slip bands in the equiaxed α phase and the direction of the β grain boundary. It changes the propagation direction away from the GB, as the main crack propagation encounters the equiaxed α phase because the slip band is not parallel to the β grain boundary. This result can be validated by Figure 4e. Excellent plastic deformation of the equiaxed α phase results in the blunting effect at the GB, which reduces the crack growth rate [30]. Figure 5d shows that the dislocation slip band occurs both in the GB α phase and the primary α lath, that is, the plastic deformation is not only confined to the GB α phase. At the trimodal β grain boundary, the plastic deformation takes place both at the GB α and the primary α phase near the grain boundary simultaneously, which reduces the local plastic deformation at the grain boundary, making the strain distribution near the grain boundary more uniform, and reducing the strain concentration at β grain boundary to some extent.

Figure 6 shows the plastic deformation and strain distribution in front of the main crack. Three β grains named grain 1, grain 2, and grain 3 are observed in Figure 6a. Several primary α_p particles are located at the grain boundary, such as α_{p1}, α_{p2}, α_{p3}, α_{p5}, and α_{p7} (Figure 6a). Two α_l colonies are observed (named colony1 and colony2). From the inverse pole figure (IPF) map of the α phase, the crystal orientations of α_{p1} to α_{p7} are (−13 −21), (−4.11 −70), (17 −8 −6), (−12 −10), (17 −80), (−13 −20) and (−12 −1 −1), respectively (the direction is perpendicular to the surface). This indicates that the anisotropy displays between these α_p particles. It displays the IPF map of the β grains in Figure 6c, which shows that the crystal orientation for grain 1, grain 2 and grain 3 is (315), (435) and (546), respectively. The Schmid factors of basal slip for α_{p1} to α_{p7} are 0.3, 0.11, 0.25, 0.14, 0.36 and 0.5, respectively, while value of colony1 and colony2 is 0.43 and 0.44, respectively (Figure 6d). However, relatively higher Schmid factor values of prismatic slip of the α_p particles are observed in Figure 6e, except for α_{p2} with 0.01 and α_{p7} with 0.25. Additionally, the Schmid factor values for colony1 and colony2 are 0.34 and 0.31, respectively. Figure 6f shows the strain distribution ahead of the crack tip, which indicates that the plastic strain is relatively inhomogeneous and forms a partition with the β grain interior instead of concentrating at the grain boundary. Even if slightly higher strain concentration is observed at the boundaries of colony1 and colony2 in Figure 6f, this strain partition effect effectively avoids the crack initiation and propagation at the grain boundary. Thus, the crack initiation and propagation mainly happens at the boundaries of α_l colonies within the β grains, as can be seen in Figure 5a.

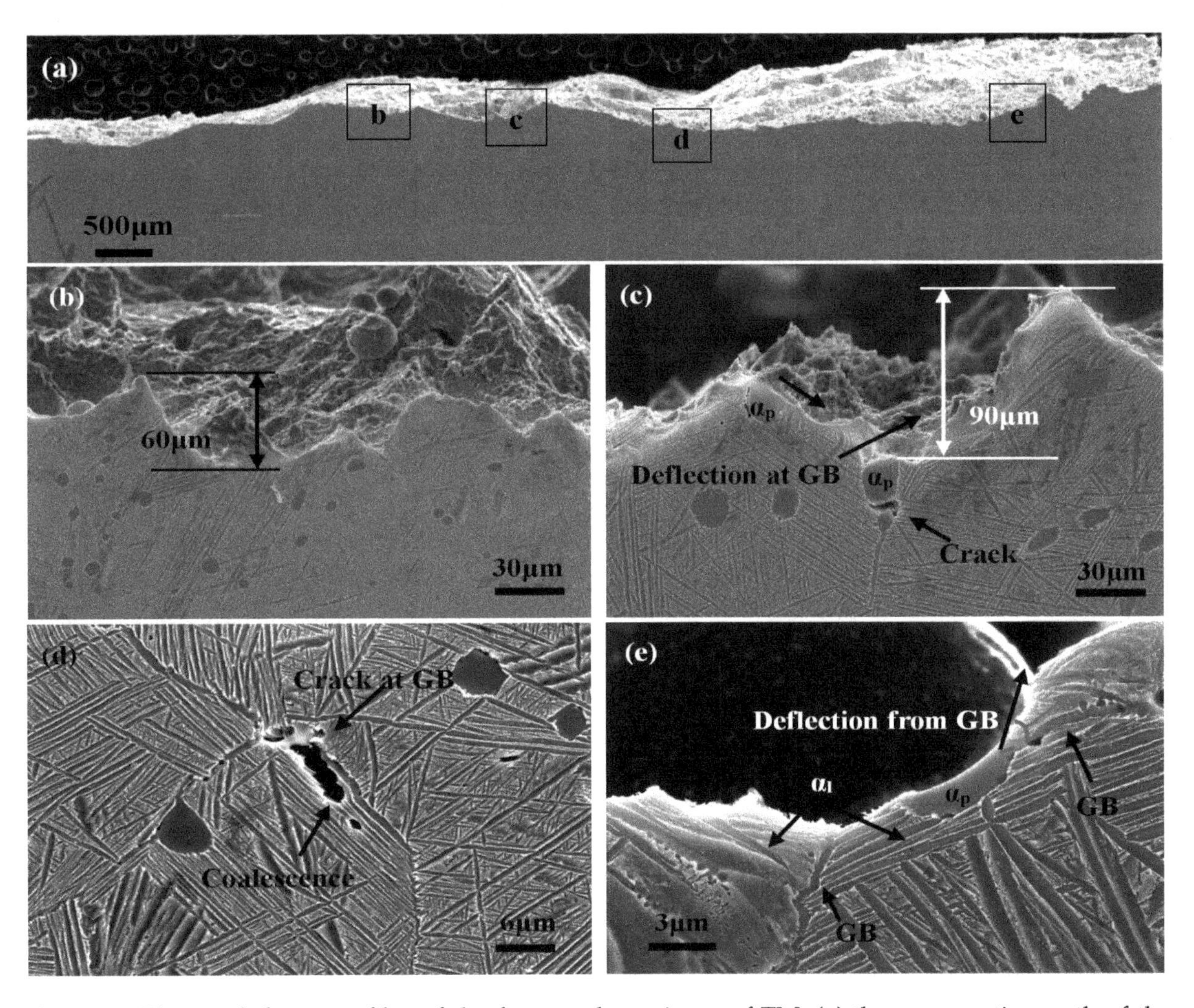

Figure 4. The crack front profiles of the fractured specimen of TM: (**a**) the propagation path of the main crack; (**b**) high fluctuation of about 60 μm of the propagation path is observed in microregion, (**c**) the crack initiates from the α_p interface but not connects with the main crack, which indicates that α_P inhibits the crack growth along the GB; (**d**) although the crack forms at the GB, it propagates into the grain interior through connecting with the microcrack in α_l, (**e**) the main cracks propagate along the grain boundaries and deviate from grain boundaries with about 58° when encounter α_l and α_p.

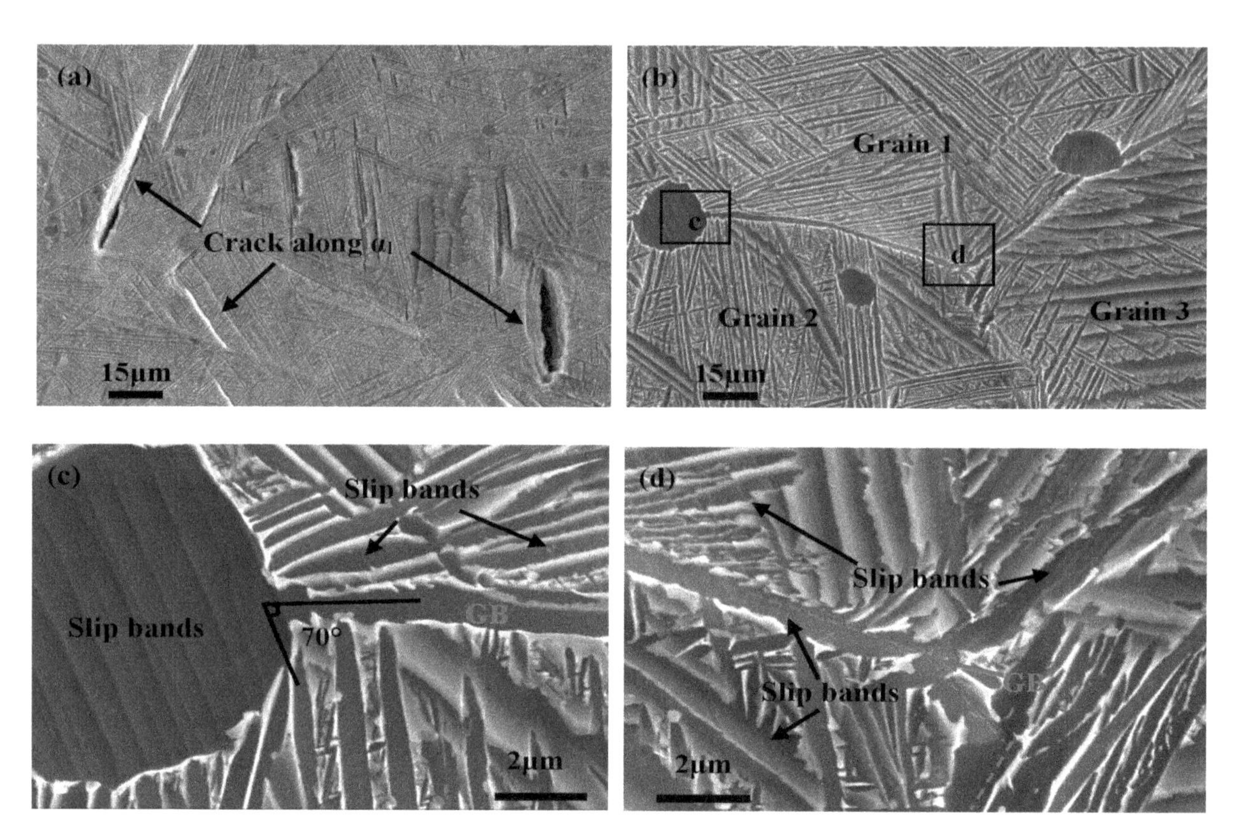

Figure 5. The plastic deformation and crack initiation behavior of the TM: (**a**) the crack initiates in α_l; (**b**) the equiaxed alpha phase (α_p) in the β GB; (**c**) the multi-slip bands in the globular α phase, (**d**) the dislocation slip bands occur both in the GB α phase and α_l; the plastic deformation is not confined to the GB α phase.

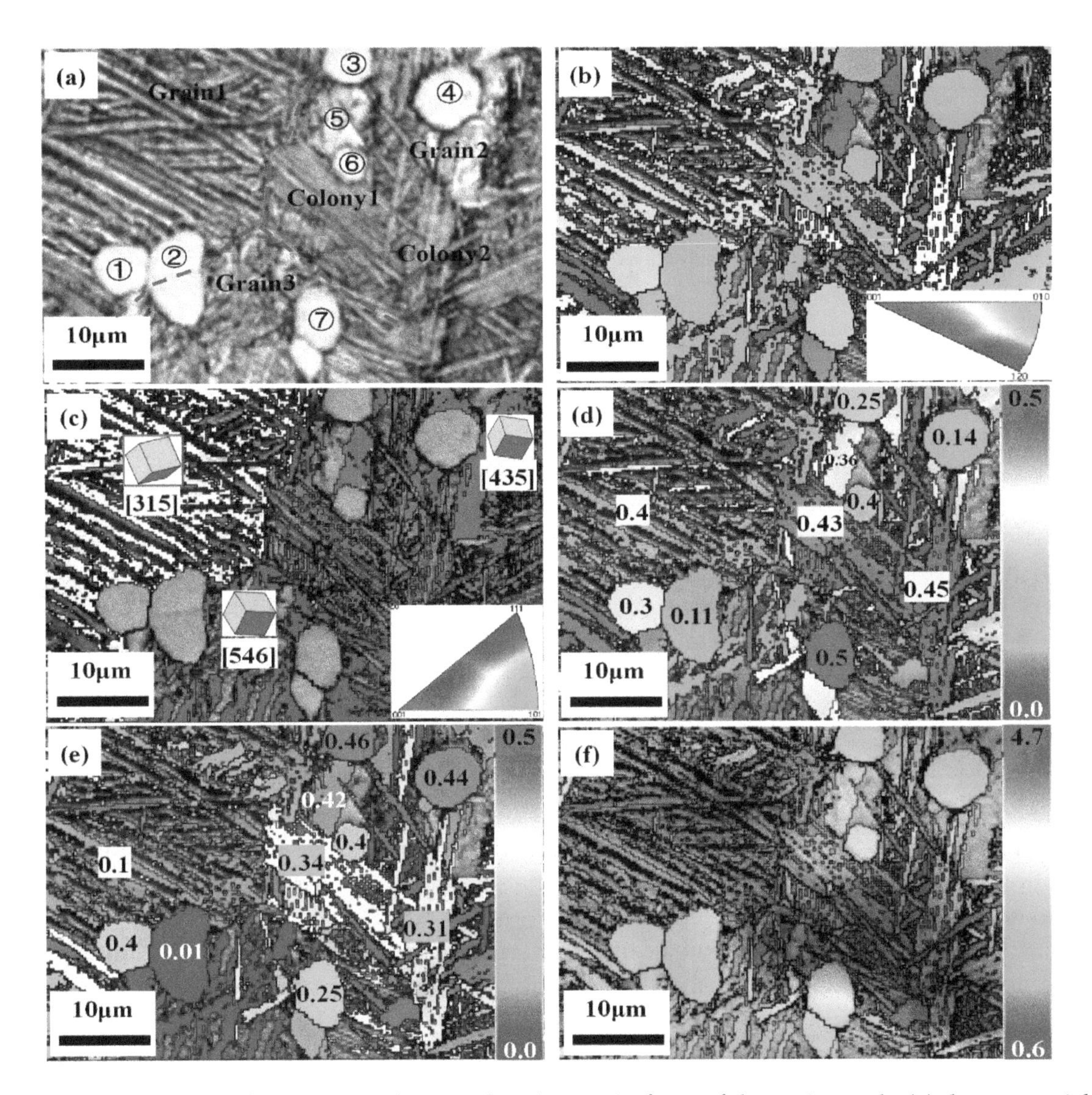

Figure 6. The plastic deformation and strain distribution in front of the main crack: (**a**) the α_p particles located at the grain boundary; (**b**) the inverse pole figure (IPF) map of the α phase; (**c**) the IPF map of the β phase; (**d**) the Schmid factor of basal slip of the α phase; (**e**) the Schmid factor of prismatic slip of the α phase; (**f**) the strain distribution near the grain boundary.

4. Discussion

The main difference of trimodal microstructure is that it contains a small amount of primary equiaxed α phases (about 5% in volume fraction) besides the primary α lath and the β_{trans} matrix. As Chan [30] reported, the softer equiaxed α phase can preferentially coordinate plastic incompatibility and cause blunting of cracks to achieve toughening. Consequently, a certain strain can be produced in the adjoin matrix, and the strain distribution near the grain boundary is more uniform and does not cause the formation of microcracks on the adjacent grain boundary or phase boundary, thus improving the fracture toughness. This can be seen in Figure 5, which shows that the plastic deformation is not only confined to the GB α phase, but occurs both in the α_p phase and the neighbor primary α lath. Based on EBSD, the strain partition within the grain interior that could reduce the stress concentration at the grain boundary to some extent was observed (Figure 6f).

Due to the thick α_l phase (32.0% in volume fraction), cracks nucleated mainly at the α_l phase within the grain interior in the trimodal microstructure instead of at the β grain boundary, which was different with the lamellar microstructure or widmannstatten microstructure [31], as shown in Figure 5a. Furthermore, it has been reported that cracks are both deviated and arrested when they reached an

α phase unfavorably oriented for prismatic slip in a two-phase titanium alloy [32]. As can be seen in Figure 6e, a very low Schmid factor value of approximately 0.01 of prismatic slip was obtained in the primary α_{p2} phases. It seems that crack arresting and crack path deviation will happen when the crack tip encounters these unfavorably oriented phases. Retardation of the crack growth occurs due to the crack arresting and deviation, as it requires more energy to expand the crack to a lower stress position, subsequently improving the crack propagation resistance of the trimodal microstructure. As the crack continues to grow, the crack tip tends to stop propagation, blunting or deviate from the grain boundary when it penetrates the equiaxed α phase, as shown in Figure 4c,e. Crack propagation will deflect along the long axis direction of the primary α lath or the direction of slip bands within the equiaxed primary α phase at the grain boundary, avoiding propagation along the β grain boundary and promoting the transgranular fracture. It can be seen that a high fluctuation (up to 90 μm) of the crack path is observed in the microregion of the TM, which increases the flexibility of crack growth and enhances the resistance of crack growth (Figure 4c).

Moreover, although the width of the GB α in the TM is about 640 nm, it displays a distinct characteristic of discontinuous and zig-zag features as marked by the blue dotted lines in Figure 7a. The width and continuity of the GB α can significantly influence the fracture toughness [33]. Researchers have reported that the thicker and continuous GB α would lead to a preferable crack propagation path and induce a detrimental influence on the fracture toughness [4,22]. However, the crack propagation is much more difficult to pass through the discontinuous grain boundary α, which is beneficial to the heightening of fracture toughness [14,33]. In contrast, in the widmannstatten microstructure, because of the lack of toughening effect of the primary equiaxed α phase, cracks are easy to initiate at grain boundaries and propagate along the β grain boundary, which results in low plasticity [31]. According to the present experiments and theoretical analysis, the schematic diagram is carried out to illustrate the effect of α morphology on the crack nucleation and growth behavior of titanium alloy, as shown in Figure 7b. It indicates that the crack mainly initiates at the primary α lath (α_l), and it avoids the initiation of the crack at the β grain boundary. Cracks will change the direction of propagation when they encounter the equiaxed α_p phase at the β grain boundary, which leads to a tortuous crack growth path. This study can provide theoretical support to tailor the microstructure and the mechanical properties of titanium alloys that contain both the α and β phase. For instance, a small number of equiaxed primary α phases are needed if high ductility and fracture toughness are required. It also indicates that a large primary equiaxed α phase is not always detrimental to fracture toughness. If the appropriate amount of the equiaxed α_p phase is obtained, the fracture toughness of the trimodal or bimodal microstructure may be higher than that of the lamellar microstructure.

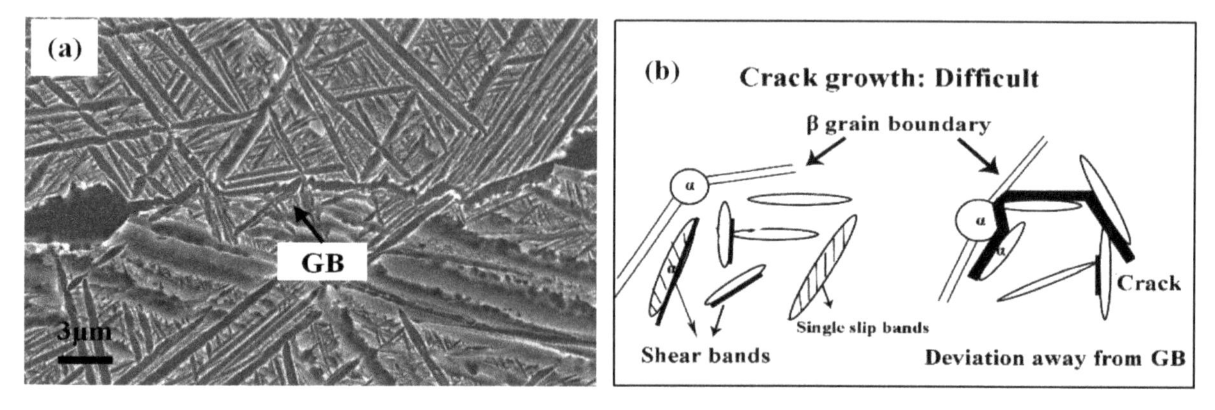

Figure 7. The features of the grain boundary α phase and the schematic diagram of crack growth behavior: (**a**) the discontinuous and "zig-zag" GB α phase, (**b**) the crack mainly initiates at α_l, and it will change the direction of propagation when it encounters the equiaxed α phase at the β grain boundary, which leads to an intergranular fracture.

5. Conclusions

The plastic deformation and crack propagation behavior of the titanium alloy with a trimodal microstructure were systematically investigated during fracture toughness tests. According to the present work, the following conclusions are drawn: A higher fracture toughness of 62 MPa·$m^{1/2}$ is obtained for the trimodal microstructure, which offers a preferable combination of strength (1186 MPa) and ductility (12.1%). In addition to dimples, a large number of tearing edges and secondary cracks are produced in the trimodal microstructure, showing transgranular fracture characteristics. The coarser and longer intragranular α lath and a small number of equiaxed α phases, as well as the discontinuous GB α, lead to a high number of branches and fluctuation of the cracks. Because of the toughening effect of the softer phase at the GB, the equiaxed α phase can preferentially coordinate plastic incompatibility and cause arresting and blunting of cracks, which improves the crack growth resistance in the trimodal microstructure. These results indicate that a small amount of the primary globular α phases located at the grain boundary will be good for improving the resistance to crack propagation. The present work can provide a theoretical support to tailor the microstructure and mechanical properties of titanium alloys in the future.

Author Contributions: Conceptualization, C.T. and G.Z.; data curation, C.T. and Q.S.; investigation, C.T. and Y.F.; project administration, G.Z.; supervision, C.T.; visualization, Q.S.; writing–original draft, C.T.; writing–review and editing, Q.S. and G.Z. All authors have read and agreed to the published version of the manuscript.

References

1. Banerjee, D.; Williams, J.C. Perspectives on titanium science and technology. *Acta Mater.* **2013**, *61*, 844–879. [CrossRef]
2. Leyens, C.; Peters, M. *Titanium and Titanium Alloys: Fundamentals and Applications*; John Wiley & Sons: Hoboken, NJ, USA, 2003.
3. Liu, Y.W.; Chen, F.W.; Xu, G.L.; Cui, Y.W.; Chang, H. Correlation between Microstructure and Mechanical Properties of Heat-Treated Ti-6Al-4V with Fe Alloying. *Metals* **2020**, *10*, 854. [CrossRef]
4. Renon, V.; Henaff, G.; Larignon, C.; Perusin, S.; Villechaise, P. Identification of Relationships between Heat Treatment and Fatigue Crack Growth of αβ Titanium Alloys. *Metals* **2019**, *9*, 512. [CrossRef]
5. He, S.T.; Zeng, W.D.; Xu, J.W.; Chen, W. The effects of microstructure evolution on the fracture toughness of BT-25 titanium alloy during isothermal forging and subsequent heat treatment. *Mater. Sci. Eng. A* **2019**, *745*, 203–211. [CrossRef]
6. Niinomi, M.; Kobayashi, T.; Inagaki, I.; Thompson, A.W. The effect of deformation-induced transformation on the fracture toughness of commercial titanium alloys. *Metall. Trans. A* **1990**, *21*, 1733–1744. [CrossRef]
7. Li, X.H.; He, J.C.; Ji, Y.J.; Zhang, T.C.; Zhang, Y.H. Study of the Microstructure and Fracture Toughness of TC17 Titanium Alloy Linear Friction Welding Joint. *Metals* **2019**, *9*, 430. [CrossRef]
8. Xu, T.W.; Zhang, S.S.; Cui, N.; Cao, L.; Wan, Y. Precipitation Behavior of Ti15Mo Alloy and Effects on Microstructure and Mechanical Performance. *J. Mater. Eng. Perform.* **2019**, *28*, 7188–7197. [CrossRef]
9. Wen, X.; Wan, M.P.; Huang, C.W.; Tan, Y.B.; Lei, M.; Liang, Y.L.; Cai, X. Effect of microstructure on tensile properties, impact toughness and fracture toughness of TC21 alloy. *Mater. Des.* **2019**, *180*, 107898. [CrossRef]
10. Shi, X.H.; Zeng, W.D.; Zhao, Q.Y. The effects of lamellar features on the fracture toughness of Ti-17 titanium alloy. *Mater. Sci. Eng. A* **2015**, *636*, 543–550. [CrossRef]
11. Wen, X.; Wan, M.P.; Huang, C.W.; Lei, M. Strength and fracture toughness of TC21 alloy with multi-level lamellar microstructure. *Mater. Sci. Eng. A* **2019**, *740*, 121–129. [CrossRef]
12. Richards, N.L. Quantitative evaluation of fracture toughness-microstructural relationships in α-β titanium alloys. *J. Mater. Eng. Perform.* **2004**, *13*, 218–225. [CrossRef]
13. Cvijović-Alagić, I.; Gubeljak, N.; Rakin, M.; Cvijović, Z.; Gerić, K. Microstructural morphology effects on fracture resistance and crack tip strain distribution in Ti-6Al-4V alloy for orthopedic implants. *Mater. Des.* **2014**, *53*, 870–880. [CrossRef]
14. Fan, J.K.; Li, J.S.; Kou, H.C.; Hua, K.; Tang, B. The interrelationship of fracture toughness and microstructure in a new near β titanium alloy Ti-7Mo-3Nb-3Cr-3Al. *Mater. Charact.* **2014**, *96*, 93–99. [CrossRef]
15. Hirth, J.P.; Froes, F.H. Interrelations between fracture toughness and other mechanical properties in titanium alloy. *Metall. Trans. A* **1977**, *8A*, 1165–1176. [CrossRef]

16. Gao, P.F.; Cai, Y.; Zhan, M.; Fan, X.G.; Lei, Z.N. Crystallographic orientation evolution during the development of tri-modal microstructure in the hot working of TA15 titanium alloy. *J. Alloys Compd.* **2018**, *741*, 734–745. [CrossRef]
17. Gao, P.F.; Fan, X.G.; Yang, H. Role of processing parameters in the development of tri-modal microstructure during isothermal local loading forming of TA15 titanium alloy. *J. Mater. Process. Technol.* **2017**, *239*, 160–171. [CrossRef]
18. Hosseini, R.; Morakabati, M.; Abbasi, S.M.; Hajari, A. Development of a trimodal microstructure with superior combined strength, ductility and creep-rupture properties in a near α titanium alloy. *Mater. Sci. Eng. A* **2017**, *696*, 155–165. [CrossRef]
19. Tan, C.S.; Sun, Q.Y.; Zhang, G.J. Role of microstructure in plastic deformation and crack propagation behaviour of an α/β titanium alloy. *Vacuum* **2020**, major revise.
20. Zhang, S.; Liang, Y.L.; Xia, Q.F.; Ou, M.G. Study on Tensile Deformation Behavior of TC21 Titanium Alloy. *J. Mater. Eng. Perform.* **2019**, *28*, 1581–1590. [CrossRef]
21. Jia, R.C.; Zeng, W.D.; He, S.T.; Gao, X.X.; Xu, J.W. The analysis of fracture toughness and fracture mechanism of Ti60 alloy under different temperatures. *J. Alloys Compd.* **2019**, *810*, 151899. [CrossRef]
22. Terlinde, G.; Rathjen, H.J.; Schwalbe, K.H. Microstructure and fracture toughness of the aged, β-Ti Alloy Ti-10V-2Fe-M. *Metall. Trans. A* **1988**, *19*, 1037–1049. [CrossRef]
23. Niinomi, M.; Kobayashi, T. Fracture characteristics analysis related to the microstructures in titanium alloys. *Mater. Sci. Eng. A* **1996**, *213*, 16–24. [CrossRef]
24. Ren, L.; Xiao, W.L.; Chang, H.; Zhao, Y.Q.; Ma, C.L.; Zhou, L. Microstructural tailoring and mechanical properties of a multi-alloyed near β titanium alloy Ti-5321 with various heat treatment. *Mater. Sci. Eng. A* **2018**, *711*, 553–561. [CrossRef]
25. Suresh, S. Fatigue crack deflection and fracture surface contact: Micromechanical models. *Metall. Trans. A.* **1985**, *16*, 249–260. [CrossRef]
26. Xu, J.W.; Zeng, W.D.; Zhou, D.D.; Ma, H.Y.; Chen, W.; He, S.T. Influence of α/β processing on fracture toughness for a two-phase titanium alloy. *Mater. Sci. Eng. A* **2018**, *731*, 85–92. [CrossRef]
27. Tan, C.S.; Sun, Q.Y.; Xiao, L.; Zhao, Y.Q.; Sun, J. Cyclic deformation and microcrack initiation during stress controlled high cycle fatigue of a titanium alloy. *Mater. Sci. Eng. A* **2018**, *711C*, 212–222. [CrossRef]
28. Huang, J.; Wang, Z.R.; Xue, K.M. Cyclic deformation response and micromechanisms of Ti alloy Ti-5Al-5V-5Mo-3Cr-0.5Fe. *Mater. Sci. Eng. A* **2011**, *528*, 8723–8732. [CrossRef]
29. Sangid, M.D. The physics of fatigue crack initiation. *Int. J. Fatigue* **2013**, *57*, 58–72. [CrossRef]
30. Chan, K.S. Toughening mechanisms in titanium aluminides. *Metall. Trans. A* **1993**, *24A*, 569–583. [CrossRef]
31. Ankem, S.; Greene, C.A. Recent developments in microstructure: Property relationships of beta titanium alloys. *Mater. Sci. Eng. A* **1999**, *263*, 127–131. [CrossRef]
32. Bantounas, I.; Lindley, T.C.; Rugg, D.; Dye, D. Effect of microtexture on fatigue cracking in Ti–6Al–4V. *Acta Mater.* **2007**, *55*, 5655–5665. [CrossRef]
33. Ghosh, A.; Sivaprasad, S.; Bhattacharjee, A.; Kar, S.K. Microstructure–fracture toughness correlation in an aircraft structural component alloy Ti-5Al-5V-5Mo-3Cr. *Mater. Sci. Eng. A* **2013**, *568*, 61–67. [CrossRef]

14

Effect of ECAP on the Microstructure and Mechanical Properties of a Rolled Mg-2Y-0.6Nd-0.6Zr Magnesium Alloy

Xiaofang Shi [1], Wei Li [1,2,*], Weiwei Hu [1], Yun Tan [1], Zhenglai Zhang [2] and Liang Tian [3]

[1] College of Material and Metallurgy, Guizhou University, Guiyang 550025, China; 18285117265@163.com (X.S.); m18786670951@163.com (W.H.); tyl19941020@163.com (Y.T.)
[2] Zhejiang Huashuo Technology Co., Ltd., Ningbo 315000, China; 522321a02@sina.com
[3] Guizhou Province Technology Innovation Service Center, Guiyang 550004, China; 13017461042@163.com
* Correspondence: wli1@gzu.edu.cn

Abstract: A fine-grained Mg-2Y-0.6Nd-0.6Zr alloy was processed by bar-rolling and equal-channel angular pressing (ECAP). The effect of ECAP on the microstructure and mechanical properties of rolled Mg-2Y-0.6Nd-0.6Zr alloy was investigated by optical microscopy, scanning electron microscopy, electron backscattered diffraction and a room temperature tensile test. The results show that the Mg-2Y-0.6Nd-0.6Zr alloy obtained high strength and poor plasticity after rolling. As the number of ECAP passes increased, the grain size of the alloy gradually reduced and the texture of the basal plane gradually weakened. The ultimate tensile strength of the alloy first increased and then decreased, the yield strength gradually decreased, and the plasticity continuously increased. After four passes of ECAP, the average grain size decreased from 11.2 μm to 1.87 μm, and the alloy obtained excellent comprehensive mechanical properties. Its strength was slightly reduced compared to the as-rolled alloy, but the plasticity was greatly increased.

Keywords: magnesium alloy; ECAP; texture; mechanical properties

---

1. Introduction

As lightweight metallic materials used for engineering applications, magnesium alloys have the advantages of low density, high specific strength, high specific stiffness, good shielding and ease of recycling. They are widely used in numerous important areas, such as military, aerospace, transportation, and electronic communications [1–3]. Mg-Y-Nd-Zr (WE) alloys are commercial high-strength rare earth magnesium alloys developed in Britain in the 1980s. They have excellent creep resistance at high temperatures and are widely used as high-strength heat-resistant engineering materials [4]. However, their application potential is limited by their low number of slip systems and their poor plasticity at room temperature. In recent years, equal-channel angular pressing (ECAP) has been widely used as a method to effectively refine grains and improve the mechanical properties of magnesium alloys [5–8].

However, under conventional conditions, the ECAP of magnesium alloys can only be carried out at higher temperatures, which leads to grain growth during the pressing process and decreases the strengthening effect of ECAP. In this context, many scholars have begun to explore ways to reduce the pressing temperature, such as through the stepwise reduction of pressing temperature [9], the reduction of pressing speed [10], the application of back pressure [11,12], and a bread jacket outside the sample [13]. These methods all reduce the pressing temperature to a certain extent and the strengthening effect of ECAP is enhanced. However, with the development of society, the requirements for materials are ever increasing and the limited strengthening effect of ECAP limits its further expansion in industrial applications. Therefore, getting rid of the single ECAP strengthening mode, combining ECAP with other strengthening methods, breaking through the traditional ECAP strengthening limit and preparing fine-grained magnesium alloys with excellent performance have become hot issues in current ECAP research. At present, the most popular method is pre-deformation before ECAP. Pre-deformation can reduce the grain size, improve the as-casted microstructure, and enhance the plastic deformation ability of the alloy, thereby effectively reducing the ECAP temperature. In addition, pre-deformation can increase the strength of the alloy. This strengthening combined with ECAP strengthening can further improve the properties of the alloy. Miyahara et al. [14] first extruded an as-cast AZ61 magnesium alloy at 437 °C, and then conducted ECAP at 200 °C. After one pass of ECAP, a submicron microstructure was obtained and the average grain size after the fourth pass was ~0.62 μm. The elongation reached 1320% in the tensile test of strain rate of 3.3×10^{-4} s^{-1} at 200 °C. Krajňák et al. [15] first extruded an as-cast AX41 magnesium alloy at 350 °C and then ECAP was carried out at 220 °C and 250 °C. After eight passes of ECAP, both obtained good plasticity; however, after ECAP at 250 °C, the average grain size was larger than that after ECAP at 220 °C, the dislocation density was lower, and the texture was not conducive to the activation of the basal plane slip systems. These factors caused the yield strength after ECAP at 250 °C to be significantly lower than that after ECAP at 220 °C. Joungsik et al. [16] studied the ECAP of a AZ31 magnesium alloy sheet. They found that after the AZ31 magnesium alloy was plate-rolled, the base surface formed a typical rolling texture, i.e., the base surface was parallel to the rolling surface, resulting in the mechanical properties of sheet showing strong anisotropy. After ECAP along the route D at 225 °C, the severe shear deformation reduced the grain size of the alloy and developed basal texture with tilted basal planes towards the pressing direction. Ultimately, the anisotropy of the mechanical properties was reduced and the hardening behavior was enhanced.

Currently, the strengthening method of extrusion or plate-rolling combined with ECAP is relatively mature, but research on magnesium alloy bar-rolling combined with ECAP has rarely been reported. In this paper, an as-cast Mg-2Y-0.6Nd-0.6Zr alloy was studied. Bar-rolling was conducted first at 400 °C and then ECAP was carried out at 340 °C. The effect of ECAP on the microstructure and properties of the rolled Mg-2Y-0.6Nd-0.6Zr alloy was investigated by microstructure observation and a mechanical properties test. The aim of this study was to fill the gap of research on magnesium alloy bar-rolling combined with ECAP and provide a theoretical basis and technical support for improving the properties of magnesium alloys.

2. Materials and Methods

A Mg-2Y-0.6Nd-0.6Zr alloy was smelted in a well-type resistance furnace (Shiyan Electric Furnace Works, Shanghai, China) and 99.9% pure magnesium (Yinguang Huasheng Magnesium Company, Shanxi, China) along with Mg-25% Y, Mg-25% Nd, and Mg-30% Zr master alloys (Xinglin Nonferrous Metals Material Co., Ltd., Shanxi, China) were used to prepare it. A quartz crucible containing pure magnesium was placed in a well-type electric resistance furnace and RJ-5 solvent (Hengfeng Chemical Co., Ltd., Henan, China), which was composed of 56% anhydrous carnallite, 30% BaCl2 and 14% CaF2, was used as the covering agent and the refining agent. The furnace was heated to 720 °C with a heating rate of 10 °C/min. After the pure magnesium was completely melted, the Mg-Y, Mg-Nd and Mg-Zr master alloys were sequentially added to the crucible and the temperature of the furnace was raised to 780 °C. The solution was stirred when the master alloys were completely melted and then the power of the furnace was turned off so that the temperature of the solution dropped as the temperature of the furnace dropped. The crucible was taken out of the furnace while the solution was lowered to 720 °C and the solution was cast into a preheated cylindrical metal mold whose size was Φ30 mm × 200 mm and then water-cooled. The cast billets were homogenized at 450 °C for 6 hours and then air-cooled. The homogenized samples were rolled on a F50-150 bar-rolling machine (Hong Feng Ji Xie, Zhejiang, China) for seven passes at 400 °C with a total strain of 0.46. The samples with the dimensions of Φ12 mm × 80 mm were machined from the as-rolled bars, and then the samples were subjected to ECAP via a mold constructed in the laboratory. The mold structure is shown in Figure 1 and the angles of Φ

and Ψwere 120° and 30°, respectively. The samples were pressed from one to six passes with a pressing velocity of 0.4 mm/s via route BC, i.e., the samples were rotated by 90° in the same direction between consecutive passes [17]. Prior to each pass, a layer of graphite and engine oil was applied to the inner wall of the mold and the surface of the sample as a lubricant and the samples were preheated together with the mold at 340 °C for 10 min. After each ECAP pass, the samples were quickly placed in water for cooling.

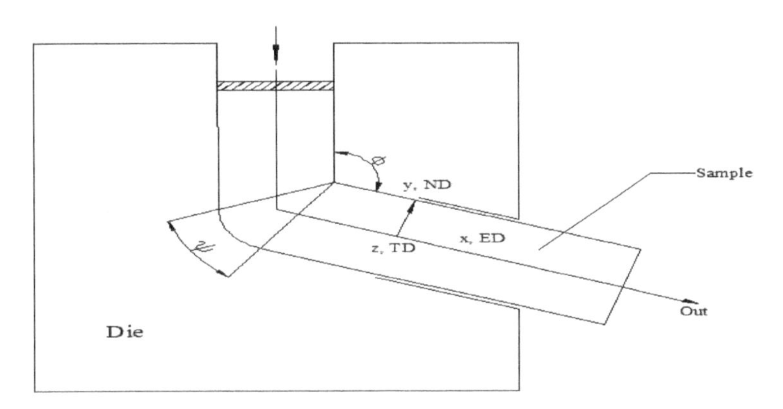

Figure 1. The schematic diagram of equal-channel angular pressing (ECAP) die (ED: extrusion direction, ND: normal direction, TD: transverse direction).

The microstructure of the samples was observed by a BH2 optical microscope (OM) (Olympus, Tokyo, Japan). Electron backscattered diffraction (EBSD) samples were prepared by a EM RES102 multi-function ion thinner (Leica, Wetzlar, Germany) and then observed the plane parallel to the extrusion direction (ED) or rolling direction (RD) on a S-3400N scanning electron microscope (Hitachi, Tokyo, Japan) and a NordlysMax3 electron backscatter diffractometer (Oxford, Abingdon, Britain) at an accelerating voltage of 20 kV and a step size of 0.2 μm. The EBSD data were analyzed by HKL Channel 5 software (Oxford, Abingdon, Britain) and the indexing rate reached 80%. The mechanical properties of the samples at room temperature were tested by an Instron 8501 universal tensile testing machine (Instron, Canton, USA). The dimensions of the tensile sample are shown in Figure 2, and were designed according to the standard of GB/T 228-2002, and the sampling direction was parallel to the ED. The tensile fracture morphology was analyzed on a SUPRA 40 scanning electron microscope (SEM) (Zeiss, Oberkochen, Germany).

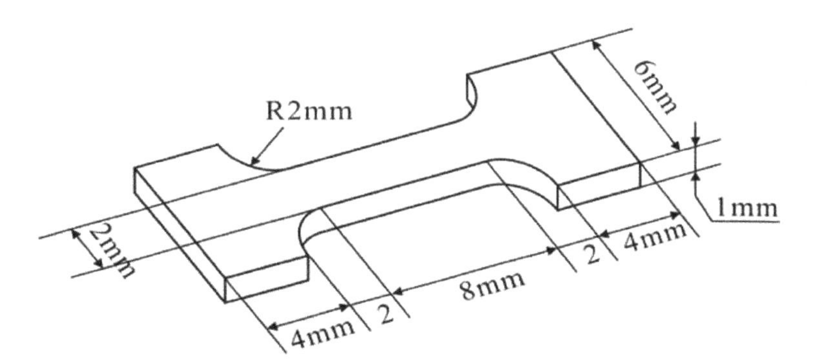

Figure 2. The dimensions of the tensile sample.

3. Results

3.1. Microstructure

Figures 3 and 4 display the microstructure and grain orientation distribution, respectively, of the Mg-2Y-0.6Nd-0.6Zr alloy subjected to rolling and after different numbers of ECAP passes. The grain size statistics and their distribution are shown in Figure 5. It can be seen in Figure 3; Figure 4 that after rolling, the grain size is relatively large, with individual large grains exceeding 30 μm and the average grain size being ~11.2 μm. After one pass of ECAP, the grain size of the alloy was remarkably reduced, some of the grains were elongated, the grain size presented a bimodal distribution, and the average grain size was ~2.43 μm. The fourth pass of ECAP led to fine equiaxed grains and the size distribution was concentrated in the range of 0–3 μm; however, at the same time, grains as large as 10 μm were also present and the average grain size was ~1.87 μm. After six passes of ECAP, the average grain size showed an increasing trend, the size of the coarse grains decreased, the grain size distribution was more homogeneous than that after four passes, and the average grain size was ~2.00 μm.

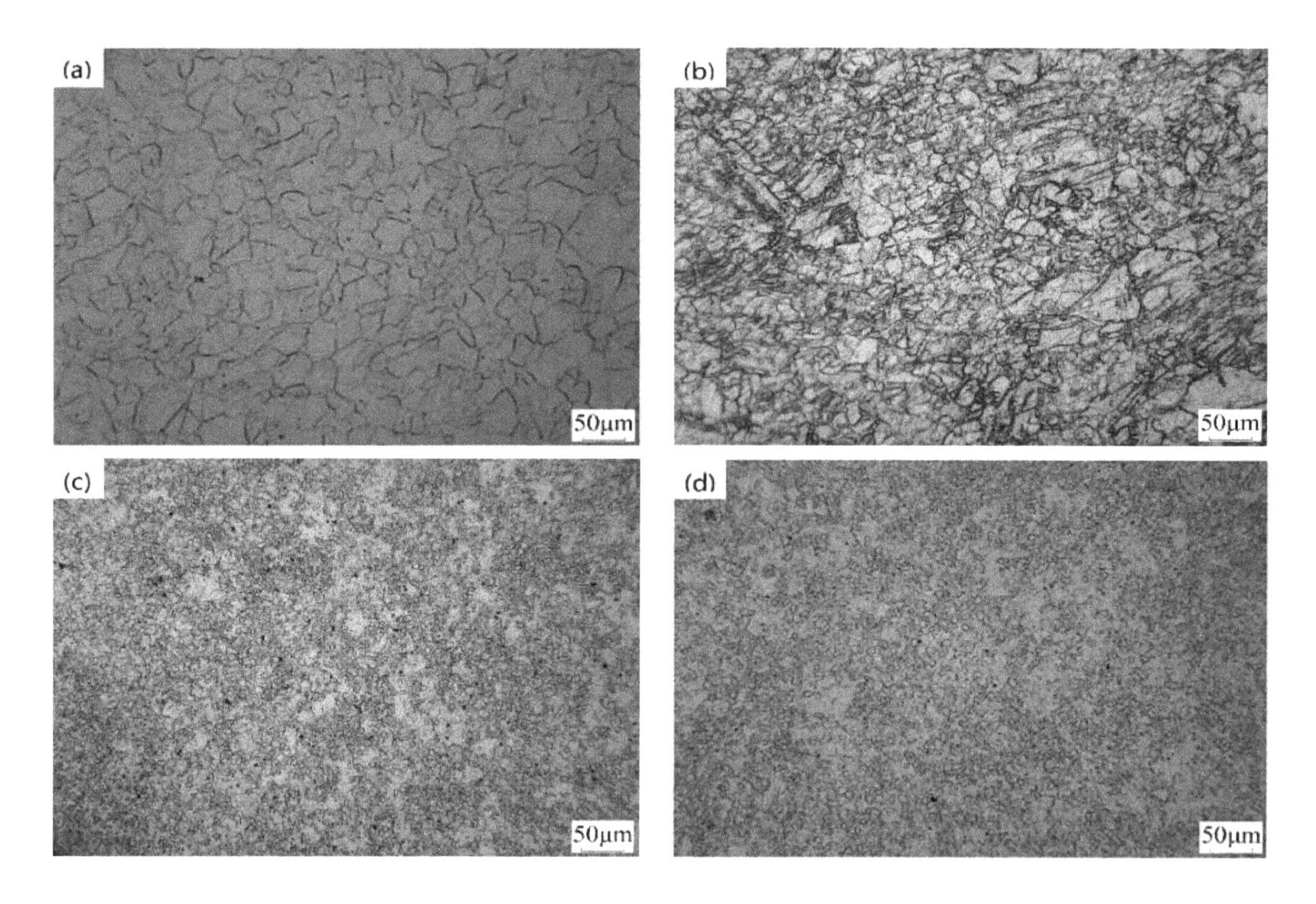

Figure 3. Microstructures of the Mg-2Y-0.6Nd-0.6Zr alloy. (**a**) as-rolled; (**b**) one pass; (**c**) four passes; (**d**) six passes.

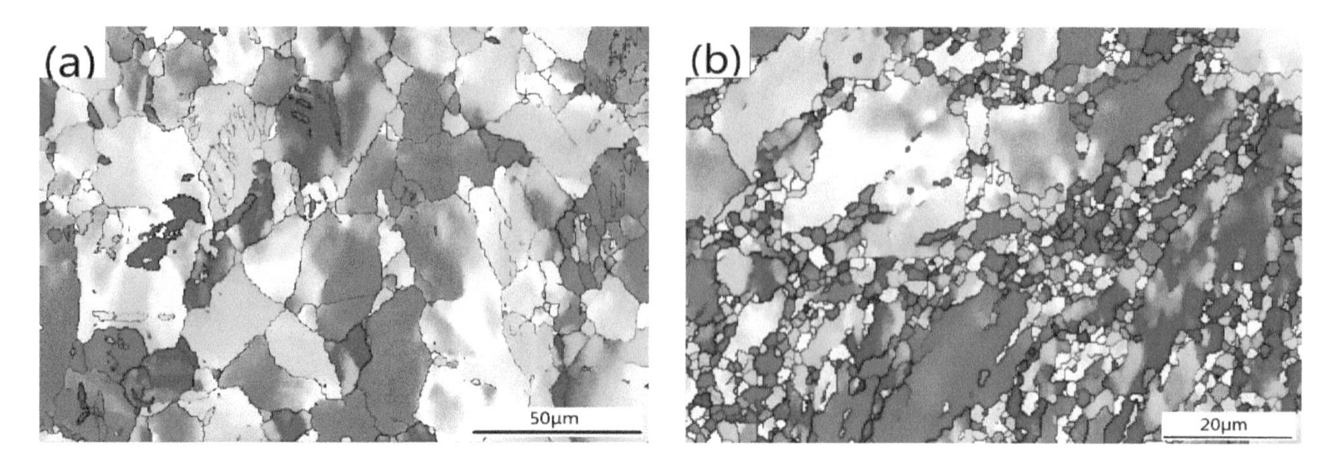

Figure 4. *Cont.*

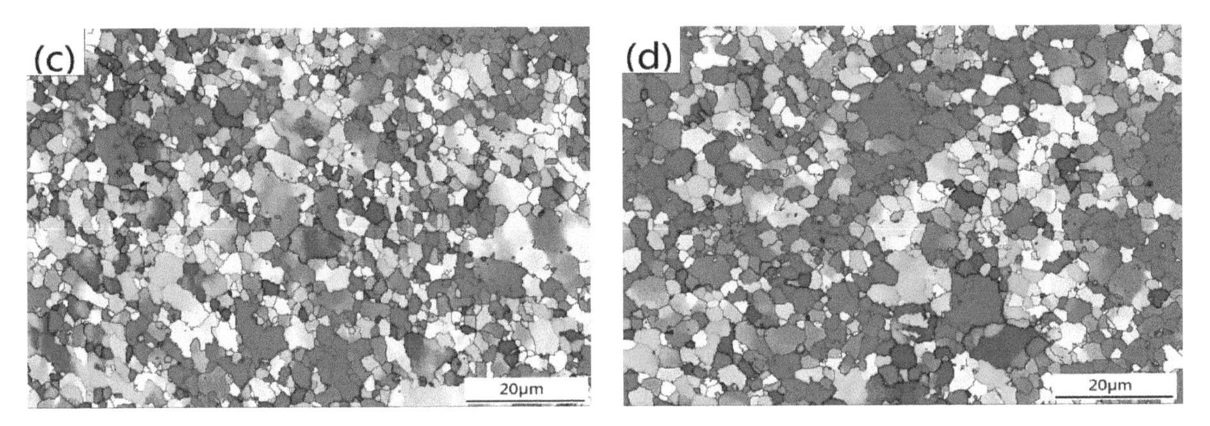

Figure 4. Grain orientation distribution of the Mg-2Y-0.6Nd-0.6Zr alloy. (**a**) as-rolled; (**b**) one pass; (**c**) four passes; (**d**) six passes.

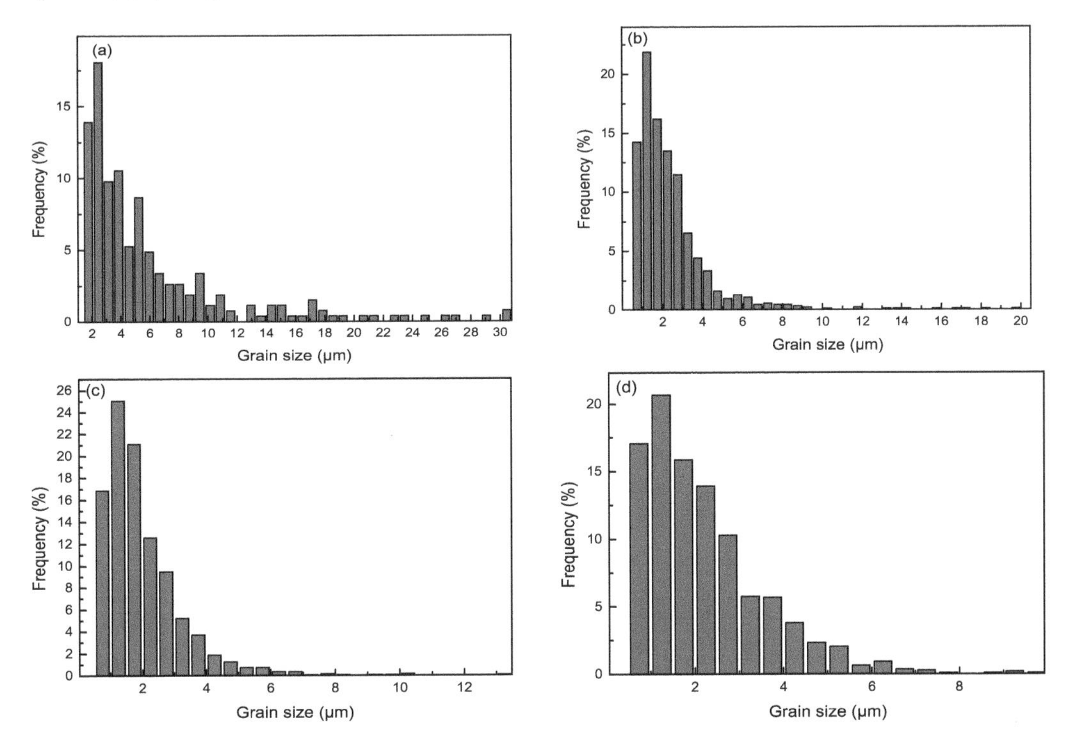

Figure 5. Grain size distribution of the Mg-2Y-0.6Nd-0.6Zr alloy. (**a**) as-rolled; (**b**) one pass; (**c**) four passes; (**d**) six passes.

3.2. Texture

Figure 6 presents the pole figure of the Mg-2Y-0.6Nd-0.6Zr alloy subjected to rolling and a different number of ECAP passes. The as-rolled alloy has a strong (0001) texture with the strongest pole density of 12.30, and the basal plane of most grains is nearly parallel to the rolling direction, as shown in Figure 6a. The pole figure of the alloy after different numbers of ECAP passes (Figure 6b–d), which indicates that the texture in the basal plane was rotated and the strength was continuously weakened with an increasing number of ECAP passes. After one pass of ECAP, the basal plane texture became dispersed, the basal plane of some grains was parallel to the extrusion direction, and the strongest pole density was decreased to 11.34. After four ECAP passes, the basal plane texture was rotated because the specimen rotated 90° along the same direction after each extrusion, the basal plane was ~30° from the extrusion direction, and the strongest pole density further decreased to 7.61. After six ECAP passes, the basal plane formed a typical inclined texture whose basal plane was parallel to the shear plane and was ~45° to the extrusion direction. The strongest pole density was 6.93, which is slightly lower than after four passes.

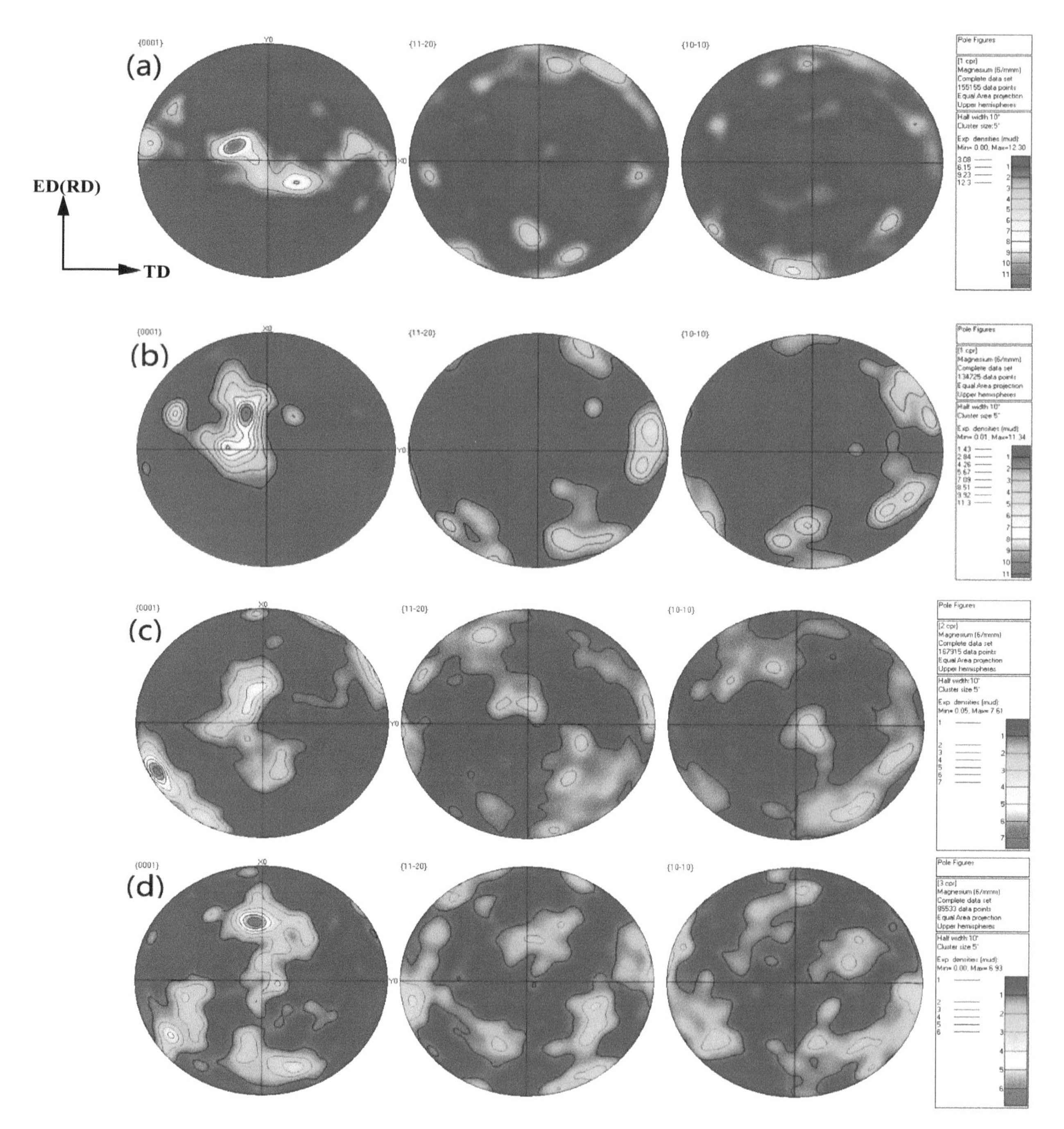

Figure 6. The pole diagram of the Mg-2Y-0.6Nd-0.6Zr alloy. (**a**) as-rolled; (**b**) one pass; (**c**) four passes; (**d**) six passes.

3.3. Mechanical Properties at Room Temperature

The samples after rolling and a different number of ECAP passes were subjected to a tensile test at room temperature. It can be seen in Figure 7 and Table 1 that the as-rolled alloy had the highest ultimate tensile strength and yield strength with values of 246 MPa and 216 MPa, respectively. However, the plasticity of the as-rolled alloy was extremely poor, and elongation was only 3.8%. With an increasing number of ECAP passes, the ultimate tensile strength first increased and then decreased, while the yield strength continuously decreased and the elongation continuously increased. After the first pass of ECAP, the ultimate tensile strength and yield strength were significantly reduced to 213 MPa and 182 MPa, respectively, but the plasticity was improved and the elongation increased to 12.3%. After four passes of ECAP, the ultimate tensile strength was greatly improved compared with the first pass of ECAP, and the yield strength did not change appreciably. The ultimate tensile strength and yield strength were 238 MPa and 180 MPa, respectively, and the elongation increased further to 19.7%. After six passes of ECAP, the alloy had the lowest ultimate tensile strength and yield strength with values of 209 MPa and 148 MPa, respectively, but the elongation reached 27.5%.

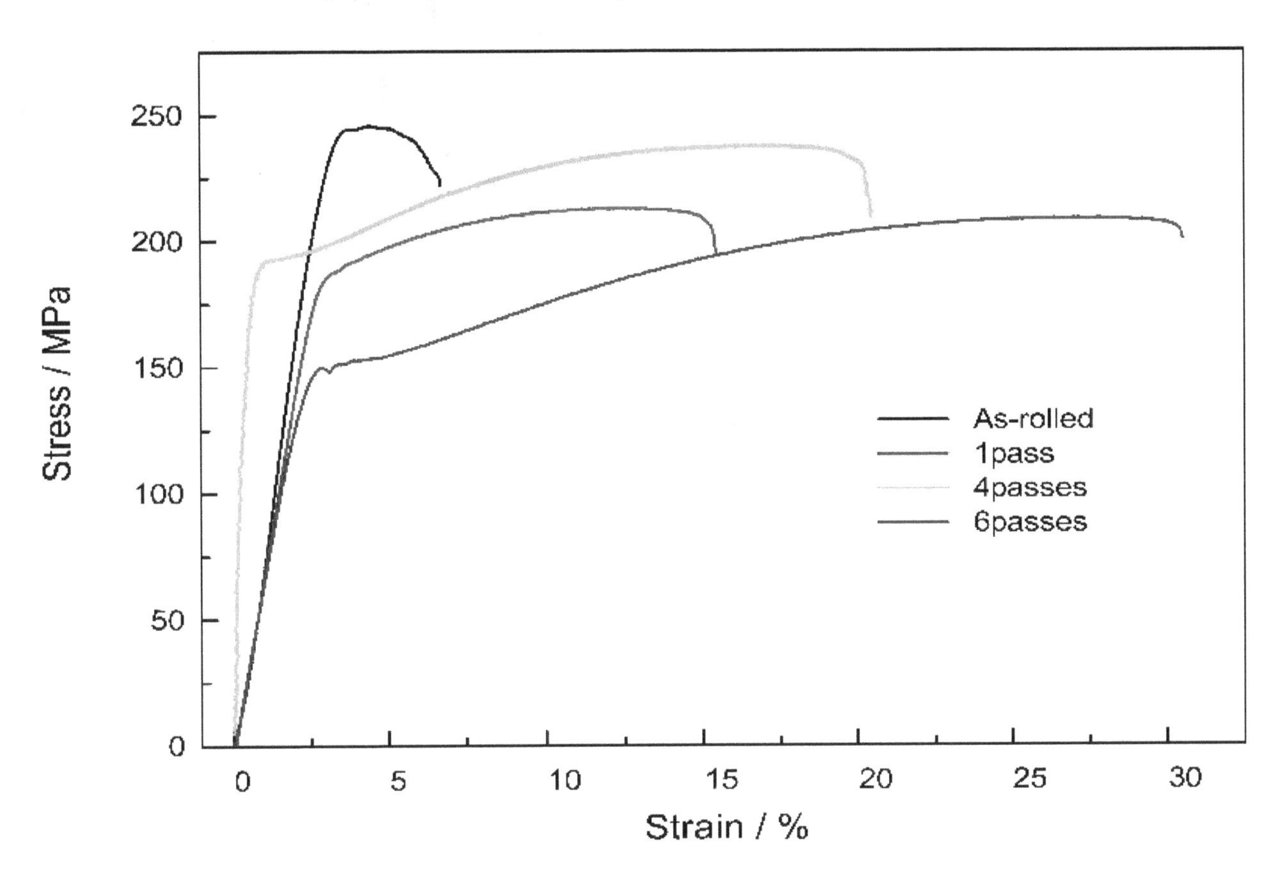

Figure 7. The tensile stress–strain curves of the Mg-2Y-0.6Nd-0.6Zr alloy.

Table 1. The mechanical properties of the Mg-2Y-0.6Nd-0.6Zr alloy.

State	Ultimate Tensile Strength/MPa	Yield Strength/MPa	Elongation/%
As-rolled	246 ± 8.3	216 ± 7.4	3.8 ± 0.12
One pass	213 ± 6.5	182 ± 5.3	12.3 ± 0.36
Four passes	238 ± 7.7	180 ± 6.2	19.7 ± 0.45
Six passes	209 ± 4.9	148 ± 5.5	27.5 ± 0.41

Figure 8 presents the tensile fracture morphology of the Mg-2Y-0.6Nd-0.6Zr alloy subjected to rolling and a different number of ECAP passes. The fracture of the as-rolled alloy (Figure 8a), is relatively flat and bright, which is a typical brittle fracture, indicating that the plasticity of the as-rolled alloy is poor. It can be seen in Figure 8b that after four passes of ECAP, a large number of dimples appeared on the fracture surface of the alloy, which indicates a typical ductile fracture, meaning that the plasticity of the alloy after ECAP was greatly improved compared with that of the as-rolled alloy. After six passes of ECAP, the dimples on the fracture surface were more uniform and deeper than those in the fourth pass, as shown in Figure 8c. This indicates that the plasticity of the alloy after six passes of ECAP is further increased compared with that after four passes of ECAP, which is consistent with the room temperature tensile test results.

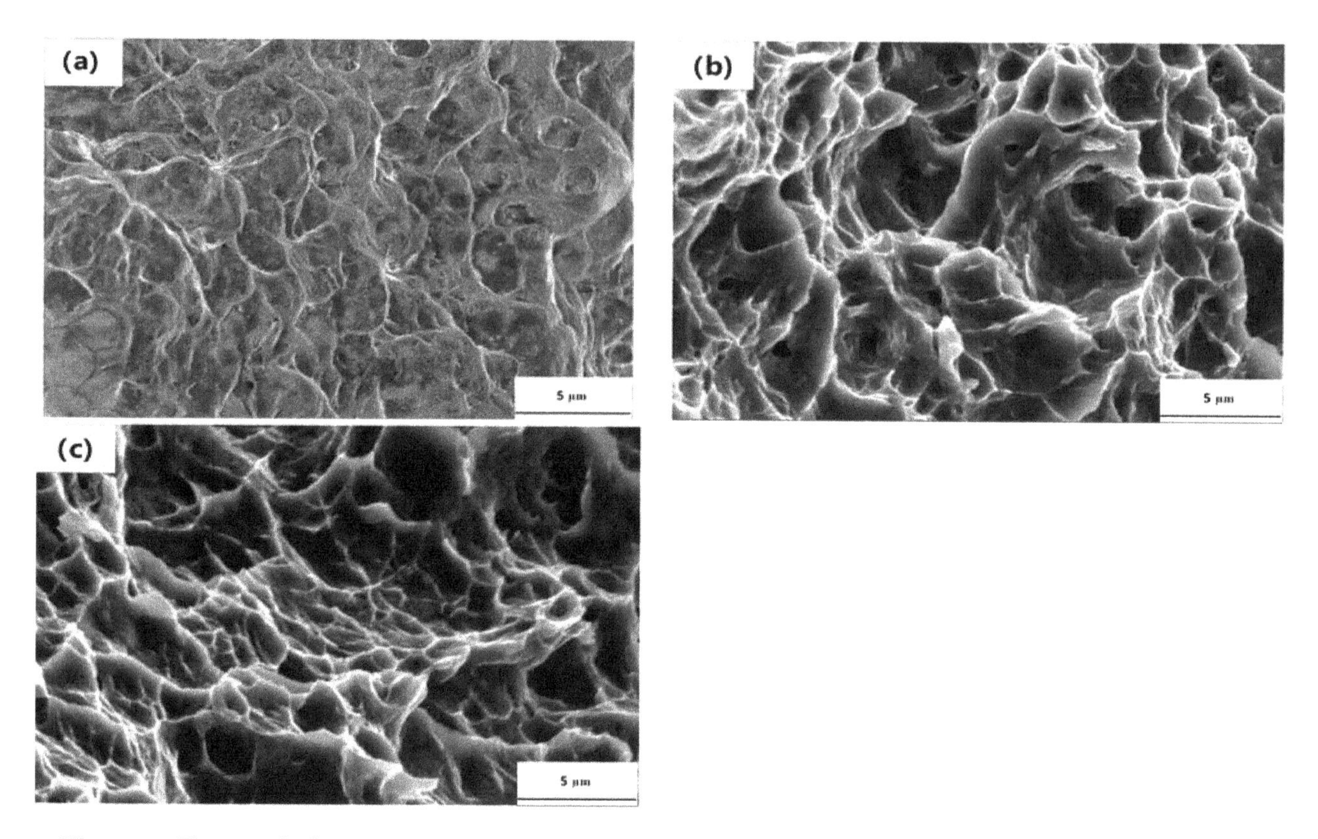

Figure 8. The tensile fracture morphology of the Mg-2Y-0.6Nd-0.6Zr alloy. (**a**) as-rolled; (**b**) four passes; (**c**) six passes.

4. Discussion

ECAP can effectively reduce the grain size of an alloy and its refining effect mainly depends on the processing temperature and total strain [18,19]. Magnesium alloys are refined by mechanical shear and dynamic recrystallization during ECAP [20]. The grains are twisted, sheared and broken when the magnesium alloy experiences shearing deformation in the mold so that the original coarse grains are divided into a plurality of fine grains; on the other hand, shear deformation can produce many dislocation tangles and shear bands which could provide nucleation sites and a driving force for dynamic recrystallization, which further reduces the grain size. It is well known that five independent slip systems are necessary for plastic deformation of polycrystalline materials. There are only two independent slip systems in the basal slip systems of magnesium alloys [21] and the non-basal slip systems of magnesium alloys are difficult to activate at low temperatures because the critical resolved shear stress (CRSS) of non-basal slip systems is much larger than that of basal slip systems [22]. Twinning is an important mechanism for low temperature plastic deformation of magnesium alloys because it can change crystal orientation and release the stress concentration caused by dislocation plugging. At high temperatures, the CRSS of the non-basal slip systems of magnesium alloys is greatly reduced and becomes easy to activate [23]. At that point, The plastic deformation of magnesium alloys does not depend on twinning but mainly depends on dislocation slipping [24]. In this study, both the rolling process and ECAP were carried out at relatively high temperatures, so twins would not be generated in the deformed microstructure, which is consistent with the results in Figures 3 and 4. In addition to the deformation temperature, the deformation mechanism of magnesium alloys is closely related to the grain size. The deformation mechanism of coarse grains is a typical slipping and twinning mechanism, while in the case of fine grains, in addition to slipping and twinning, grains can also coordinate deformation through grain boundary sliding and rotation [25]. These mechanisms work together to improve the deformation capacity of the alloys. When the Mg-2Y-0.6Nd-0.6Zr alloy was rolled at 400 °C, the grain size of the alloy was relatively large, the main deformation mechanism of the alloy was dislocation slipping, and the non-basal slip played an important role in order to satisfy the conditions of the five independent slip systems. The seven-pass rolling treatment caused the dislocation to proliferate in the grains, forming a large number of dislocation tangles and storing a lot of energy. In the subsequent ECAP process, these dislocation tangles and the stored energy can provide favorable conditions for dynamic recrystallization. With an increasing number of ECAP passes, the grain size is continuously reduced under the combined action of mechanical shearing and dynamic recrystallization, the deformation mechanism of the alloy is transformed from solely dislocation slipping to a combination of dislocation slipping and grain boundary sliding and rotation, and the plastic deformation capacity of the alloy is continuously enhanced. After four passes of ECAP, the dynamic recrystallization of the alloy was almost complete, the effect of mechanical shearing and dynamic recrystallization was greatly weakened, and the grain size would have increased if the extrusion passes has been further increased. Feng et al. [26,27] also found in the ECAP studies of an AZ31 magnesium alloy and Al-Mg-Si alloy that the grain size has a tendency to grow after the ECAP reaches a certain number of passes.

It can be seen in Table 1 that the as-rolled alloy kept the highest strength and the worst plasticity. This is because the as-rolled alloy accumulated a large amount of strain after seven passes of rolling, which led to the increase in dislocation density and the increase in the plastic deformation resistance of the alloy. In addition, water-cooling produces greater internal stress in the alloy after rolling, which accelerates the crack extension rate. Finally, the as-rolled alloy obtained a high strength and a poor plasticity due to work-hardening. After one pass of ECAP, the average grain size of the alloy decreased from ~11.2 μm to ~2.43 μm. It is well known that the finer the grain size, the better the strength and plasticity of the alloy, but the strength of the alloy decreased after one pass of ECAP. This is due to the fact that the dislocations were annihilated during the process of thermal insulation and subsequent ECAP so that the work-hardening effect was substantially weakened compared with the as-rolled alloy. Finally, the strength of the alloy decreased and the plasticity increased after one pass of ECAP.

With an increasing number of ECAP passes, the grain size of the alloy was gradually reduced, and the elongation increased, but the yield strength gradually decreased. In particular, after the sixth pass of ECAP, the yield strength decreased from a value of 180 MPa for the fourth pass to 148 MPa, which is contrary to the traditional Hall–Petch relationship. Kim et al. [28] concluded that the mechanical properties of magnesium alloys are closely related to texture and grain size after ECAP. When the base plane is 45° from the extrusion direction, the Schmid factor (SF) tends to be 0.5, the alloy is in a soft orientation, and the yield strength decreases. When the base plane is parallel or perpendicular to the extrusion direction, the SF tends to be 0, the alloy is in a hard orientation, and the yield strength increases. Figure 9 presents the SF distribution of the Mg-2Y-0.6Nd-0.6Zr alloy after different number of ECAP passes. Figure 9a shows that the SF of the as-rolled alloy is low with an average value of 0.282. After one pass of ECAP, because the basal plane of some grains was parallel to the extrusion direction (Figure 6a), the ratio of SF approaching 0 was higher, but the average value, 0.285, was close to that of the as-rolled. At that time, the alloy was in a hard orientation so the yield strength was improved. After four passes of ECAP, the average grain size of the alloy decreased from ~2.43 μm to ~1.87 μm, while the strength and plasticity of the alloy increased. However, the average value of the SF increased from 0.285 to 0.317; the softening effect of texture is equivalent to the strengthening effect of grain refinement, so the yield strength did not change significantly. After six passes of ECAP, the alloy formed a typical inclined texture whose basal plane was parallel to the shear plane and was ~45° to the extrusion direction. Here, the ratio of SF factor approaching 0.5 was higher, with an average value of 0.354, and the alloy was in a soft orientation. Moreover, the grain size of the alloy after six passes of ECAP was slightly larger than that after four passes of ECAP, so the yield strength was greatly reduced compared with the fourth pass of ECAP. However, due to the uniform distribution of the grain size, the deformation compatibility of the alloy increased. Therefore, the plasticity of the alloy after six ECAP passes was higher than that after the fourth ECAP pass and the elongation reached

27.5%. Muralidhar et al. [29,30] also obtained similar conclusions in their ECAP studies of AZ31 and AZ80 magnesium alloys.

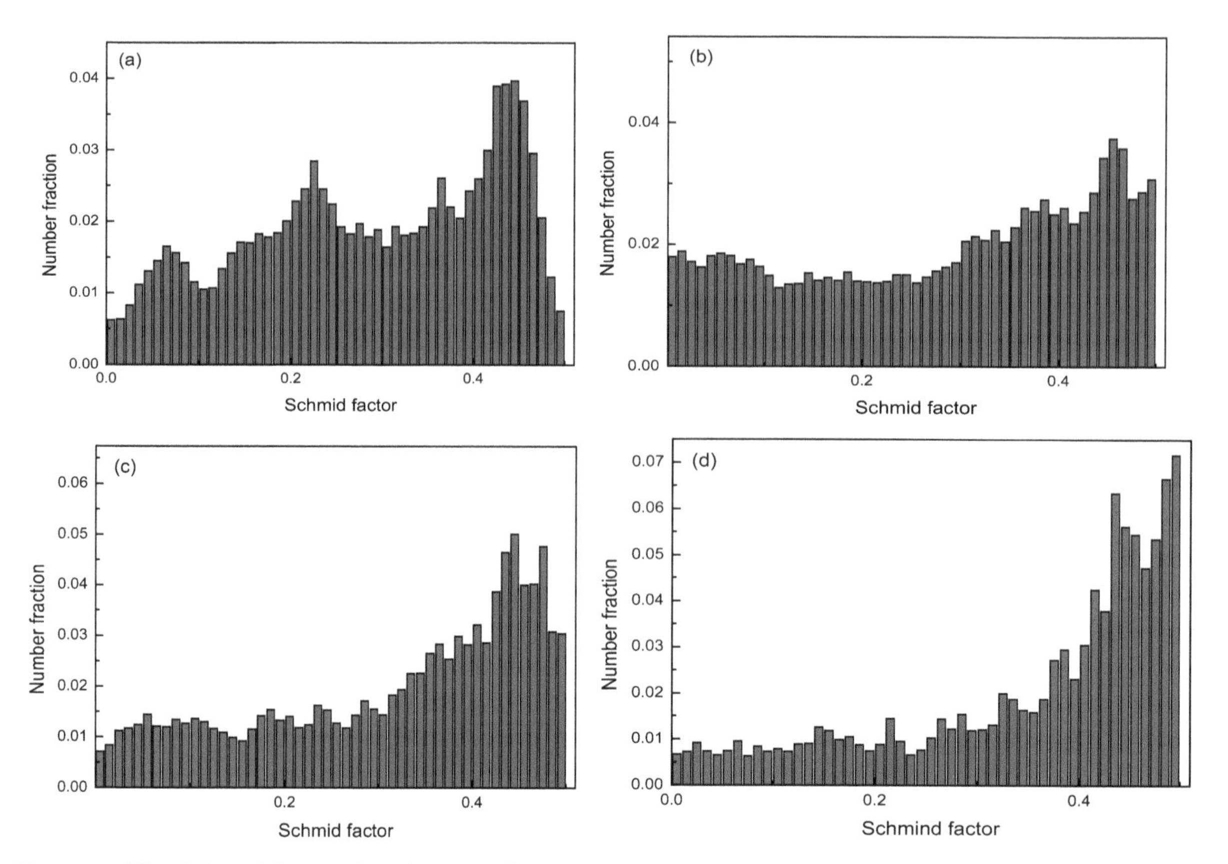

Figure 9. The Schmid factor distribution of the Mg-2Y-0.6Nd-0.6Zr alloy. (**a**) as-rolled; (**b**) one pass; (**c**) four passes; (**d**) six passes.

5. Conclusions

In this paper, the ECAP of a rolled Mg-2Y-0.6Nd-0.6Zr alloy was successfully processed at 340 °C. The microstructure and mechanical properties of the Mg-2Y-0.6Nd-0.6Zr alloy before and after ECAP were investigated. The following are the main conclusions:

(1) After the Mg-2Y-0.6Nd-0.6Zr alloy was rolled for seven passes at 400 °C, a high strength was obtained due to work-hardening. The ultimate tensile strength and yield strength were 216 MPa and 246 MPa, respectively, but the plasticity was extremely poor and the elongation was only 3.8%.

(2) ECAP of the Mg-2Y-0.6Nd-0.6Zr alloy was carried out at 340 °C. With an increasing number of ECAP passes, the grain size of the alloy gradually decreased under the combined action of mechanical shearing and dynamic recrystallization. After four passes of ECAP, the average grain size of the alloy decreased from 11.2 μm to 1.87 μm and the grain size no longer decreased as the number of ECAP passes increased.

(3) With an increasing number of ECAP passes, the plasticity of the Mg-2Y-0.6Nd-0.6Zr alloy increased continuously and the yield strength and tensile strength first increased and then decreased due to a combination of the fine-grain strengthening and texture softening. After four passes of ECAP, good comprehensive mechanical properties were obtained and the strength was slightly decreased compared with the as-rolled alloy, but the plasticity was greatly increased.

Author Contributions: W.L. and L.T. designed the experiment; Y.T. and Z.Z. conducted the experiment; testing, W.H.; X.S., collected data, analyzed and wrote the paper; checking and modifying, W.L.

Acknowledgments: This research was funded by the National Natural Science Foundation of China (51661007), China Postdoctoral Science Foundation (2018M633416), Guizhou Science and Technology Fund (qian ke heJ [2017] 1022), Guizhou Science and Technology Platform Project (qian ke he talent platform [2019]5053).

References

1. Aghion, E.; Bronfin, B.; Eliezer, D. The role of the magnesium industry in protecting the environment. *J. Mater. Proc. Technol.* **2001**, *117*, 381–385. [CrossRef]
2. Mordike, B.L.; Ebert, T. Magnesium: Properties—applications—potential. *Mater. Sci. Eng. A* **2001**, *302*, 37–45. [CrossRef]
3. Luo, A.A.; Mishra, R.K.; Powell, B.R.; Sachdev, A.K. Magnesium Alloy Development for Automotive Applications. *Mater. Sci. Forum* **2012**, *706*, 69–82. [CrossRef]
4. Wang, X.; Liu, C.; Xu, L.; Xiao, H.; Zheng, L. Microstructure and mechanical properties of the hot-rolled Mg–Y–Nd–Zr alloy. *J. Mater. Res.* **2013**, *28*, 1386–1393. [CrossRef]
5. Yamashita, A.; Horita, Z.; Langdon, T.G. Improving the mechanical properties of magnesium and a magnesium alloy through severe plastic deformation. *Mater. Sci. Eng. A* **2001**, *300*, 142–147. [CrossRef]
6. Chen, B.; Lin, D.L.; Jin, L.; Zeng, X.Q.; Lu, C. Equal-channel angular pressing of magnesium alloy AZ91 and its effects on microstructure and mechanical properties. *Mater. Sci. Eng. A* **2008**, *483*, 113–116. [CrossRef]
7. Ma, A.; Jiang, J.; Saito, N.; Shigematsu, I.; Yuan, Y.; Yang, D.; Nishida, Y. Improving both strength and ductility of a Mg alloy through a large number of ECAP passes. *Mater. Sci. Eng. A* **2009**, *513*, 122–127. [CrossRef]
8. Figueiredo, R.B.; Langdon, T.G. Grain refinement and mechanical behavior of a magnesium alloy processed by ECAP. *J. Mater. Sci.* **2010**, *45*, 4827–4836. [CrossRef]
9. Biswas, S.; Dhinwal, S.S.; Suwas, S. Room-temperature equal channel angular extrusion of pure magnesium. *Acta Mater.* **2010**, *58*, 3247–3261. [CrossRef]
10. Ding, S.X.; Lee, W.T.; Chang, C.P.; Chang, L.W.; Kao, P.W. Improvement of strength of magnesium alloy processed by equal channel angular extrusion. *Scr. Mater.* **2008**, *59*, 1006–1009. [CrossRef]
11. Xu, C.; Xia, K.; Langdon, T.G. Processing of a magnesium alloy by equal-channel angular pressing using a back-pressure. *Mater. Sci. Eng. A* **2009**, *527*, 205–211. [CrossRef]
12. Zhang, N.X.; Ding, H.; Li, J.Z.; Wu, X.L.; Li, Y.L.; Xia, K. Microstructure and Mechanical Properties of Ultra-Fine Grain AZ80 Alloy Processed by Back Pressure Equal Channel Angular Pressing. *Mater. Sci. Forum* **2011**, *667*, 547–552. [CrossRef]
13. Lei, W.; Wei, L.; Wang, H.; Sun, Y. Effect of annealing on the texture and mechanical properties of pure Mg by ECAP at room temperature. *Vacuum* **2017**, *144*, 281–285. [CrossRef]
14. Miyahara, Y.; Horita, Z.; Langdon, T.G. Exceptional superplasticity in an AZ61 magnesium alloy processed by extrusion and ECAP. *Mater. Sci. Eng. A* **2006**, *420*, 240–244. [CrossRef]
15. Krajňák, T.; Minárik, P.; Stráská, J.; Gubicza, J.; Máthis, K.; Janeček, M. Influence of equal channel angular pressing temperature on texture, microstructure and mechanical properties of extruded AX41 magnesium. *J. Alloys Compd.* **2017**, *705*, 273–282. [CrossRef]
16. Suh, J.; Victoria-Hernández, J.; Letzig, D.; Golle, R.; Volk, W. Enhanced mechanical behavior and reduced mechanical anisotropy of AZ31 Mg alloy sheet processed by ECAP. *Mater. Sci. Eng. A* **2016**, *650*, 523–529. [CrossRef]
17. Furukawa, M.; Iwahashi, Y.; Horita, Z.; Nemoto, M.; Langdon, T.G. The shearing characteristics associated with equal-channel angular pressing. *Mater. Sci. Eng. A* **1998**, *257*, 328–332. [CrossRef]
18. Akihiro, Y.; Daisuke, Y.; Zenji, H.; Terence, G. Langdon Influence of pressing temperature on microstructural development in equal-channel angular pressing. *Mater. Sci. Eng. A* **2000**, *287*, 100–106.
19. Ramin, J.; Mohammad, S.; Hamid, J. ECAP effect on the micro-structure and mechanical properties of AM30 magnesium alloy. *Mater. Sci. Eng. A* **2014**, *593*, 178–184.
20. Su, C.W.; Lu, L.; Lai, M.O. A model for the grain refinement mechanism in equal channel angular pressing of Mg alloy from microstructural studies. *Mater. Sci. Eng. A* **2006**, *434*, 227–236. [CrossRef]
21. Yoo, M.H. Slip, Twinning, and Fracture in Hexagonal Close-Packed Metals. *Metall. Trans. A* **1981**, *12*, 409–418. [CrossRef]
22. Koike, J. Enhanced deformation mechanisms by anisotropic plasticity in polycrystalline Mg alloys at room temperature. *Metall. Mater. Trans. A* **2005**, *36*, 1689–1696. [CrossRef]
23. Xin, R.L.; Wang, B.S.; Zhou, Z.; Huang, G.J.; Liu, Q. Effects of strain rate and temperature on microstructure and texture for AZ31 during uniaxial compression. *Trans. Nonferrous Met. Soc. China* **2010**, *20*, s594–s598. [CrossRef]
24. Chino, Y.; Kimura, K.; Mabuchi, M. Twinning behavior and deformation mechanisms of extruded AZ31 Mg alloy. *Mater. Sci. Eng. A* **2008**, *486*, 481–488. [CrossRef]

25. Partridge, P.G. The crystallography and deformation modes of hexagonal close-packed metals. *Metall. Rev.* **1967**, *12*, 169–194.
26. Feng, X.M.; Tao-Tao, A.I. Microstructure evolution and mechanical behavior of AZ31 Mg alloy processed by equal-channel angular pressing. *Trans. Nonferrous Met. Soc. China* **2009**, *19*, 293–298. [CrossRef]
27. Khelfa, T.; Rekik, M.A.; Khitouni, M.; Cabrera-Marrero, J.M. Structure and microstructure evolution of Al–Mg–Si alloy processed by equal-channel angular pressing. *Int. J. Adv. Manuf. Technol.* **2017**, *92*, 1731–1740. [CrossRef]
28. Kim, W.J.; Hong, S.I.; Kim, Y.S.; Min, S.H.; Jeong, H.T.; Lee, J.D. Texture development and its effect on mechanical properties of an AZ61 Mg alloy fabricated by equal channel angular pressing. *Acta Mater.* **2003**, *51*, 3293–3307. [CrossRef]
29. Muralidhar, A.; Narendranath, S.; Nayaka, H.S. Effect of equal channel angular pressing on AZ31 wrought magnesium alloys. *J. Magnes. Alloys* **2013**, *1*, 336–340. [CrossRef]
30. Wang, L.; Mostaed, E.; Cao, X.; Huang, G.; Fabrizi, A.; Bonollo, F.; Chi, C.; Vedani, M. Effects of texture and grain size on mechanical properties of AZ80 magnesium alloys at lower temperatures. *Mater. Des.* **2016**, *89*, 1–8. [CrossRef]

15

Emergence and Progression of Abnormal Grain Growth in Minimally Strained Nickel-200

Olivia D. Underwood [1,*], Jonathan D. Madison [2] and Gregory B. Thompson [3]

1 Connector & LAC Technology, Sandia National Laboratories, Albuquerque, NM 87185, USA
2 Materials Mechanics, Sandia National Laboratories, Albuquerque, NM 87185, USA; jdmadis@sandia.gov
3 Metallurgical and Materials Engineering, The University of Alabama, Tuscaloosa, AL 35487, USA; gthompson@eng.ua.edu
* Correspondence: odunder@sandia.gov

Abstract: Grain boundary engineering (GBE) is a thermomechanical processing technique used to control the distribution, arrangement, and identity of grain boundary networks, thereby improving their mechanical properties. In both GBE and non-GBE metals, the phenomena of abnormal grain growth (AGG) and its contributing factors is still a subject of much interest and research. In a previous study, GBE was performed on minimally strained (ε < 10%), commercially pure Nickel-200 via cyclic annealing, wherein unique onset temperature and induced strain pairings were identified for the emergence of AGG. In this study, crystallographic segmentation of grain orientations from said experiments are leveraged in tandem with image processing to quantify growth rates for abnormal grains within the minimally strained regime. Advances in growth rates are shown to vary directly with initial strain content but inversely with initiating AGG onset temperature. A numeric estimator for advancement rates associated with AGG is also derived and presented.

Keywords: abnormal grain growth; grain boundary engineering; electron backscattered diffraction; growth rate

---

1. Introduction

Abnormal grain growth (AGG) is a mechanism by which a subset of grains grow at a rate faster than others. AGG is of significance because, like many other grain-scale and sub-grain-scale features such as freckles [1,2], precipitates [2], or dendrite arm spacing [3], AGG has been shown to have a significant effect on properties across many material systems under various loading conditions [4–7]. The exact mechanisms underpinning the occurrence of AGG are still unclear and remain a subject of much interest and research. Though debate on specific mechanisms exists, most authors agree that the formation of AGG is related to low-energy, high-mobility grain boundaries such as $\sum3$ and its variants (e.g., $\sum9$ and $\sum27$) [8–16]. This is of note as Grain Boundary Engineering (GBE) is a specific type of thermomechanical process used to improve material properties by altering their grain boundary network and often resulting in increased frequency of $\sum3$ boundaries [17,18].

In a previous study by the authors, GBE was performed on commercially pure Nickel-200 cold-worked to plastic strains (ε) of 3%, 6%, and 9%, respectively. This pre-strained material was then subjected to cyclic annealing schedules in which the dwell temperature was increased with each cycle. Among these cases, greater initial cold-work resulted in lower AGG onset temperatures. Specifically, for ε = 0%, 3%, 6%, and 9%, AGG was initially observed at 780 °C, 760 °C, 740 °C, and 720 °C, respectively. Furthermore, in the vicinity of AGG, $\sum3$, $\sum9$, and $\sum27$, boundaries exhibited local maxima either at or in the thermal cycle following the observed onset temperature for AGG [19]. While the kinetics of AGG are hypothesized analytically [8,20–23], it is rather difficult to reliably ascertain abnormal grain growth rates experimentally for a variety of reasons. These challenges include fluctuations in relative size needed to resolve AGG within a given region of interest [24], full discretization of AGG amidst a population of continuously growing grains within a changing global grain size distribution [25], and the practicality of acquiring a reasonable amount of observations within a limited thermal range to support quantification [20].

Additionally, while texture is not a central emphasis in this work, it is noteworthy that other studies have shown that both grain growth and recrystallization kinetics can be additionally influenced by grain orientation and strain path [26–29]. In this work, the effect of varying thermal and mechanical strain routes are examined specifically in relation to grain growth. To that end, one of the aforementioned studies [19] is revisited and careful electron backscatter diffraction (EBSD) experiments, similar to those reported in [30], are performed. In these experiments, however, EBSD is combined with image segmentation based on crystallographic orientation to extract advancement rates associated with initial onset (emergence) and continuing growth (progression) of abnormally large grains in a simple Ni system.

2. Materials and Methods

Commercially pure Ni-200 bars (with dimensions of 152 mm × 6.35 mm × 6.35 mm) were thermomechanically processed using a Stanat Model: TA 215 rolling mill (Stanat Mfg. Co. Inc., Westbury, NY, USA) and a Lindberg/Blue M tube furnace (Lindberg/Mph, Riverside, MI, USA), see Figure 1. Samples were cold rolled using multiple rolling passes to achieve a desired reduction of 0%, 3%, 6%, or 9%, respectively. Once rolled, samples were metallographically polished to a 0.05 μm colloidal silica finish. Focus ion beam (FIB) and micro-hardness indent fiducial markers were then placed on each sample so specific regions of interest (ROIs) could be tracked throughout each stage of the experiment while not impeding the field of view associated with any ROI. At room temperature, EBSD was performed on the rolling direction (RD) plane of the pre-selected ROIs for the 0%, 3%, 6%, and 9% strained cases. Samples were then cyclically and progressively annealed for 30 min at 700 °C, 720 °C, 740 °C, 760 °C, 780 °C, and 800 °C, under a flowing argon-rich or hydrogen-rich atmosphere to prevent oxidation.

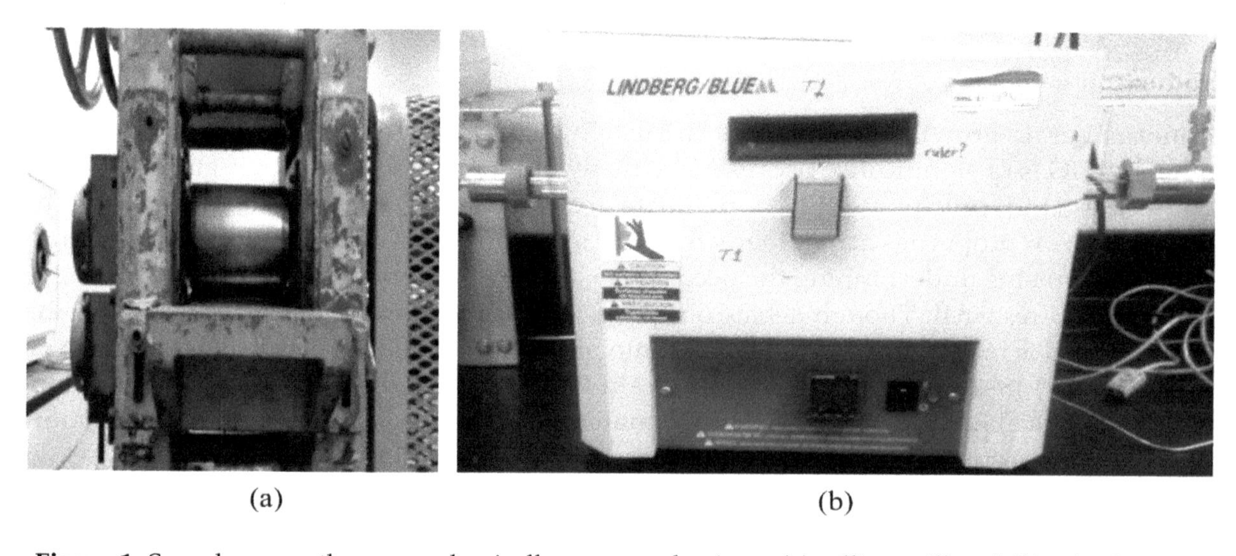

(a) (b)

Figure 1. Samples were thermomechanically processed using a (**a**) rolling mill and (**b**) tube furnace.

After each annealing cycle, see Figure 2, samples were returned to room temperature and EBSD was performed on the same pre-selected ROI to document any changes in the grain population. Operating parameters for the collection of EBSD data utilized the following arrangement: A nominal beam current of 4 nA, a camera binning size of 4 × 4 or 6 × 6, and an indexing step size of 0.5 or 1 μm over multiple cross-sectional areas of 500 × 500 square μm or larger. TexSEM Laboratories (TSL) orientation imaging microscopy analysis by EDAX, Inc. (Mahwah, NJ, USA). was used to analyze

EBSD data and all EBSD maps were filtered using the Neighbor Confidence Index (CI) Correlation followed by Grain CI Standardization cleanup processes, where an average confidence index of >0.1 was used. The reader is directed to reference [19] for additional information regarding the experimental procedure should further detail be desired.

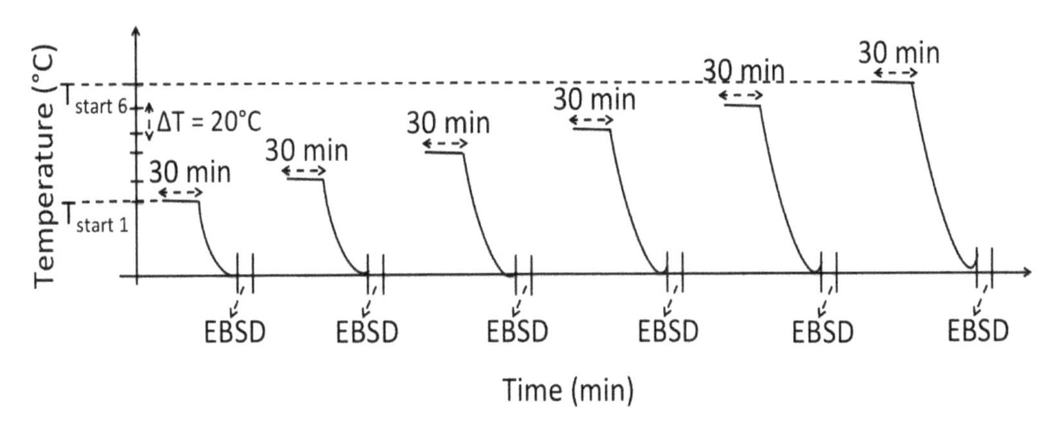

Figure 2. Cyclic annealing schedule with increasing isothermal dwelling anneals used in this study.(Reproduced with permission from [19].)

Grain boundary characterization for the unstrained "as-received" Nickel-200 material is shown in Figure 3. The mean grain size diameter, including twin boundary distributions, displayed a near Gaussian distribution with a mean grain size of 7 μm and an area fraction mode of 20% at a grain size of 15 μm, Figure 3a. The polycrystalline material also bore no specific bias in texture or anisotropy in grain morphology but presented a twin-containing equiaxed microstructure, see Figure 3b–e. The orientation maps and pole figure in Figure 3c–e are shown with respect to the RD.

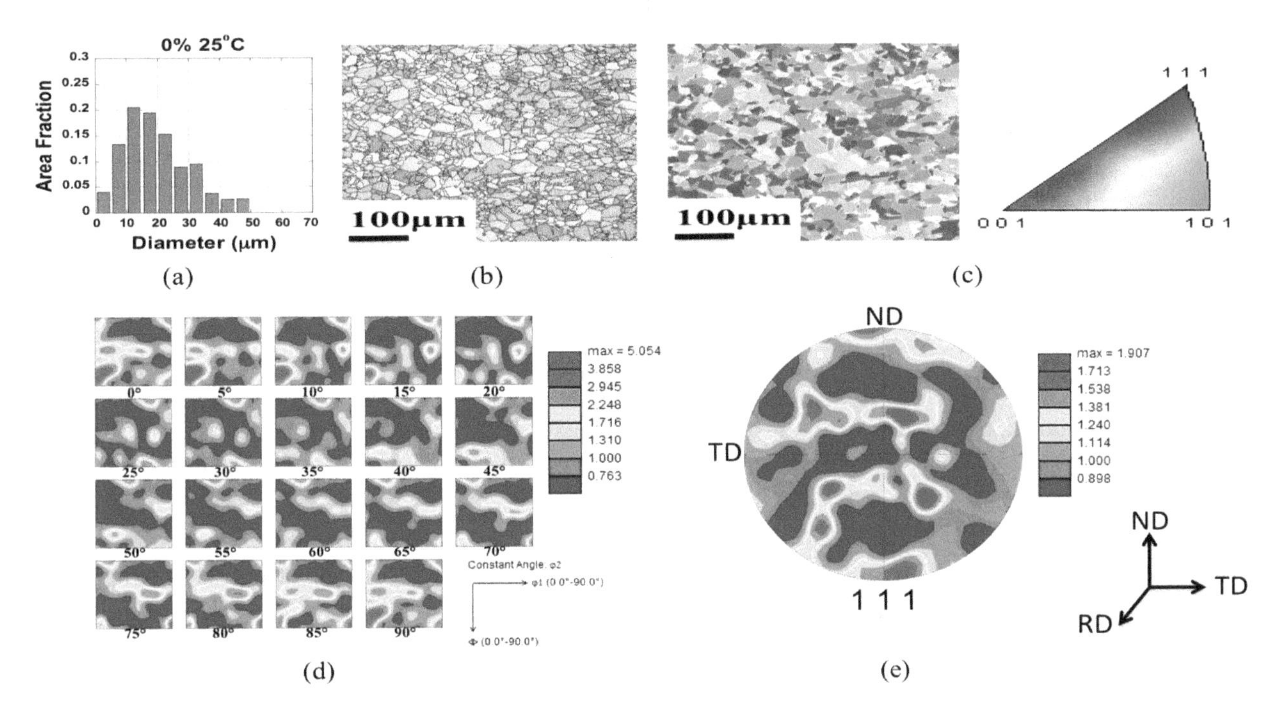

Figure 3. Baseline grain boundary metrics for unstrained, commercially pure Ni-200. (**a**) Grain area fraction distribution; (**b**) electron backscatter diffraction (EBSD) grain contrast map; (**c**) grain orientation map with inverse pole figure (IPF); (**d**) orientation distribution function map and (**e**) pole figure. (Orientations reported with respect to the rolling direction (RD), normal direction (ND) and transverse direction (TD) for (**c**–**e**); Adapted with permission from [19].)

3. Results

For initial induced strains of 0%, 3%, 6%, and 9%, a series of EBSD maps are shown in Figure 4 where a given ROI is maintained along each row. Each cell reveals the EBSD map acquired following a specific cyclic annealing dwell temperature. As can be seen, AGG was observed to initiate at 780 °C for 0% strain; 760 °C for 3% strain; 740 °C for 6% strain; and 720 °C for 9% strain. For the cases in which abnormal grain growth was observed, bimodal distributions for grain size were also seen. Grains occupying the secondary peak of these bimodal distributions, which corresponded to higher mean grain sizes, were considered abnormally large [19]. Quantitatively, these grains corresponded to the upper 30% of their specific grain-size distributions across nearly all cases. So in this way, a consistent numeric threshold was applied for delineation of abnormally large grains. In Figure 4, these grains are outlined in black to assist in clearly identifying their locations and presence.

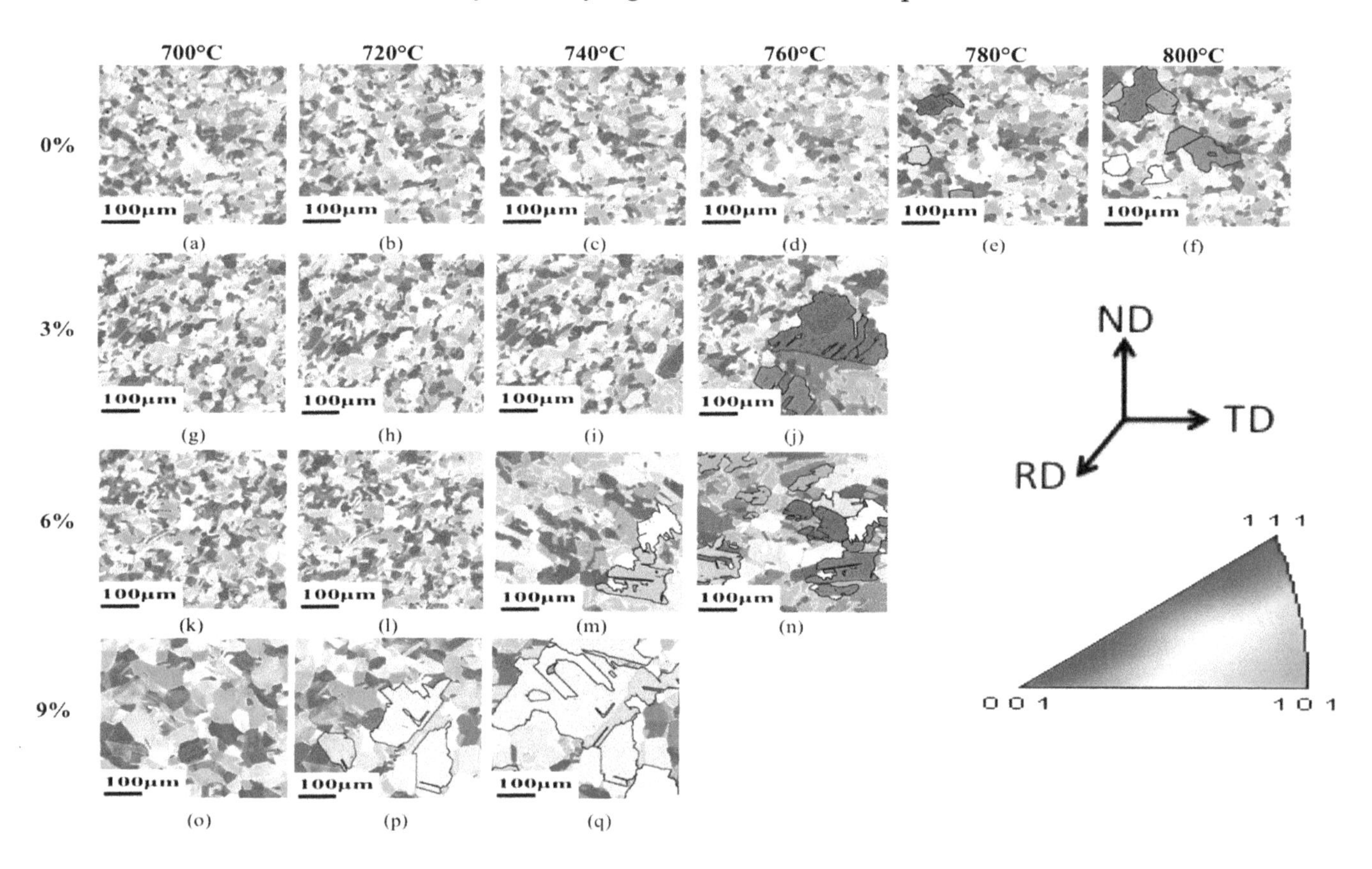

Figure 4. EBSD progression maps for 0% (**a–f**), 3% (**g–j**), 6% (**k–n**), and 9% (**o–q**) induced strain following annealing treatments ranging from 700 °C to 800 °C. Abnormally large grains are outlined in black. (Orientation is shown with respect to the rolling direction (RD); Adapted with permission from [19].)

Inverse pole figure (IPF) maps with respect to the rolling direction (RD) for the ROIs presented in Figure 4 are shown below in Figure 5. The white circles indicate the orientations associated with the abnormally large grains. Minor changes in the texture populations contained within the ROIs are observable at or immediately before the onset of AGG. As might be expected, the population shifts are clearly indicative of the abnormally large grains increasingly occupying significant portions of the ROIs. However, the IPF maps also reveal that among the fields of view investigated, there appears to be no singularly preferred orientation among abnormally large grains in this study.

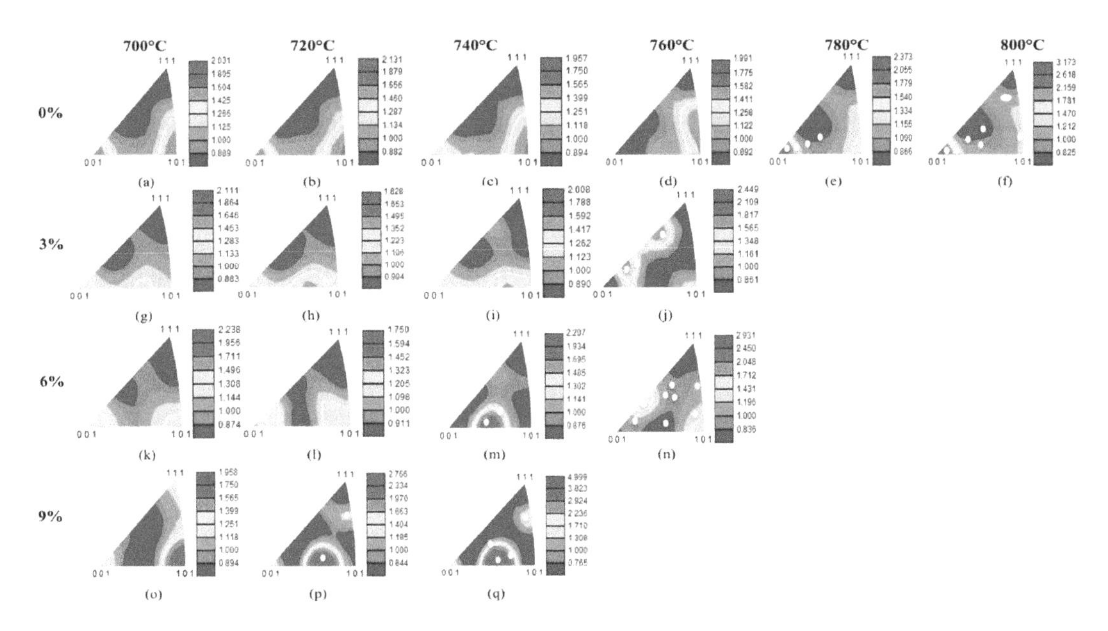

Figure 5. Inverse pole figure maps for 0% (**a–f**), 3% (**g–j**), 6% (**k–n**), and 9% (**o–q**) induced strain following annealing treatments ranging from 700 °C to 800 °C. White circles indicate the location of abnormally large grains. Orientation is shown with respect to the rolling direction (RD).

By utilizing the inherent segmentation available through the differentiation of crystallographic orientations, abnormally large grains were identified, as shown in Figure 4, and then isolated within their fields of view, see Figure 6. These sequestered grains were then used to calculate a local AGG area fraction measurement. By utilizing each successive area fraction, the rates of AGG were determined for both their onset and continued growth behavior. While differences in growth rates for twin and non-twinned grains are not captured here, the experimental quantification of the aggregate development of abnormal grain growth is notable.

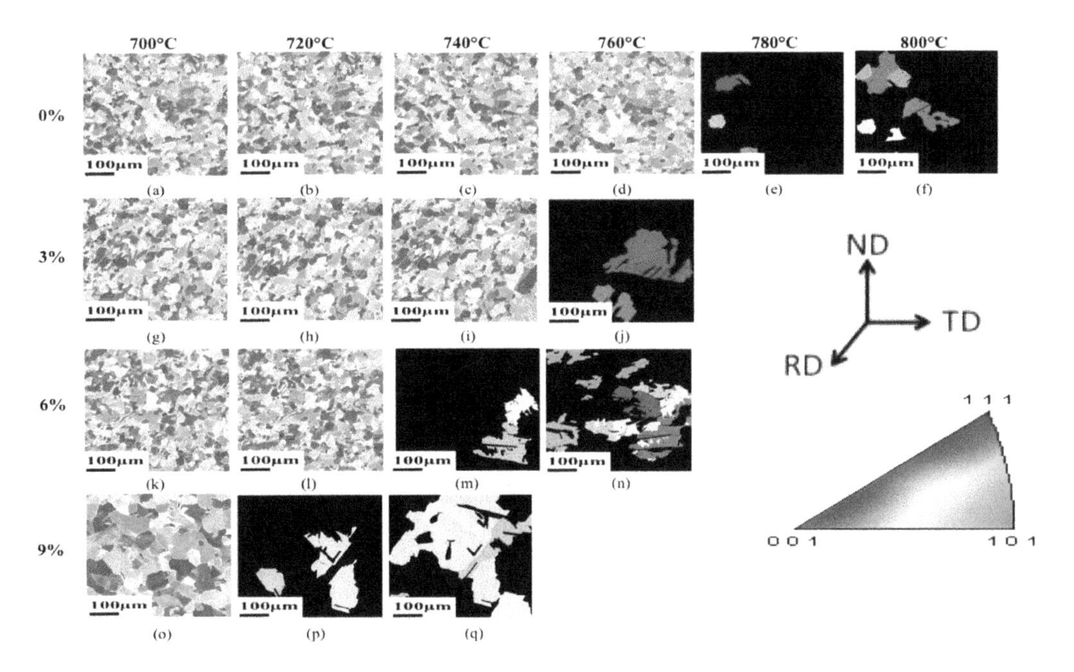

Figure 6. AGG (abnormal grain growth) identification progression maps for 0% (**a–f**), 3% (**g–j**), 6% (**k–n**), and 9% (**o–q**) induced strain following annealing treatments ranging from 700 °C t o 8 0 0 °C.(Orientation is shown with respect to the rolling direction (RD); Adapted with permission from [19].)

The measured area fractions of abnormally large grains are plotted as a function of annealing temperature associated with their AGG onset, see Figure 7a. These area fractions are also shown as a function of the number of thermal cycles (n) beginning with the cycle prior to AGG, see Figure 7b. The error bars shown depict the area fraction variability associated with a $\pm$10% adjustment to the upper 30% grain size threshold value imposed. As can be observed, the growth rates for all strain levels exhibit a rather consistent two-stage behavior. Where stage 1 appears relatively slow and coincides with the initial observation of AGG, and stage 2 continues rapidly once AGG has clearly presented itself. This more rapid stage is seen to advance at a rate roughly 3 times that of stage 1 for the 0%, 3%, and 9% cases, see Table 1. For convenience, the authors will henceforth refer to stage 1 growth as "emergence" and stage 2 as "progression". Due to challenges associated with reliably identifying the emergence stage for the 3% case, emergence rates are not reported for 3% strain.

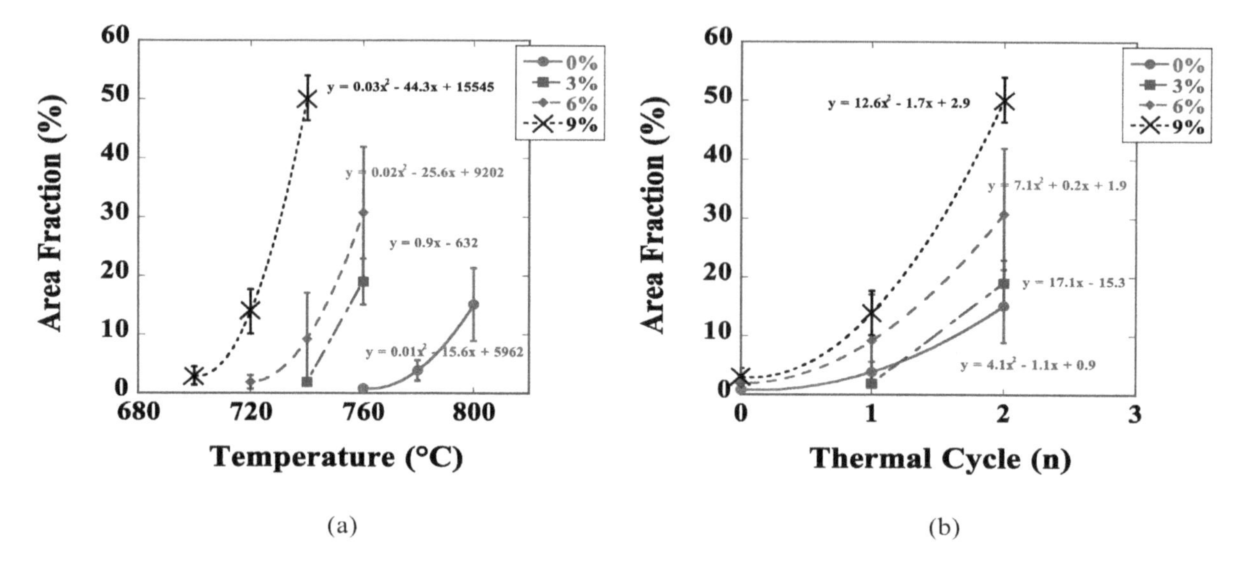

Figure 7. Area fraction of abnormally growing grains for 0%, 3%, 6%, and 9% induced strain as a function of (**a**) temperature and (**b**) thermal cycle.

Table 1. Emergence and progression rates for abnormally growing grains.

ε (%)	Emergence	Progression	Progression/Emergence
	Δ (Area Fraction $_{n=1\text{–}0}$)	Δ (Area Fraction $_{n=2\text{–}1}$)	
0	3.03	11.23	3.71
3	-	17.12	-
6	7.34	21.55	2.93
9	10.97	36.25	3.30

As seen in Figure 7, the greater the initial strain content, the higher the rate of abnormal grain growth in both the emergence and progression stages. Alternatively, the higher the AGG onset temperature, the slower the overall AGG advancement. While twinned and non-twinned grains were not differentiated in this analysis, the evolution of grains occupying large regions of cross-sectional area within their physical neighborhood were tracked rather successfully. To provide a more generalized extrapolation of these relative rates, the derivative of the parabolic curve fits in Figure 7b are provided in Figure 8. Here, each data series corresponds to a specific strain content and is denoted by color. The data markers provide the exact values of the calculated differential for up to 10 thermal cycles. Again, no calculated differential values for 3% are included due to the unavailability of a second order polynomial to differentiate.

Figure 8 illustrates a few informative trends. First, AGG in Ni-200 manifests unique growth rates for a given strain irrespective of the number of cyclic annealing exposures it experiences.

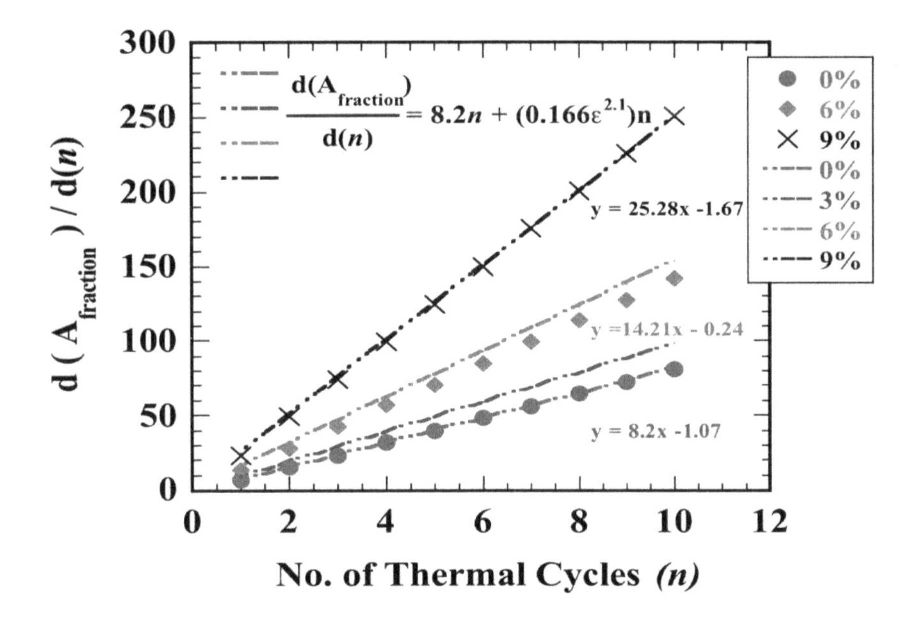

Figure 8. Rate of abnormal grain growth for low strained Ni-200 as a function of the number of thermal annealing cycles.

Stated another way, the initial strain content appears to strongly influence the emergence rate, and this imprint is preserved as abnormal grain growth continues. This estimation can be further simplified and approximated mathematically for all observed strains. This empirical relationship is provided in Equation (1) where:

$$\frac{\mathrm{d}\,(A_{\mathrm{fraction}})}{\mathrm{d}(n)} = 8.2n + (\frac{\varepsilon^{2.1}}{6})n. \quad (1)$$

In this instance, n is the quantity of cyclic thermal anneals and ε is the initial plastic strain. These predictive estimates are also included in Figure 8 as dashed lines to assist with their identification in comparison to the calculated differentials. Please note: A numerical prediction provided by Equation (1) is included for an initial 3% strain case by a dashed line, despite the current work's inability to experimentally resolve the emergence rate for 3% strain. However, as can be seen, the predictive agreement is quite reasonable across all strain series shown. Furthermore, Equation (1) also illustrates that higher strain AGG rates can be estimated with a nominal (0%) AGG rate, when further incentivized by the added contribution of the higher strains as provided in the second term of Equation (1). As shown, the numeric approximation must also be accompanied by an appropriate set of prefactors. These values are provided in Equation (1) based on the experiments performed here. This suggests that in the case of consistent AGG advancement from repeated cyclic anneals, at least within the low strain regime, a methodology for the prediction of AGG advancement may be reasonably devised by clearly understanding or merely quantifying the unstrained AGG rates and the relevant scaling factors associated with additional strain content.

4. Discussion

The authors acknowledge that the findings of this work are derived from a limited number of observations within a single and specific lot of a commercially pure Ni-200 material, thermomechanically processed by rolling and exposure to a specific, consistent, and repeated cyclic annealing heat treatment. It should be anticipated that the specific area fraction measurements may be influenced by a number of these factors. Specifically, the most influential factors may be the cross-sectional areas selected for ROIs within each EBSD map, the 20 °C thermal jump between dwells, the 30 min dwell-time interval, and even the utilization of rolling to impart initial plastic strain. While these factors may introduce quantitative variation into specific measures, the authors assert these differences are likely to be insignificant to the fundamental trends observed. It has been shown that differing average grain size and cumulative area fractions can be obtained within the same pre-strained material under cyclic or isothermal annealing by introducing variations in dwell temperature, ramp rate, or hold times. However, once normalized by the local mean grain size, these varying distributions converge to a singular, invariant, cumulative fraction distribution [31]. Likewise, AGG observations could reasonably be expected to follow a similar convergence behavior when multiple observations are normalized to the local mean grain size present. The authors would further assert that the true benefit of this study is held in the observation of self-similar AGG rates being clearly and quantifiably related to initial plastic strain when exposed to equivalent, consistent, and repeated thermal exposures, not the specific area fractions reported in themselves.

Furthermore, the systematic observations across a breadth of strains and temperatures are notable in that AGG is shown to be predictively related to initial strain content. As seen here, initial strain bears a definite and significant influence on not only the emergence rate for AGG, but also, by extension, the continued progression rate. These findings are supported by the work of Decker et al. [32], He et al. [33–35], and Cho et al. [36], who all showed that among instances of self-similar AGG, initial strain content can serve as a significant driving force for grain boundary migration among low-energy, high-mobility grain boundaries [9,36,37]. In this work, this driving force is incorporated into the predictive estimate for AGG advancement as a superposition of a nominal rate (where $\varepsilon = 0$) combined with a weighted contribution from the additional plastic strain present. While good agreement has been achieved here, further experiments, along the lines of those reported herein, could additionally refine these observations and/or develop similar descriptions for other material systems. Such experiments would, however, require identification of self-similar AGG across a collection of strains and temperatures as a pre-requisite.

Two specific studies to further elucidate the AGG advancement rates reported here could include: (1) Capturing regions of interest larger than 500×500 square µm to acquire greater grain populations for measurement, and thereby increase the potential for identification of additional AGG events; or (2) decrease the cyclic annealing thermal or dwell time intervals to values smaller than 20 °C or 30 min, respectively. This would allow the examination of AGG emergence and progression across smaller observation spans. Further studies of this type would either grow the range of instances available for data collection or increase the resolution and granularity over which the AGG phenomena could be observed.

5. Conclusions

(1) Emergence and progression rates for abnormally large grains in minimally strained, commercially pure Ni-200 were observed experimentally by EBSD and quantified using image segmentation based on crystallographic orientation and grain size distributions.

(2) Emergence and progression growth rates were shown to scale directly with increased initial strain content and inversely with onset AGG temperature.

(3) A predictive estimate for the rate of AGG area fraction advancement, as a function of repeated thermal cycles, was determined based on the derivative of the experimental area fraction measures. This estimate has the following form:

$$\frac{\mathrm{d}\,(A_{\text{fraction}})}{\mathrm{d}(n)} = C_1 n + (C_2 \varepsilon^{\alpha})n, \tag{2}$$

where for the experiments performed herein, $C_1 = 8.2$, $C_2 = 1/6$ and $\alpha = 2.1$.

(4) This numeric estimate indicates two notable implications:

i. Area fraction advancement rates for AGG in minimally strained, commercially pure Ni-200 proceed at unique rates for a given initial strain content;

ii. This advancement rate can be reasonably approximated across the low strain regime by the superposition of a nominal rate (where $\varepsilon = 0$) combined with the contribution of additional plastic strain (ε), modified by some prefactor value (C_2), and scaled by an appropriate exponential (α).

Acknowledgments: Sandia National Laboratories is a multi-mission laboratory managed and operated by National Technology and Engineering Solutions of Sandia, Limited Liability Company (LLC), a wholly owned subsidiary of Honeywell International, Inc. for the U.S. Department of Energy's National Nuclear Security Administration under contract DE-NA0003525. A subset of experiments for this work were supported by the National Science Foundation under Grant no. DMR-1151109. Compressive rolling was conducted at the Redstone Arsenal in Huntsville, AL, USA and the authors would like to thank Daniel Renner for initial metallographic preparation. Significant electron backscatter diffraction and annealing studies were conducted at the Center for Nanophase Materials Sciences, a U.S. Department of Energy, Office of Science and User Facility in Oak Ridge, TN, USA. The authors benefited greatly from the technical assistance of Donovan Leonard and James Kiggans and, the authors would also like to thank Rodney McCabe from Los Alamos National Laboratory for guidance in EBSD data processing. All other work was performed at Sandia National Laboratories and the authors would like to thank Alice Kilgo for sample preparation, Charles Walker for annealing studies, and Bonnie McKenzie and Joseph Michael for additional EBSD experiments. The late Professor Jeffrey L. Evans is also recognized for his initial motivation of these studies.

Author Contributions: Olivia D. Underwood and Jonathan D. Madison conceived and designed the experiments; Olivia D. Underwood performed the experiments and analyzed the data; Olivia D. Underwood, Jonathan D. Madison, and Gregory B. Thompson wrote the paper.

References

1. Madison, J.D.; Spowart, J.E.; Rowenhorst, D.J.; Aagesen, L.K.; Thorton, K.; Pollock, T.M. Fluid flow and defect formation in the three-dimensional dendritic structure of nickel-based single crystals. *Metall. Mater. Trans. A* **2012**, *43*, 369–380. [CrossRef]
2. Pollock, T.M.; Tin, S. Nickel-Based Superalloys for Advanced Turbine Engines: Chemistry, Microstructure, and Properties. *J. Propuls. Power* **2006**, *22*, 361–374. [CrossRef]
3. Osorio, W.R.; Goulart, P.R.; Santos, G.A.; Neto, C.M.; Garcia, A. Effect of dendritic arm spacing on mechanical properties and corrosion resistance of Al 9 Wt Pct Si and Zn 27 Wt Pct Al alloys. *Metall. Mater. Trans. A* **2006**, *37*, 2525–2538. [CrossRef]
4. Gabb, T.P.; Kantzos, P.T.; Palsa, B.; Telesman, J.; Gayda, J.; Sudbrack, C.K. *Fatigue Failure Modes of the Grain Size Transition Zone in a Dual Microstructure Disk*; John Wiley & Sons: New York, NY, USA, 2012; pp. 63–72.
5. Flageolet, B.; Yousfi, O.; Dahan, Y.; Villechaise, P.; Cormier, J. *Characterization of Microstructures Containing Abnormal Grain Growth Zones in Alloy 718*; John Wiley & Sons: New York, NY, USA, 2012; pp. 594–606.
6. Gabb, T.P.; Kantzos, P.T.; Gayda, J.; Sudbrack, C.K.; Palsa, B. Fatigue resistance of the grain size transition zone in a dual microstructure superalloy disk. *Int. J. Fatigue* **2011**, *33*, 414–426. [CrossRef]
7. Randle, V.; Coleman, M. Grain growth control in grain boundary engineered microstructures. *Mater. Sci. Forum* **2012**, *715–716*, 103–108. [CrossRef]
8. Holm, E.A.; Miodownik, M.A.; Rollett, A.D. On abnormal subgrain growth and the origin of recrystallization nuclei. *Acta Mater.* **2003**, *51*, 2701–2716. [CrossRef]
9. Fang, S.; Dong, Y.P.; Wang, S. The abnormal grain growth of P/M nickel-base superalloy: Strain storage and CSL boundaries. *Adv. Mater. Res.* **2015**, *1064*, 49–54. [CrossRef]
10. Watson, R.; Preuss, M.; Da Fonseca, J.Q.; Witulski, T.; Terlinde, G.; Buscher, M. Characterization of Abnormal Grain Coarsening in Alloy 718. *MATEC Web Conf.* **2014**, *14*, 1–5. [CrossRef]
11. Brons, J.G.; Thompson, G.B. A comparison of grain boundary evolution during grain growth in FCC metals. *Acta Mater.* **2013**, *61*, 3936–3944. [CrossRef]
12. Randle, V.; Booth, M.; Owen, G. Time evolution of sigma 3 annealing twins in secondary recrystallized nickel. *J. Microsc.* **2005**, *217*, 162–166.
13. Kazaryan, A.; Wang, Y.; Dregia, S.A.; Patton, B.R. Grain growth in anisotropic systems: Comparison of effects of energy and mobility. *Acta Mater.* **2002**, *50*, 2491–2502. [CrossRef]

14. Upmanyu, M.; Hassold, G.N.; Kazaryan, A.; Holm, E.A.; Wang, Y.; Patton, B.; Srolovitz, D.J. Boundary mobility and energy anisotropy effects on microstructural evolution during grain growth. *Int. Sci.* **2002**, *10*, 201–216.
15. Homma, H.; Hutchinson, B. Orientation dependence of secondary recrystallization in silicon-iron. *Acta Mater.* **2003**, *51*, 3795–3805. [CrossRef]
16. Lin, P.; Palumbo, G.; Harase, J.; Aust, K.T. Coincident site lattice (CSL) grain boundaries and goss texture development in Fe-3% Si Alloy. *Acta Mater.* **1996**, *44*, 4677–4683. [CrossRef]
17. Watanabe, T. An approach to grain boundary design for strong and ductile polycrystals. *J. Glob.* **1984**, *11*, 47–84.
18. Randle, V. Twinning-related grain boundary engineering. *Acta Mater.* **2004**, *52*, 4067–4081. [CrossRef]
19. Underwood, O.D.; Madison, J.D.; Martens, R.L.; Thompson, G.B.; Welsh, S.L.; Evans, J.L. An examination of abnormal grain growth in low strain nickel-200. *Metallogr. Microstruct. Anal.* **2016**, *5*, 302–312. [CrossRef]
20. Randle, V.; Horton, D. Grain growth phenomena in nickel. *Scr. Metall. Mater.* **1994**, *31*, 891–895. [CrossRef]
21. Hillert, M. On the theory of normal and abnormal grain growth. *Acta Metall.* **1965**, *13*, 227–238. [CrossRef]
22. Holm, E.A.; Hassold, G.N.; Miodownik, M.A. On misorientation distribution evolution during anisotropic grain growth. *Acta Mater.* **2001**, *49*, 2981–2991. [CrossRef]
23. Rollett, A.D.; Srolovitz, D.J.; Anderson, M.P. Simulation and theory of abnormal grain growth-anisotropic grain boundary energies and mobilities. *Acta Metall.* **1989**, *37*, 1227–1240. [CrossRef]
24. Rollett, A.D.; Mulins, W.W. On the growth of abnormal grains. *Scr. Mater.* **1997**, *36*, 975–980. [CrossRef]
25. Thompson, C.V.; Frost, H.J.; Spaepen, F. The relative rates of secondary and normal grain growth. *Acta Metall.* **1987**, *35*, 887–890. [CrossRef]
26. Sztwiertnia, K. Recrystallization textures and the concept of oriented growth revisited. *Mater. Lett.* **2014**, *123*, 41–43. [CrossRef]
27. Jensen, D.J. Growth rates and misorientation relationships between growing nuclei/grains and the surrounding deformed matrix during recrstalization. *Acta Metall. Mater.* **1995**, *43*, 4117–4129. [CrossRef]
28. Akhiani, H.; Nezakat, M.; Sonboli, A.; Szpunar, J. The origin of annealing texture in a cold-rolled incoloy 800H/Ht after different strain paths. *Mater. Sci. Eng.* **2014**, *619*, 334–344. [CrossRef]
29. Akhiani, H.; Nezakat, M.; Szpunar, J.A. Evolution of deformation and annealing textures in incoloy 800H/HT via different rolling paths and strains. *Mater. Sci. Eng.* **2014**, *614*, 250–263. [CrossRef]
30. Wilson, A.W.; Madison, J.D.; Spanos, G. Determining phase volume fraction in steels by electron backscattered diffraction. *Sci. Mater.* **2001**, *45*, 1335–1340. [CrossRef]
31. Sahay, S.S.; Malhotra, C.P.; Kolkhede, A.M. Accelerated grain growth behavior during cyclic annealing. *Acta Mater.* **2003**, *51*, 339–346. [CrossRef]
32. Decker, R.F.; Rush, A.I.; Dano, A.G.; Freeman, J.W. *Abnormal Grain Growth in Nickel-Base Heat-Resistant Alloys*; University of North Texas: Denton, TX, USA, 1957.
33. He, G.; Tan, L.; Liu, F.; Huang, L.; Huang, Z.; Jiang, L. Unraveling the formation mechanism of abnormally large grains in an advanced polycrystalline nickel base superalloy. *J. Alloys Compd.* **2017**, *718*, 405–413. [CrossRef]
34. He, G.; Liu, F.; Huang, L.; Huang, Z.; Jiang, L. Controlling grain size via dynamic recrystallization in an advanced polycrystalline nickel base superalloy. *J. Alloys Compd.* **2017**, *701*, 909–919. [CrossRef]
35. He, G.; Tan, L.; Liu, F.; Huang, L.; Huang, Z.; Jiang, L. Revealing the role of strain rate during multi-pass compression in an advanced polycrystalline nickel base superalloy. *Mater. Charact.* **2017**, *128*, 123–133. [CrossRef]
36. Cho, Y.K.; Yoon, D.Y.; Henry, M.F. The Effects of deformation and pre-heat treatment on abnormal grain growth in RENÉ 88 superalloy. *Metall. Mater. Trans. A* **2001**, *32*, 3077–3090. [CrossRef]
37. Bozzolo, N.; Agnoli, A.; Souai, N.; Bernacki, M.; Loge, R.E. Strain induced abnormal grain growth in nickel base superalloys. *Mater. Sci. Forum* **2013**, *753*, 321–324. [CrossRef]

16

Effect of Mo Addition on the Chemical Corrosion Process of SiMo Cast Iron

Marcin Stawarz [1,*] and Paweł M. Nuckowski [2]

[1] Department of Foundry Engineering, Silesian University of Technology, 7 Towarowa Street, 44-100 Gliwice, Poland

[2] Department of Engineering Materials and Biomaterials, Silesian University of Technology, 18A Konarskiego Street, 44-100 Gliwice, Poland; pawel.nuckowski@polsl.pl

* Correspondence: marcin.stawarz@polsl.pl

Abstract: The study was carried out to evaluate five SiMo cast iron grades and their resistance to chemical corrosion at elevated temperature. Corrosion tests were carried out under conditions of an actual cyclic operation of a retort coal-fired boiler. The duration of the study was 3840 h. The range of temperature changes during one cycle was in the range of 300–650 °C. Samples of SiMo cast iron with Si content at the level of 5% and variable Mo content in the range 0%–2.5% were used as the material for the study. The examined material was subjected to preliminary metallographic analysis using scanning microscopy and an Energy dispersive spectroscopy (EDS) system. The chemical composition was determined on the basis of a Leco spectrometer and a Leco carbon and sulfur analyzer. The examination of the oxide layer was carried out with the use of Scanning electron microscope (SEM), EDS, and X-ray diffraction (XRD) methods. It was discovered that, in the analyzed alloys, oxide layers consisting of Fe_2O_3, Fe_3O_4, SO_2, and Fe_2SiO_4 were formed. The analyzed oxide layers were characterized by high adhesion to the substrate material, and their total thickness was about 20 μm.

Keywords: chemical corrosion; SiMo cast iron; fayalite; hematite; magnetite; maghemite; sulfur dioxide

---

1. Introduction

Corrosive behavior of SiMo cast iron in the air and flue gases was presented in the works [1–5]. When we subject pure iron to the corrosion process at elevated temperatures and in the ambient atmospheric air, a multilayer oxide structure composed of FeO, Fe_3O_4, and Fe_2O_3 is formed [5]. For ductile cast iron, an oxide layer is formed, located both in the material and in the surface layer as a result of migration of Fe atoms [5]. After introducing an alloying element in the form of Si into cast iron, a SiO_2 compound is formed at the metal–oxide layer point of contact, which constitutes a barrier to further oxidation processes [5]. Simultaneously, SiO_2 can react with O, Fe, and FeO. The result of these reactions may be the formation of fayalite, Fe_2SiO_4 [6–8]. In the paper [9], the author writes that the oxide layer on the surface of SiMo cast iron is composed of the following sub-layers situated from the outside to the inside of the material: Fe_2O_3, Fe_3O_4, FeO, FeO + Fe_2SiO_4.

The oxide layer adheres well to the base material and the inner layer consisting of FeO + Fe_2SiO_4 [9]. The higher the silicon content in the base material, the faster the oxide layer forms. A number of studies concerning the corrosion resistance of SiMo cast iron focus on a relatively short time of exposure to oxidation (500 h on average) [10]. These studies are conducted mainly in terms of the use of SiMo cast iron in automotive castings, as described by Rouczka [11] and many other authors [12–17]. SiMo cast iron is an increasingly popular material, and research on this material is also conducted with a focus on optimizing the manufacturing process. In their work, Guzik et al. [18] write about the method of introducing two flexible hoses with the diameter of Ø 9 mm; one filled with a FeSi + Mg mixture, and the other with a graphitizing modifier for the treatment drum ladle. Guzik et al. [18] describe it as a new method of secondary treatment of ferritic cast iron production of SiMo type. This method can be used for the production of ductile iron melted in an induction furnace [18,19].

SiMo cast iron can also be successfully used in other areas of industry: exhaust parts for combustion engines, turbocharger housings and rotors, gas turbine components, molds for the glass industry, molds for aluminum alloys, zinc, forging dies, heat treatment furnace components, aluminum melting furnace components, and waste incineration furnaces. This happens wherever elevated operating temperatures and gases resulting from the combustion, e.g., of solid fuels are involved. A good example of such a system is a coal-fired retort furnace. Nyashina and her team write about the problems related to the emission of pollutants during the combustion process [20]. Released into the atmosphere with exhaust gases, nitrogen oxides (mainly NO and NO_2) are the main reason why photochemical smog appears, which reaches the stratosphere to act as a catalyst for ozone layer depletion. Rapid oxidation of NOx and SOx and their interaction with water vapor in the atmosphere generates tiny droplets of sulfuric (H_2SO_4) and nitric (HNO_3) acids [20]. Sulfuric acid causes significant losses in the ecosystem, which has been mentioned by many authors [21]. It is important to optimize the combustion process of solid fuels in boilers by improving the materials from which these boilers are built. For the above reasons, in this work, studies of resistance to chemical corrosion of SiMo cast iron were carried out during actual operation of a retort boiler. The duration of the study was 3840 hours. To date, the corrosion resistance of SiMo cast iron during the operation of a retort boiler has not been described in the literature. Due to its properties, it can be successfully used for manufacturing furnace elements fired with solid, liquid, or gaseous fuels.

2. Methods and Materials

Experimental melts were conducted in an induction furnace (PI25, ELKON Sp. z o.o., Rybnik, Poland) with medium frequency and a capacity of 25 kg. The charge consisted of steel scrap with low sulphur content. Other ingredients added during the melting were ferrosilicon FeSi75, synthetic graphite of carbon content above 99.35%, and FeMo65-rich alloy. The spheroidization process of cast iron was conducted at the bottom of the ladle after covering the nodulizing agent with pieces of steel scrap. The magnesium-rich alloy used in the studies was FeSiMg5RE. The studies were carried out under conditions of cyclic temperature changes in the range of 300–650 °C. The duration of the study was 3840 h. The full cycle time of heating and cooling was 12 min 30 s. A total of 18,432 full cycles of heat load were carried out. The length of the test cycle was selected so that the samples would reach the assumed minimum and maximum temperatures. Temperature measurement of the samples was performed with a NiCr-Ni thermocouple, with no recording of temperature changes in time, and the measurement of surface temperature of the samples was performed to determine the minimum and maximum temperature for the test cycle. The tests were carried out under conditions of a reverberatory furnace (the scheme of a single retort stoker is shown in Figure 1).

The fuel used was bituminous coal with a calorific value of 26–28 MJ/kg, a combustion heat of 29 MJ/kg, a granulation of 5–25 mm, a humidity of <10%, a maximum ash content of 7%, and a maximum sulfur content of 0.6%. The fuel used was certified by Główny Instytut Górnictwa (Central Mining Institute).

In the studies, samples of SiMo cast iron with Si content of 5% and Mo content of 0%–2.5% were used. The chemical composition was determined on the basis of a Leco spectrometer (Model No 607-500, Leco Corporation, 3000 Lakeview Ave, St. Joseph, MI, USA) and a CS-125 Leco carbon and sulfur analyzer (Leco Corporation, 3000 Lakeview Ave, St. Joseph, MI, USA). The chemical composition of the tested samples is presented in Table 1.

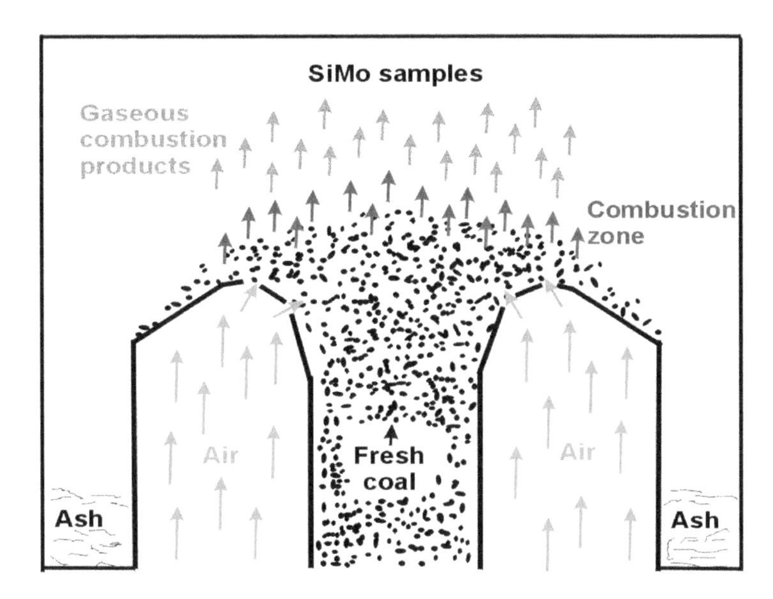

Figure 1. Scheme of a single retort with (yellow) SiMo samples placed above the stoker.

Table 1. Chemical composition of the tested SiMo cast iron.

Melt Number	Chemical Composition, % of Weight						
	C	Si	Mo	P	S	Mg	Fe (Balance)
SiMo 1	3.02	5.03	0.01	0.021	0.009	0.032	91.878
SiMo 2	2.94	5.14	0.47	0.020	0.005	0.029	91.396
SiMo 3	3.04	4.94	1.09	0.022	0.005	0.031	90.872
SiMo 4	2.71	5.17	1.92	0.018	0.009	0.062	90.111
SiMo 5	2.74	4.42	2.51	0.022	0.007	0.033	90.268

In order to determine the phase composition of the studied material, X-ray diffraction analyses were carried out with the use of an X'Pert Pro multipurpose x-ray diffractometer by Panalytical (Almelo, The Netherlands). The measurements were conducted utilizing filtered radiation of a cobalt anode lamp ($\lambda K\alpha = 0.179$ nm) as well as a PIXcel 3D detector on the diffracted beam axis. The diffraction lines were recorded in the Bragg–Brentano geometry in the angular scope of 10°–120° (2θ), with the step of 0.026° and the step time of 100 s. Furthermore, to obtain more precise information from the surface oxide layer, grazing incidence diffraction (GID) geometry with a proportional detector on the diffracted beam axis was used. In this geometry, a primary X-ray beam was set at a constant, low angle (1.5°) related to the sample plane, which affected the corresponding slope of diffraction vectors related to the normal to surface. This allowed us to obtain during the measurement a constant penetration depth of the X-ray beam, limited mainly to the surface oxide layer. The analysis of the obtained diffraction patterns was made in the Panalytical High Score Plus software (ver.: 3.0e), with the dedicated Panalytical Inorganic Crystal Structure Database (PAN-ICSD).

The analysis of the structure and the chemical composition was performed on a Phenom ProX scanning microscope (Phenom-World Eindhoven, Noord-Brabant, Netherlands) equipped with an energy-dispersive X-ray spectrometer (EDS).

3. Results

3.1. SEM Analysis

Figure 2 shows photos of SiMo cast iron samples after the chemical corrosion resistance test cycle. The microstructure of the cast iron consisted of a ferritic matrix, graphite nodules, molybdenum carbide (Mo_2C), a carburized zone (Figure 2e), and the passive layer and the loose oxide layer on the top surface of the samples. Microstructure components are highlighted in Figure 2.

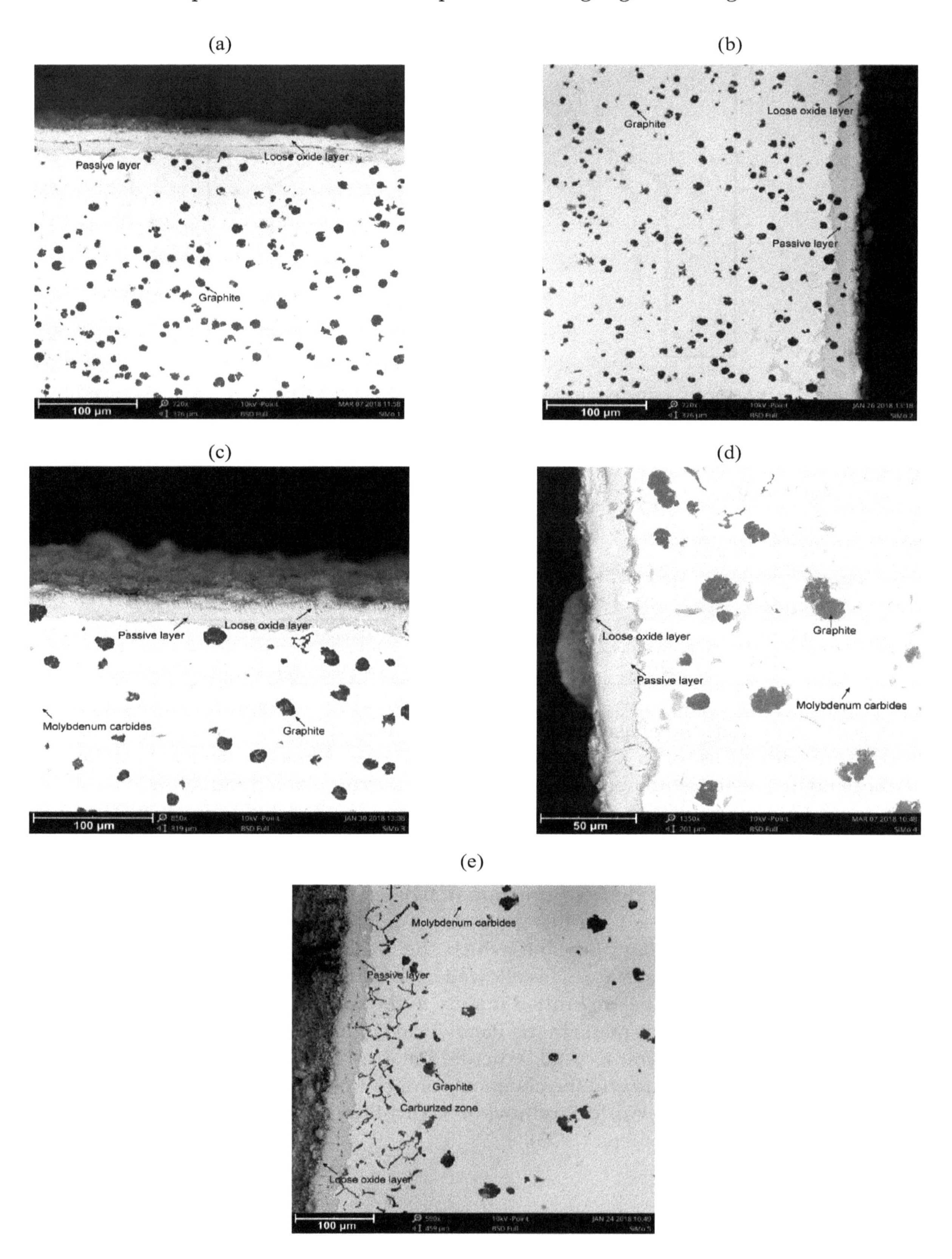

Figure 2. Microstructure of SiMo cast iron after the corrosion test cycle. SiMo cast iron with addition of: (**a**) 0.01% Mo, (**b**) 0.47% Mo, (**c**) 1.09% Mo, (**d**) 1.92% Mo, (**e**) 2.51% Mo. SEM.

In all cases (SiMo samples 1–5), the thickness of passive layers is around 10 μm. For elevated molybdenum content (Figure 2e, SiMo melt 5, 2.51% Mo) in the near-surface layer, the carburized zone in the form of black inclusions distributed in the vicinity of Mo_2C carbide precipitates is clearly visible. All the cases considered are characterized by a cohesive passive layer tightly adhering to the sample. No cracks nor defects in the passive layer were observed, even after the process of preparing metallographic sections (cutting, grinding, and polishing).

3.2. EDS Analysis

Figures 3 and 4 show the collective results of metallographic studies for selected alloys using the EDS system. The presented results indicate the presence of a loose oxide layer (blue), which is adjacent to the passive layer. The passive layer is an area also marked in blue, where the area has an increased silicon content (intense yellow bands on the maps—see Figure 3e, Figure 4e, Figure 5e). The increased Si content in the passive layer results from the diffusion of iron atoms from this layer to the loose oxide layer, which makes the area richer in Si. Penetrating oxygen reaches the area enriched with silicon and forms compounds with it (e.g., SiO_2) creating a tight barrier to the propagation of corrosive phenomena. Of course, a SiO_2 compound can react with Fe, O, and FeO, forming a fayalite Fe_2SiO_4 [8,17].

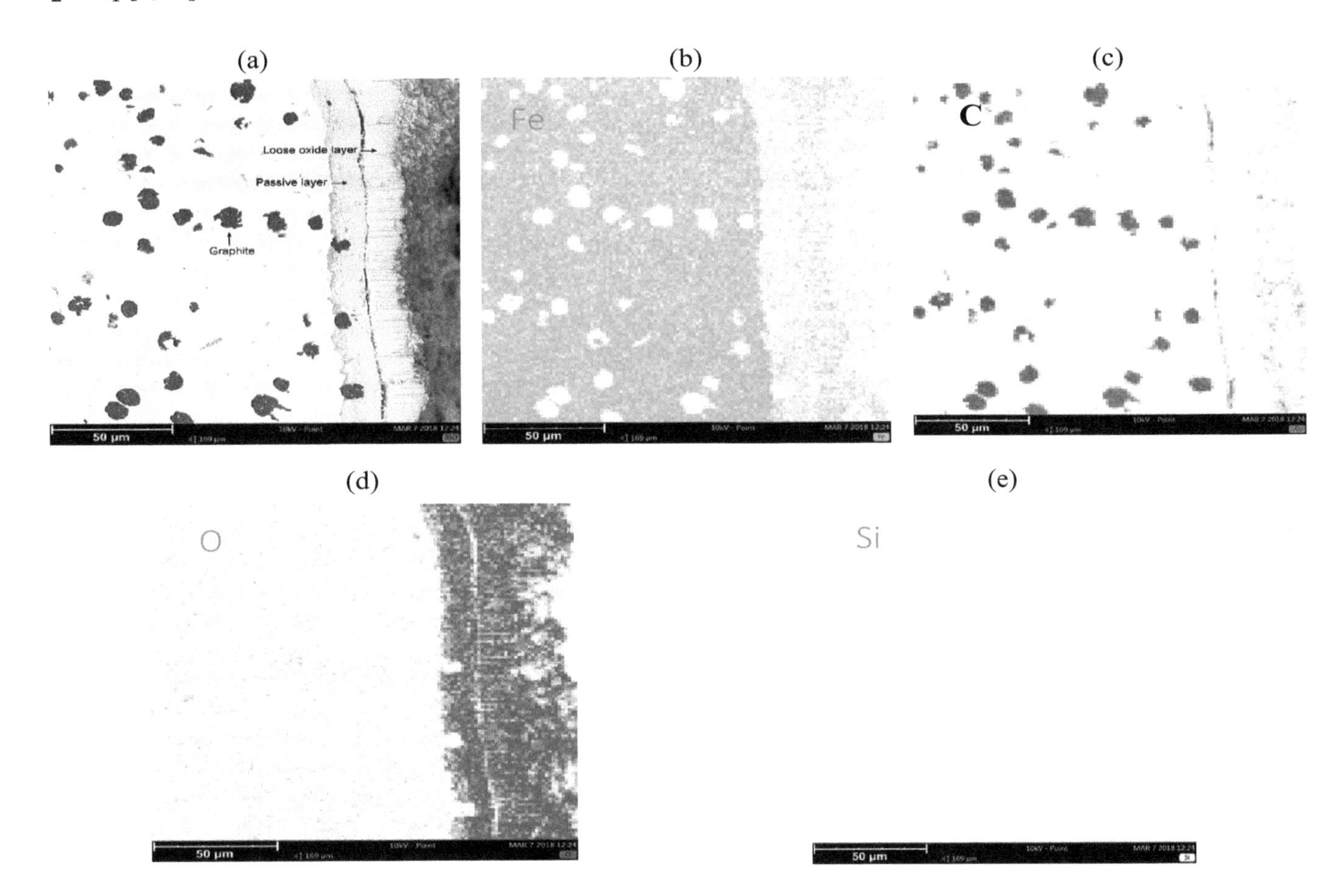

Figure 3. Microstructure of SiMo cast iron, 0.01% Mo (**a**). Elements decomposition maps. Map for (**b**) Fe, (**c**) C, (**d**) O, and (**e**) Si.

The characteristic phenomenon of these alloys is their ability to perform so-called "self-healing". Removing the passive layer creates a new one in its place. The average thickness of the loose oxide layer was 10 μm for the tested samples, while the thickness of the passive layer was also around 10 μm. Of course, the thickness of this layer depends on the local conditions on the surface of the sample exposed to the corrosive agents.

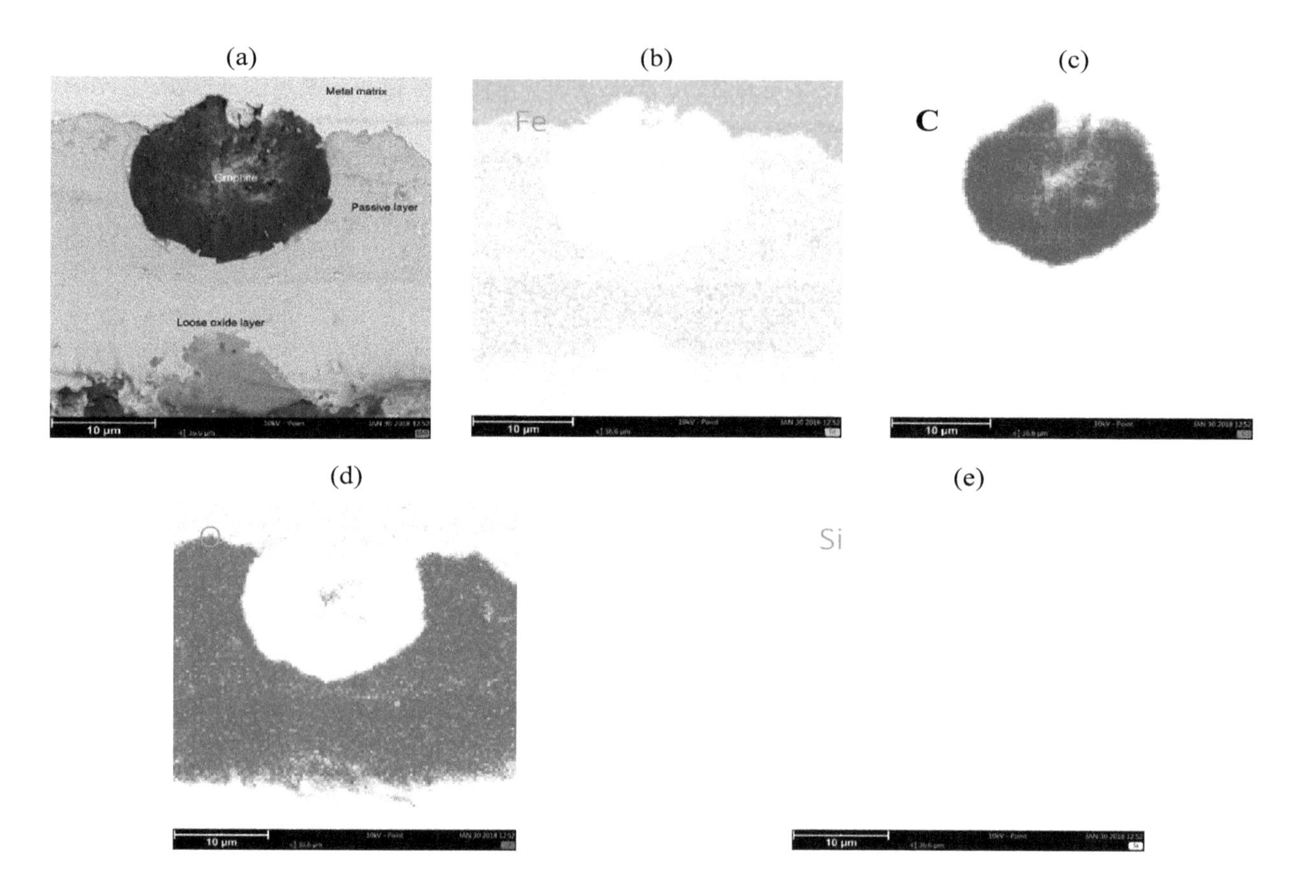

Figure 4. Microstructure of SiMo 3 cast iron, 1.09% Mo (**a**). Map for (**b**) Fe, (**c**) C, (**d**) O, and (**e**) Si.

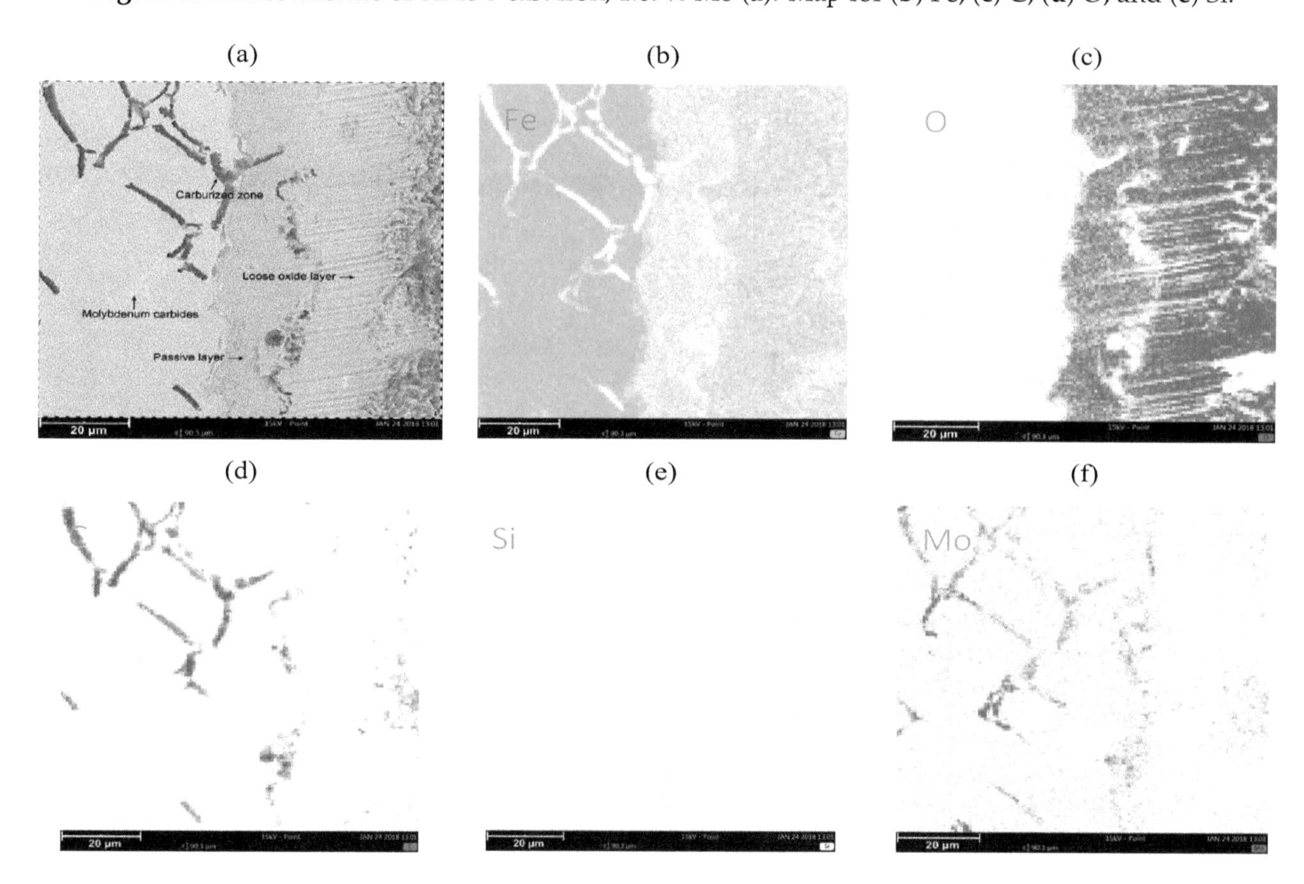

Figure 5. View of the carburized layer. SiMo sample 5. 2.51% Mo (**a**). Map for (**b**) Fe, (**c**) O, (**d**) C, (**e**) Si, and (**f**) Mo.

Figure 5 shows the analysis of the sample area (for the SiMo melt 5), where a significant degree of carburation of the casting material layer was observed. The layer is placed under the passive layer. This can be explained by the high concentration of carbon in the corrosive atmosphere of the furnace, and the penetration of carbon atoms through the passive layer towards the center of the casting.

3.3. XRD Analysis

The above SEM and EDS results were complemented with X-ray diffraction. Two measurements of the surface area of a selected sample were made (Figure 6), one with Bragg–Brentano geometry and the other one using the grazing incidence diffraction (GID) geometry. The GID geometry allows us to limit the penetration of the X-ray beam, so that diffractive information can be obtained mainly from the surface layer (in this particular case, the evidence is the almost complete disappearance of the diffraction lines from the substrate). In the figure below Figure 7, two diffractograms are superimposed on each other. The blue diffractogram was obtained using Bragg–Brentano geometry and the red diffractogram using GID geometry at a 1.5-degree tilt of the primary beam in relation to the measurement plane.

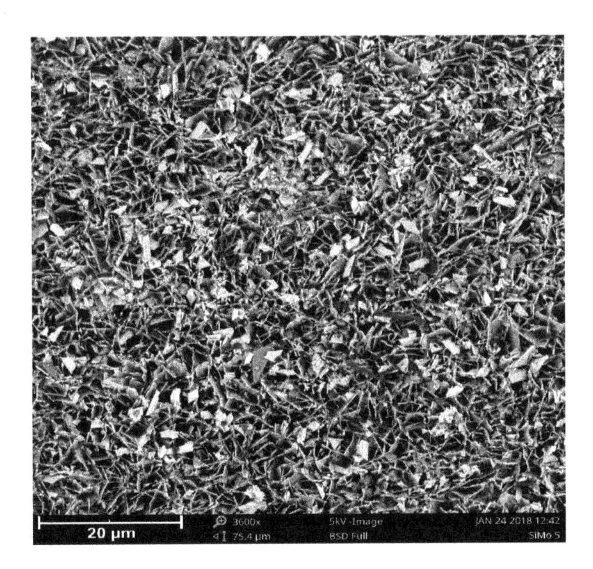

Figure 6. View of the loose oxide layer on a SiMo 5 sample. SEM.

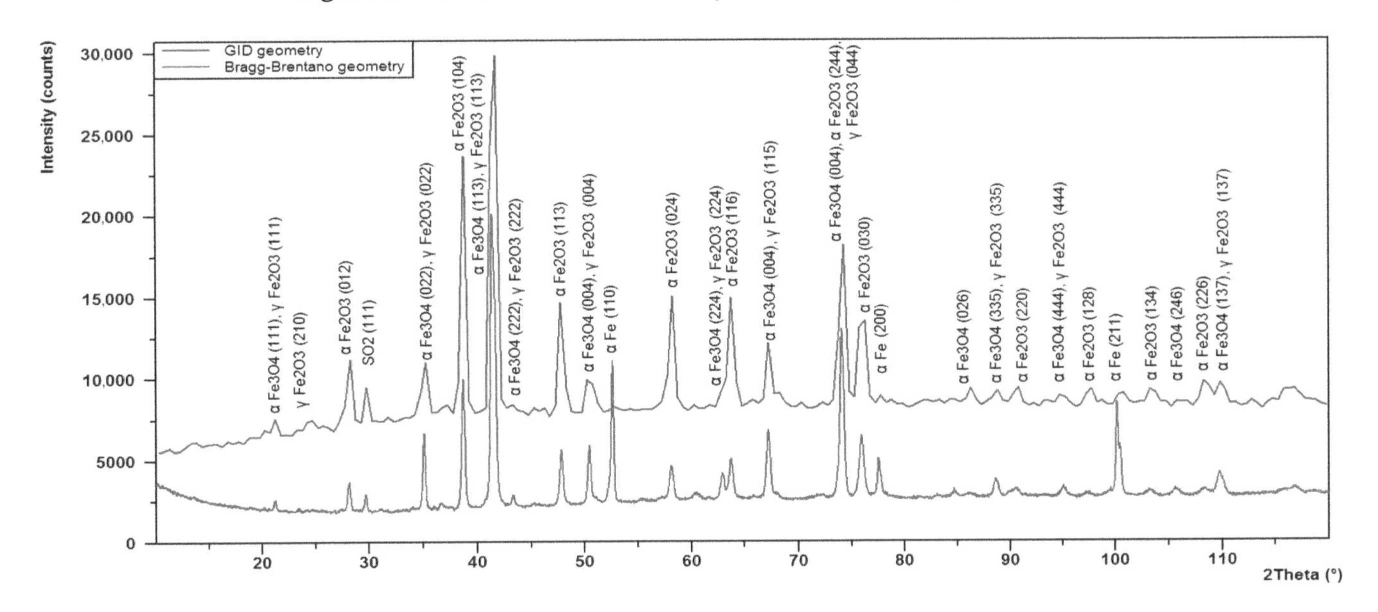

Figure 7. Diffractogram from a sample. Blue chart, Bragg–Brentano geometry; red chart, grazing incidence diffraction (GID) geometry. SiMo 3. 1.09% Mo.

From the results of the analysis presented in Figure 7, the following components of the surface oxide layer were identified. The compounds identified were, among others: α - Fe_2O_3 (hematite) with a hexagonal lattice R-3c (98-002-2505), α - Fe_3O_4 (magnetite) with a cubic lattice Fd-3m (98-026-3007), γ - Fe_2O_3 (maghemite) with a cubic lattice Fd-3m (98-017-2905), and a small share of SO_2 (an orthorhombic lattice).

4. Discussion

The presented photos of SiMo cast iron samples (Figure 2), after a series of chemical corrosion resistance tests, clearly show the microstructure of the cast iron with the passive layer and the loose oxide layer on the surface of the samples. The obtained results are analogous to the results presented by M. Ekström in his paper [5]. The results obtained by M. Ekström were obtained under different test conditions in a flue gas atmosphere. For all cases analyzed in the study, the thickness of the oxide and passive layers oscillates around the level of 20 μm. For elevated molybdenum content (SiMo melt 5, 2.51% Mo) in the near-surface layer, the carburized zone, in the form of black inclusions distributed in the vicinity of Mo_2C carbide precipitates, is clearly visible. The layer is placed under the passive layer, which is clearly visible in Figure 5. We found no description of a similar case in the available literature. The obtained effect can be explained by the high concentration of carbon in the corrosive atmosphere of the furnace; penetration of carbon atoms through the passive layer in the direction of the casting center, combined with the high concentration of molybdenum in the casting, resulted in the accumulation of carbon atoms around the precipitates rich in molybdenum (Mo_2C carbides).

The results obtained from the XRD analysis allowed for the following components of the surface oxide layer to be described, among others: α - Fe_2O_3 (hematite), α - Fe_3O_4 (magnetite), γ - Fe_2O_3 (maghemite), and SO_2 [22–24].

The X-ray diffraction patterns of magnetite and maghemite are very similar. This is related to their similar structures. Both of these two oxide phases crystallize in the cubic system and their lattice parameters are very close. For this reason, it is difficult to differentiate these structures. However, some works [25,26] report that the maghemite phase gives two additional small diffraction lines at 23.77° (210) and 26.10° (211). One of these, line (210), was identified even on a diffractogram obtained in Bragg–Brentano geometry. The formed magnetite and maghemite oxide layers do not show the much higher intensity of line (113) in relation to the highest hematite line (104). This result can be explained by the similar crystal growth of these oxide phases. Also, some authors report that the formation of maghemite (γ - Fe_2O_3) is a result of oxidation of the magnetite (α - Fe_3O_4) [27,28]. One of the variables deciding the quantitative share of hematite, magnetite, and maghemite is the range of oxidation temperature described by M. Marciuš et al. [29].

5. Conclusions

Based on the conducted study, it can be stated that SiMo cast iron is fully resistant to chemical corrosion during retort furnace operations. All SiMo cast iron samples were characterized by a cohesive oxidized layer, consisting of a passive layer on the casting material side and an oxide surface layer.

The two layers adhered quite well to one another. No cracks nor defects in any of the elements of the oxidized layer were observed. The surface oxide layer was found to consist of the following compounds: Fe_2O_3, Fe_3O_4, and SO_2.

For the sample with increased Mo content, a significant carburization of the near-surface layer of the sample was observed, especially in the areas adjacent to molybdenum carbide. The carburized edge zone of the sample did not affect the corrosion resistance of SiMo cast iron.

The areas strongly enriched with silicon are the fayalite Fe_2SiO_4 resulting from the reaction of SiO_2 with O, Fe, and FeO.

Due to the relatively low operating temperature range, we suggest that the Mo content in the alloy be reduced to the range of 0%–0.5% Mo.

Author Contributions: research concept, M.S.; conducting the experiment, M.S.; microstructure studies, M.S. and P.M.N.; diffractive analysis, P.M.N.; writing—original manuscript preparation, M.S.; writing—review and edition, M.S. and P.M.N. All authors have read and agreed to the published version of the manuscript.

Acknowledgments: This publication was financed by the statutory subsidy of the Faculty of Mechanical Engineering of the Silesian University of Technology in 2019.

References

1. Tholence, F.; Norell, M. High-Temperature Corrosion of Cast Irons and Cast Steels in Dry Air. *Mater. Sci. Forum* **2001**, *369*, 197–204. [CrossRef]
2. Tholence, F.; Norell, M. AES characterization of oxide grains formed on ductile cast irons in exhaust environments. *Surf. Interface Anal.* **2002**, *34*, 535–539. [CrossRef]
3. Tholence, F.; Norell, M. Nitride precipitation during high temperature corrosion of ductile cast irons in synthetic exhaust gases. *J. Phys. Chem. Solids* **2005**, *66*, 530–534. [CrossRef]
4. Tholence, F.; Norell, M. High Temperature Corrosion of Cast Alloys in Exhaust Environments I-Ductile Cast Irons. *Oxid. Met.* **2007**, *69*, 13–36. [CrossRef]
5. Ekström, M.; Szakálos, P.; Jonsson, S. Influence of Cr and Ni on High-Temperature Corrosion Behavior of Ferritic Ductile Cast Iron in Air and Exhaust Gases. *Oxid. Met.* **2013**, *80*, 455–466. [CrossRef]
6. Yang, Y.L.; Cao, Z.Y.; Qi, Y.; Liu, Y.B. The Study on Oxidation Resistance Properties of Ductile Cast Irons for Exhaust Manifold at High Temperatures. *Adv. Mater. Res.* **2010**, *97*, 530–533. [CrossRef]
7. Choe, K.H.; Lee, S.M.; Lee, K.W. High Temperature Oxidation Behavior of Si-Mo Ferritic Ductile Cast Iron. *Mater. Sci. Forum* **2010**, *654*, 542–545. [CrossRef]
8. Stawarz, M.; Janerka, K.; Dojka, M. Selected Phenomena of the In-Mold Nodularization Process of Cast Iron That Influence the Quality of Cast Machine Parts. *Processes* **2017**, *5*, 68. [CrossRef]
9. Henderieckx, G.D. *Silicon Cast Iron*; Gietech BV: Terneuzen, The Netherlands, 2009.
10. Brady, M.P.; Muralidharan, G.; Leonard, D.; Haynes, J.A.; Weldon, R.G.; England, R.D. Long-Term Oxidation of Candidate Cast Iron and Stainless Steel Exhaust System Alloys from 650 to 800 °C in Air with Water Vapor. *Oxid. Met.* **2014**, *82*, 359–381. [CrossRef]
11. Roucka, J.; Abramova, E.; Kana, V. Properties of type SiMo ductile irons at high temperatures. *Arch. Metall. Mater.* **2018**, *63*, 601–607.
12. Zeytin, H.K.; Kubilay, C.; Aydin, H.; Ebrinc, A.A.; Aydemir, B. Effect of microstructure on exhaust manifold cracks produced from SiMo ductile iron. *J. Iron Steel Res. Int.* **2009**, *16*, 32–36. [CrossRef]
13. Mohammadi, F.; Alfantazi, A. Corrosion of ductile iron exhaust brake housing in heavy diesel engines. *Eng. Fail. Anal.* **2013**, *31*, 248–254. [CrossRef]
14. Cygan, B.; Stawarz, M.; Jezierski, J. Heat treatment of the SiMo iron castings—Case study in the automotive foundry. *Arch. Foundry Eng.* **2018**, *18*, 103–109.
15. Li, D. Mixed graphite cast iron for automotive exhaust component applications. *China Foundry* **2017**, *14*, 519–524. [CrossRef]
16. Unkić, F.; Glavaš, Z.; Terzić, K. The influence of elevated temperatures on microstructure of cast irons for automotive engine thermo-mechanical loaded parts. *Mater. Geoenvironment* **2009**, *56*, 9–23.
17. Ibrahim, M.; Nofal, A.; Mourad, M.M. Microstructure and Hot Oxidation Resistance of SiMo Ductile Cast Irons Containing Si-Mo-Al. *Met. Mater. Trans. A* **2016**, *48*, 1149–1157. [CrossRef]
18. Guzik, E.; Kopyciński, D.; Wierzchowski, D. Manufacturing of Ferritic Low-Silicon and Molybdenum Ductile Cast Iron with the Innovative 2PE-9 Technique. *Arch. Met. Mater.* **2014**, *59*, 687–691. [CrossRef]
19. Vaško, A.; Belan, J.; Tillová, E. Static and Dynamic Mechanical Properties of Nodular Cast Irons. *Arch. Metall. Mater.* **2019**, *64*, 185–190.
20. Nyashina, G.S.; Kuznetsov, G.V.; Strizhak, P. Energy efficiency and environmental aspects of the combustion of coal-water slurries with and without petrochemicals. *J. Clean. Prod.* **2018**, *172*, 1730–1738. [CrossRef]
21. Bhadra, B.N.; Jhung, S.H. Oxidative desulfurization and denitrogenation of fuels using metal-organic framework-based/-derived catalysts. *Appl. Catal. B Environ.* **2019**, *259*, 118021. [CrossRef]
22. Avvakumov, E.G.; Golovin, A.V.; Paukshtis, E.A.; Ivanov, V.P.; Kolomiichuk, V.N.; Kustova, G.N.; Burgina, E.B.; Litvak, G.S.; Cherepanova, S.V.; Tsybulya, S.V.; et al. *Golden Book of Phase Transitions*, 1st ed.; Tomaszewski P.E.: Wroclaw, Poland, 2002; pp. 1–123.

23. Ju, S.; Cai, T.; Lu, H.-S.; Gong, C.-D. Pressure-Induced Crystal Structure and Spin-State Transitions in Magnetite (Fe3O4). *J. Am. Chem. Soc.* **2012**, *134*, 13780–13786. [CrossRef] [PubMed]
24. Post, B.; Schwartz, R.S.; Fankuchen, I. The crystal structure of sulfur dioxide. *Acta Crystallogr.* **1952**, *5*, 372–374. [CrossRef]
25. Kim, W.; Suh, C.-Y.; Cho, S.-W.; Roh, K.-M.; Kwon, H.; Song, K.; Shon, I.-J. A new method for the identificationand quantification of magnetite-maghemite mixtureusing conventional X-ray diffraction technique. *Talanta* **2012**, *94*, 348–352. [CrossRef] [PubMed]
26. Darezereshki, E.; Ranjbar, M.; Bakhtiari, F. One-step synthesis of maghemite (γ-Fe2O3) nano-particles by wet chemical method. *J. Alloy. Compd.* **2010**, *502*, 257–260. [CrossRef]
27. Long, R.; Zhou, S.; Wiley, B.J.; Xiong, Y. Oxidative etching for controlled synthesis of metal nanocrystals: Atomic addition and subtraction. *Chem. Soc. Rev.* **2014**, *43*, 6288–6310. [CrossRef]
28. Li, C.; Wei, Y.; Liivat, A.; Zhu, Y.; Zhu, J. Microwave-solvothermal synthesis of Fe3O4 magnetic nanoparticles. *Mater. Lett.* **2013**, *107*, 23–26. [CrossRef]
29. Marciuš, M.; Ristić, M.; Ivanda, M.; Music, S. Formation of Iron Oxides by Surface Oxidation of Iron Plate. *Croat. Chem. Acta* **2012**, *85*, 117–124. [CrossRef]

17

Deformation Behavior of High-Mn TWIP Steels Processed by Warm-to-Hot Working

Vladimir Torganchuk [1], Aleksandr M. Glezer [2,*], Andrey Belyakov [1] and Rustam Kaibyshev [1]

1 Belgorod State University, Belgorod 308015, Russia; torganchuk@bsu.edu.ru (V.T.); belyakov@bsu.edu.ru (A.B.); rustam_kaibyshev@bsu.edu.ru (R.K.)
2 National University of Science & Technology (MISIS), Moscow 119049, Russia
* Correspondence: a.glezer@mail.ru

Abstract: The deformation behavior of 18%Mn TWIP steels (upon tensile tests) subjected to warm-to-hot rolling was analyzed in terms of Ludwigson-type relationship, i.e., $\sigma = K_1 \cdot \varepsilon^{n1} + \exp(K_2 - n_2 \cdot \varepsilon)$. Parameters of K_i and n_i depend on material and processing conditions and can be expressed by unique functions of inverse temperature. A decrease in the rolling temperature from 1373 K to 773 K results in a decrease in K_1 concurrently with n_1. Correspondingly, true stress approached a level of about 1750 MPa during tensile tests, irrespective of the previous warm-to-hot rolling conditions. On the other hand, an increase in both K_2 and n_2 with a decrease in the rolling temperature corresponds to an almost threefold increase in the yield strength and threefold shortening of the stage of transient plastic flow, which governs the duration of strain hardening and, therefore, manages plasticity. The change in deformation behavior with variation in the rolling temperature is associated with the effect of the processing conditions on the dislocation substructure, which, in turn, depends on the development of dynamic recovery and recrystallization during warm-to-hot rolling.

Keywords: high-Mn steel; deformation twinning; dynamic recrystallization; grain refinement; work hardening

---

1. Introduction

High-manganese austenitic steels with low stacking fault energy (SFE) are currently considered as promising materials for various structural/engineering applications because of their outstanding mechanical properties [1–3]. Owing to their low SFE, these steels are highly susceptible to deformation twinning, which results in the twinning-induced plasticity (TWIP) effect. Austenitic TWIP steels are characterized by pronounced strain hardening, which retards the strain localization and cracking during plastic deformation and, therefore, provides a beneficial combination of high strength with high ductility [1]. Deformation twinning, therefore, is the most crucial deformation mechanism governing the mechanical properties of high-Mn TWIP austenitic steels [4]. The deformation twins appear as bundles of closely spaced twins with thickness of tens nanometers, crossing over the original grains [5,6]. The deformation twins prevent the dislocation motion and promote an increase in dislocation density, leading to strain hardening. Frequent deformation twinning develops in steels with SFE in the range of approx. 20 mJ/m^2 to 50 mJ/m^2, which can be adjusted by manganese and carbon content [4].

The exact values of mechanical properties of austenitic steels, e.g., yield strength, ultimate tensile strength, total elongation, etc., depends on processing conditions. Hot rolling is frequently used as a processing technology for various structural steels and alloys. Final mechanical properties of processed steels and alloys depend on their microstructures that develop during hot working. Metallic materials with low SFE like high-Mn austenitic steels experience discontinuous dynamic recrystallization (DRX) during hot plastic deformation [7]. The developed microstructures depend on the deformation temperature and/or strain rate. Namely, the DRX grain size decreases with a decrease in temperature and/or an increase in strain rate and can be expressed by a power law function of temperature compensated strain rate (Z) [8]. A decrease in the deformation temperature to warm deformation conditions results in a change in the DRX mechanism from discontinuous to continuous, leading to a decrease in the grain size exponent in the relationship between the grain size and the deformation conditions, although this relationship remains qualitatively the same as that for hot working conditions [9]. The grain refinement with an increase in Z is accompanied with an increase in the dislocation density in DRX microstructures, irrespective of the DRX mechanisms [10]. Thus, the yield strength of the warm to hot worked semi-products can be evaluated by using various structural parameters. This approach has been successfully applied for strength evaluation of a range of structural steels and alloys, including high-Mn TWIP steels subjected to various thermo-mechanical treatments [5,6,10–12]. On the other hand, the effect of processing conditions on the deformation behavior of high-Mn TWIP steels has not been qualitatively evaluated, although, particularly for these steels, the deformation behavior is one of the most important properties, which manages the practical applications of the steels. It should be noted that the stress-strain behavior of austenitic steels with low SFE cannot be described by any well elaborated models like Hollomon or Swift equations, especially, at relatively small strains because of exceptional strain hardening [13]. Ludwigson modified the Hollomon-type relationship with an additional term to compensate the large difference between experimental and predicted flow stresses at small strains for such metals and alloys [14]. In spite of certain achievements in the application of the Ludwigson-type equation for the stress-strain behavior prediction, the selection of suitable parameters in this equation is still arbitrary in many ways. The aim of the present study, therefore, is to obtain the relationships between the processing conditions, the developed microstructures, and the stress-strain equation parameters in order to predict the tensile deformation behavior of advanced high-Mn TWIP steels processed by warm-to-hot rolling.

2. Materials and Methods

Two steels, Fe-18%Mn-0.4%C and Fe-18%Mn-0.6%C, have been selected in the present study as typical representatives of high-Mn TWIP steels. The steel melts were annealed at 1423 K, followed by hot rolling with about 60% reduction. The steels were characterized by uniform microstructures consisting of equiaxed grains with average sizes of 60 μm (18Mn-0.4C) and 50 μm (18Mn-0.6C). The steels were subjected to plate rolling at various temperatures from 773 K to 1373 K to a total rolling reduction of 60%. After each 10% rolling reductions, the samples were re-heated to the designated rolling temperature. Structural investigations were carried out using a Quanta 600 scanning electron microscope (SEM), equipped with an electron backscattering diffraction (EBSD) analyzer incorporating orientation imaging microscopy (OIM). The SEM samples were electro-polished at a voltage of 20 V at room temperature using an electrolyte containing 10% perchloric acid and 90% acetic acid. The OIM images were subjected to a clean-up procedure, setting the minimal confidence index of 0.1. The tensile tests were carried out using Instron 5882 testing machine with tensile specimens with a gauge length of 12 mm and a cross section of 3×1.5 mm^2 at an initial strain rate of 10^{-3} s^{-1}. The tensile axis was parallel to the rolling axis.

3. Results and Discussion

3.1. Developed Microstructures

Typical deformation microstructures that develop in the high-Mn steels during warm to hot rolling are shown in Figure 1. The mechanisms of microstructure evolution operating in austenitic steels during warm to hot working and the developed microstructures have been considered in detail in previous studies [10]. The deformation microstructures in the present high-Mn steels subjected to warm to hot rolling at temperatures of 773–1323 K can be briefly characterized here as follows. The temperature range above 1073 K corresponds to hot deformation conditions. Therefore, the deformation microstructures evolved during deformation in this temperature range result from the development of discontinuous DRX. The uniform microstructures consisting of almost equiaxed grains with numerous annealing twins are clearly seen in the samples hot rolled at temperatures above 1073 K (Figure 1a,b). The transverse DRX grain size decreases from 50–80 μm to 5–10 μm with a decrease in the rolling temperature from 1323 K to 1073 K.

In contrast, DRX hardly develops during warm rolling at temperatures below 1073 K. The deformation microstructures composed of flattened original grains evolve during warm rolling (Figure 1c,d). It is worth noting that the transverse grain size in the deformation microstructures developed during warm rolling does not remarkably depend on the rolling temperature. Relatively low deformation temperature suppresses discontinuous DRX. In this case, the structural changes are controlled by dynamic recovery. Under conditions of warm working, continuous DRX, which is assisted by dynamic recovery, can be expected after sufficiently large strains [7]. The present steels, however, are characterized by low SFE of 20–30 mJ/m^2 [4]. Such a low SFE makes the dislocation rearrangements during plastic deformation difficult and slows down the recovery kinetics. Therefore, 60% rolling reduction as applied in the present study is not enough for continuous DRX development in high-Mn steels. The final grain size, therefore, seems to be dependent on the original grain size and the total rolling reduction.

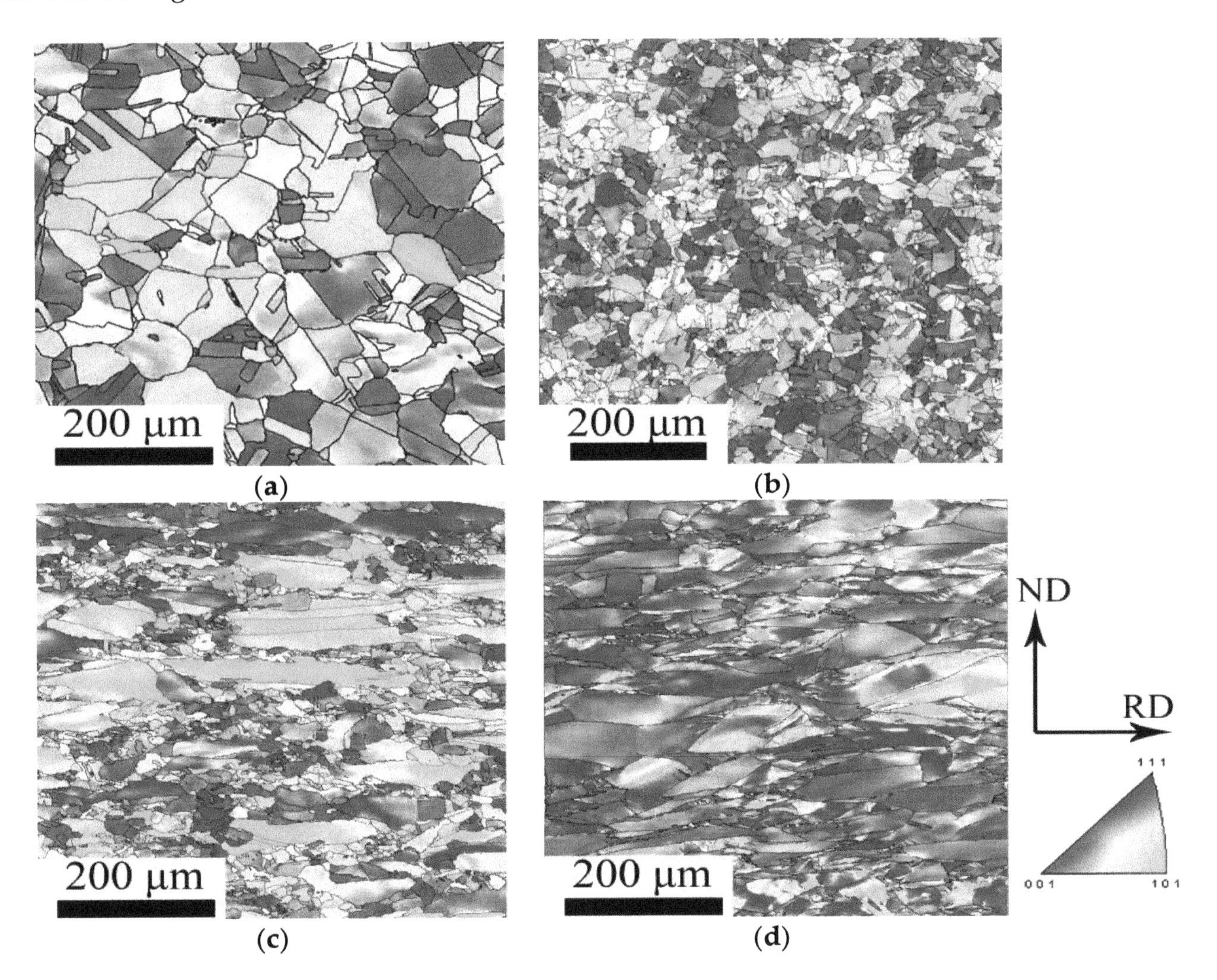

Figure 1. Typical OIM (orientation imaging microscopy) micrographs for deformation microstructures evolved in the Fe-18%Mn-0.4%C steel during hot-to-warm rolling at 1273 K (**a**), 1173 K (**b**), 1073 K (**c**) and 973 K (**d**). Colored orientations are shown for the transverse direction (TD).

3.2. Mechanical Properties

The stress-elongation curves obtained by tensile tests of the high-Mn steels processed by warm-to-hot rolling at different temperatures in the range of 773–1373 K are shown in Figure 2. A decrease in the rolling temperature results commonly in an increase in the strength and a decrease in the plasticity. The effect of the rolling temperature on the tensile tests properties is more pronounced for the warm working domain, i.e., rolling at temperatures below 1073 K, than that for hot working conditions, i.e., rolling at temperatures above 1073 K. A decrease in the temperature from 1373 K to 1073 K results in an increase in the yield strength ($\sigma_{0.2}$) by about 130 MPa while the ultimate tensile strength (UTS) does not change remarkably. In contrast, further decrease in the rolling temperature from 1073 K to 773 K leads to almost twofold increase in $\sigma_{0.2}$, which approaches about 900 MPa, and increases UTS by about 200 MPa. Correspondingly, the strengthening by warm to hot rolling is accompanied by a degradation of plasticity. It is interesting to note that total elongation gradually decreases with a decrease in the rolling temperature for Fe-18%Mn-0.6%C steel, where as that in Fe-18%Mn-0.4%C steel exhibits a kind of bimodal temperature dependence. The total elongation in the Fe-18%Mn-0.4%C steel tends to saturate at a level of 60–65% as the rolling temperature increases above 1073 K, following a rapid increase from 30% to 55% with an increase in the rolling temperature from 773 K to 1073 K.

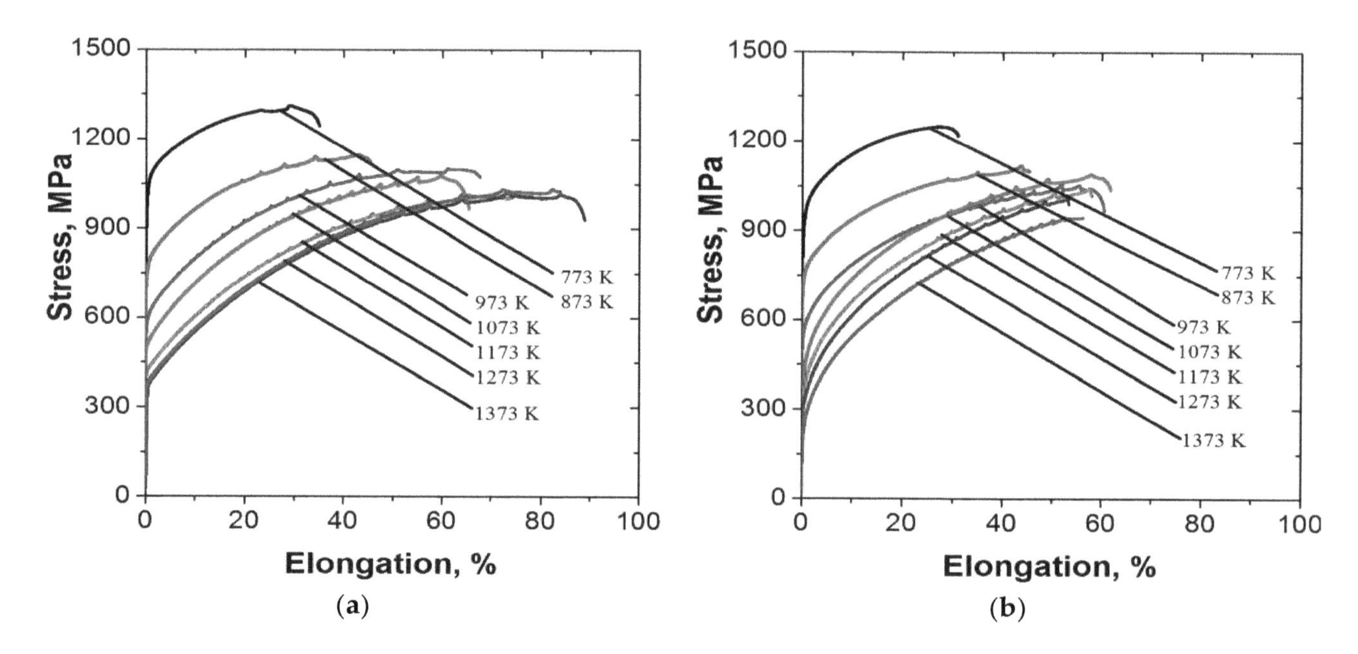

Figure 2. Engineering stress vs elongation curves of the Fe-18%Mn-0.6%C (**a**) and Fe-18%Mn-0.4%C (**b**) steels subjected to rolling at the indicated temperatures.

An apparent saturation for the total elongation of the Fe-18%Mn-0.4%C steel with an increase in the rolling temperature above 1073 K can be associated with a variation in the deformation mechanisms operating during tensile tests. The steel with lower carbon content has somewhat lower SFE [4] and, thus, may involve the strain-induced martensite upon tensile tests at room temperature. This difference in deformation mechanisms has been considered as a reason for the difference in plasticity [10]. The Fe-18%Mn-0.4%C steel subjected to hot rolling at temperatures above 1073 K exhibits the maximal plasticity, which can be obtained in the case of partial ε-martensitic transformation, whereas the Fe-18%Mn-0.6%C steel demonstrates increasing plasticity, which is improved by deformation twinning, with an increase in the rolling temperature. On the other hand, the strength properties, which depend on the grain size and dislocation density, are certainly affected by the rolling temperature, even in the range of hot working.

The Fe-18%Mn-0.6%C steel exhibits higher strength and elongation than the Fe-18%Mn-0.4%C steel after rolling warm to hot rolling in the studied temperature range. This additional strengthening of the Fe-18%Mn-0.6%C steel can be attributed to the difference in carbon content, which has been considered as the contributor to the yield strength of high-Mn TWIP steels [15]. The difference in 0.2 wt % carbon should result in about 85 MPa difference in the yield strength [15].

3.3. Tensile Behavior

Generally, the strength and plasticity during tensile tests depends on strain hardening, which, in turn, depends on the operating deformation mechanisms [1]. The plastic deformation of austenitic steels with low SFE at an ambient temperature is commonly expressed by the Ludwigson relation [14]:

$$\sigma = K_1 \cdot \varepsilon^{n1} + \exp(K_2 - n_2 \cdot \varepsilon), \quad (1)$$

where the first term with the strength factor of K_1 and strain hardening exponent n_1 represents Hollomon equation and the second term has been introduced by Ludwigson to incorporate the transient deformation stage, which differentiates the deformation behavior of fcc-metals/alloys with low-to medium SFE from other materials at relatively small strains. The stress of $\sigma = \exp(K_2)$ is close to the stress of plastic deformation onset and an inverse value of n_2 corresponds to the transient stage duration.

The parameters of K_1, K_2, n_1, n_2 providing the best correspondence between Equation (1) and experimental stress-strain curves are listed in Table 1. Note here, similar values for parameters of Ludwigson relation have been reported in other studies on low SFE austenitic steels [14,16–18]. The larger values of K_2 and K_1 for the present 0.6%C steel as compared to those for the 0.4%C steel reflect the higher stress levels of the former at early deformations and at large tensile strains, respectively. On the other hand, the n_1 values for both steels are close, suggesting similar strain hardening at large tensile strains, irrespective of some differences in the carbon content and SFE. The n_2 values are also close for both steels, indicating the same effect of the rolling temperature on the transient deformation stage during subsequent tensile tests.

Table 1. Parameters of the Ludwigson equation for the Fe-18%Mn-0.4%C and Fe-18%Mn-0.6%C steels processed by warm-to-hot rolling.

Steel	Rolling Temperature (K)	K_1 (MPa)	n_1	K_2	n_2
Fe-18%Mn-0.4%C	773	2248	0.24	6.0	37.8
Fe-18%Mn-0.4%C	873	2260	0.35	6.2	30.8
Fe-18%Mn-0.4%C	973	2342	0.46	6.0	20.3
Fe-18%Mn-0.4%C	1073	2410	0.48	5.8	25.2
Fe-18%Mn-0.4%C	1173	2556	0.58	5.7	12.8
Fe-18%Mn-0.4%C	1273	2575	0.63	5.5	11.4
Fe-18%Mn-0.4%C	1373	2598	0.67	5.5	12.3
Fe-18%Mn-0.6%C	773	2368	0.25	6.2	33.5
Fe-18%Mn-0.6%C	873	2410	0.38	6.3	24.5
Fe-18%Mn-0.6%C	973	2554	0.49	6.2	20.7
Fe-18%Mn-0.6%C	1073	2600	0.55	6.1	20.7
Fe-18%Mn-0.6%C	1173	2608	0.64	5.9	17.5
Fe-18%Mn-0.6%C	1273	2608	0.68	5.9	16.0
Fe-18%Mn-0.6%C	1373	2617	0.73	5.8	12.5

The monotonous changes of obtained parameters with rolling temperature suggest unique relationships between all parameters and deformation conditions. The effects of processing temperature on the parameters of Equation (1) are represented in Figure 3. Except for K_1, all parameters can be expressed by unique linear functions of the inverse rolling temperatures (Figure 3). The bimodal temperature dependencies obtained for K_1 with inflection points at 1073 K in Figure 3 are associated with the transition from warm to hot rolling conditions at this temperature, which reflects clearly on the deformation microstructures (see Figure 1). Using the indicated (Figure 3) linear relationships between the parameters of Ludwigson relation and the inverse rolling temperatures, the true stress vs strain curves calculated by Equation (1) are shown in Figure 4, along with the experimental curves obtained by tensile tests. Figure 4a,c show a general view of the stress-strain curves to validate the first term of Equation (1), whereas Figure 4b,d are plotted in double logarithmic scale to display the deformation behaviors at relatively small strains, which are described by the second term of Equation (1). The clear correspondence between the calculated and experimental plots testifies to the proposed treatments above.

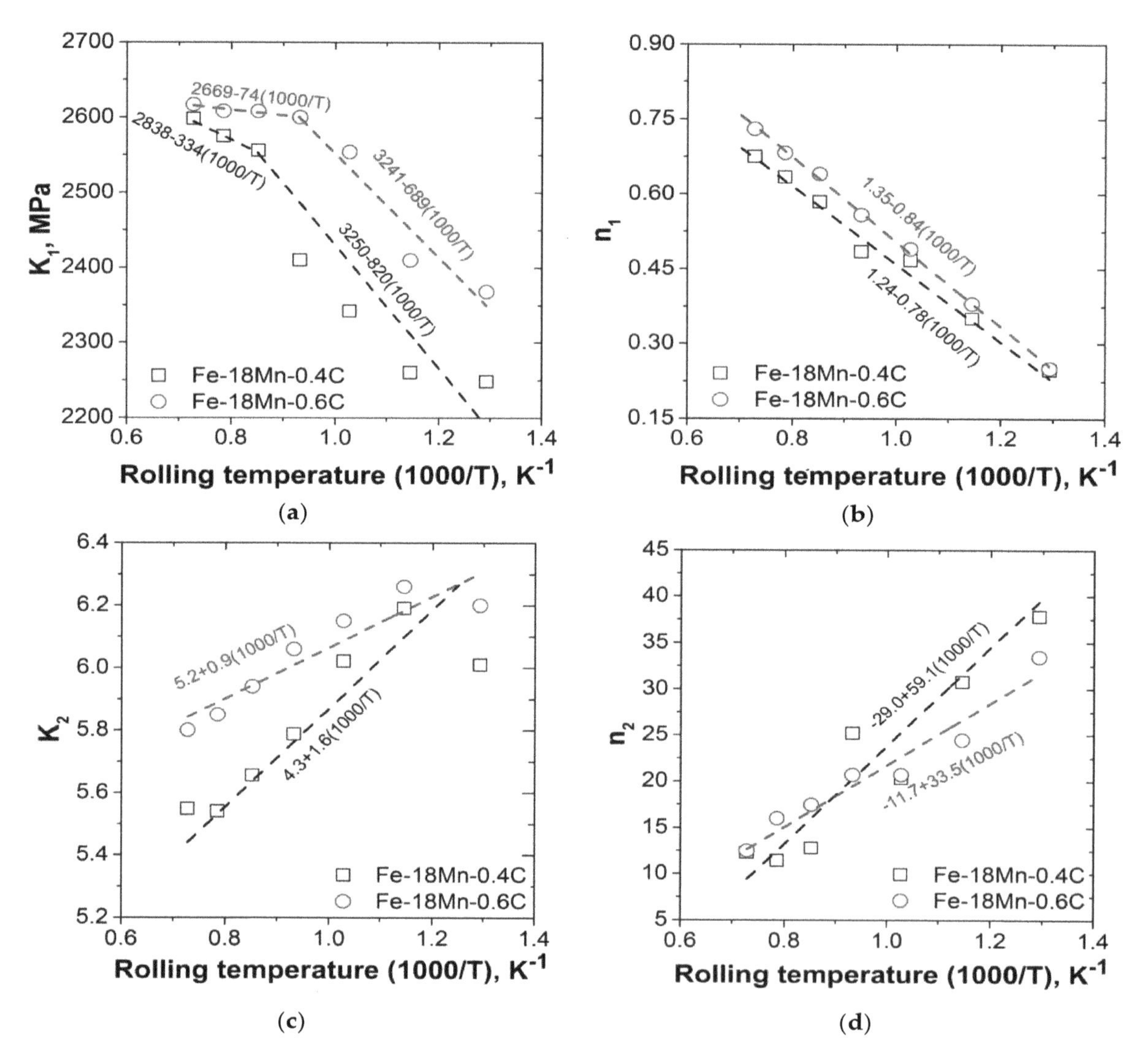

Figure 3. Effect of the rolling temperatures on the parameters of Ludwigson equation, K_1 (**a**), n_1 (**b**), K_2 (**c**), and n_2 (**d**).

The tensile deformation behavior of the steel samples should indeed be closely related to the steel microstructures, which were evolved by previous thermo-mechanical treatment. In turn, the developed microstructures depend on the processing conditions, i.e., rolling temperature, as the main processing variable in the present study. Generally, the deformation microstructures including the mean grain size and dislocation density that develop in metallic materials during warm-to-hot working can be expressed by power law functions of Zener-Hollomon parameter (temperature-compensated strain rate); $Z = \varepsilon \cdot \exp(Q/RT)$, where Q and R are the activation energy and universal gas constant, respectively [7,12]. Such microstructural changes are associated with thermally activated mechanisms of microstructure evolution in metallic materials. Therefore, the unique linear relationships in Figure 3 between the parameters of the flow stress predicting equation and the inverse rolling temperature suggest exponential relationships between the flow stress and the microstructures developed by warm-to-hot rolling. The second (exponential) term in Equation (1) predicts the flow stresses at relatively small strains (transient deformation), when the deformation behavior is associated with the dislocation ability to planar glide [14]. Thus, the stress-strain relationship of the high-Mn TWIP steels depends on their microstructures, namely, dislocation densities, evolved by previous thermo-mechanical treatments. Similar conclusions about a dominant role of dislocation density in the yield strength [10] and the work-hardening rate [19] were drawn in other studies on TWIP steels.

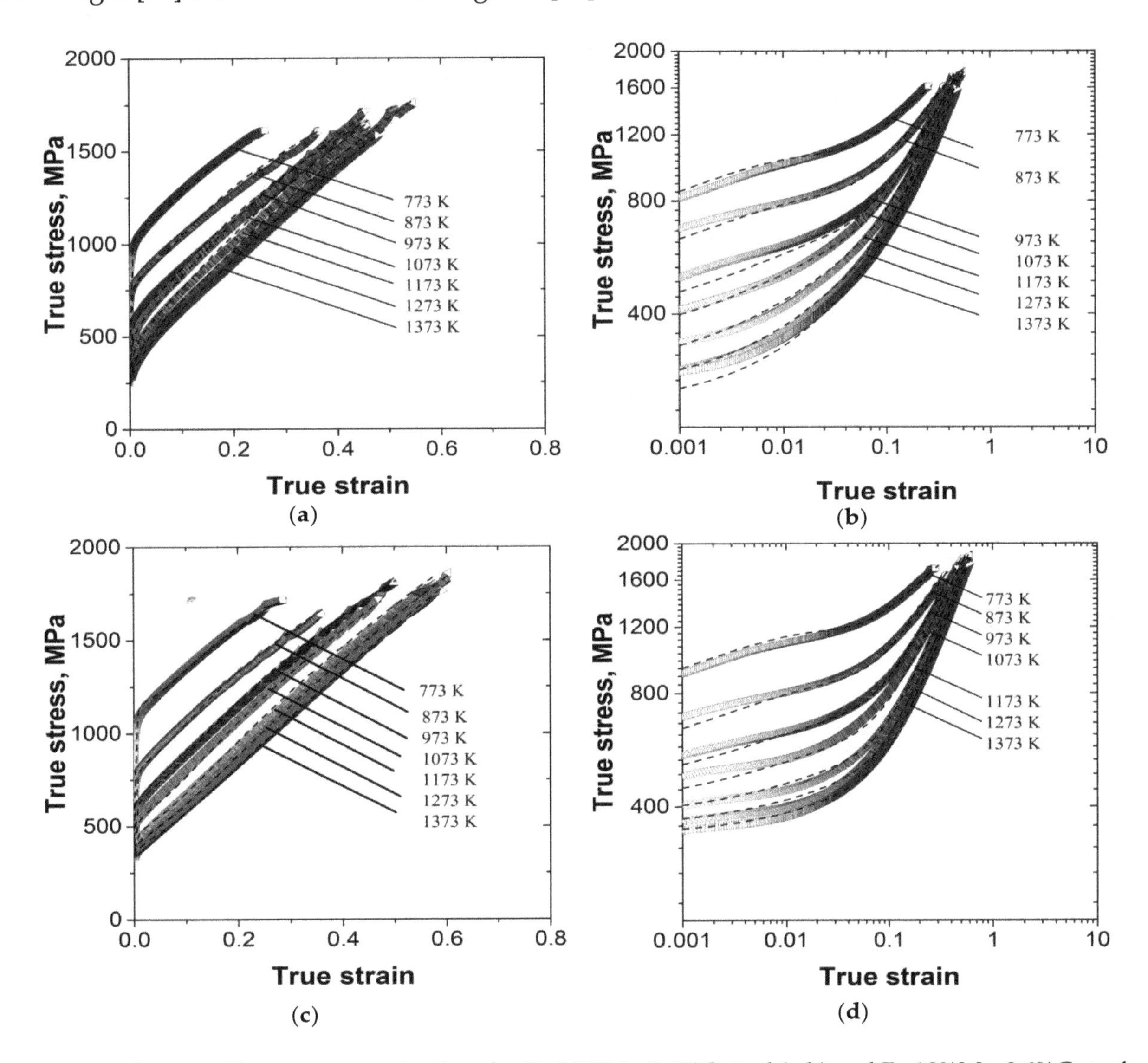

Figure 4. True tensile stress vs strain plots for Fe-18%Mn-0.4%C steel (**a**,**b**) and Fe-18%Mn-0.6%C steel (**c**,**d**) subjected to warm-to-hot rolling at indicated temperatures. The stress-strain curves obtained by tensile tests are shown by thick gray-scaled lines and those calculated by Equation (1) are shown by dashed lines.

It is worth noting in Figure 4 that maximal true stresses during tensile tests comprise about 1750 MPa for all steel samples, irrespective of the previous rolling conditions. Such gradual change in the altitude and slope of the true stress-strain curves can be represented by a gradual decrease in K_1 concurrently with n_1. An apparent saturation for the true stresses can be attributed to the strain-hardening ability owing to dislocation accumulation.

The strain, at which the transient deformation stage decays (ε_L) can be evaluated from the following relation [14]:

$$\exp(K_2 - n_2 \cdot \varepsilon_L)/(K_1 \varepsilon_L{}^{n1}) = r, \quad (2)$$

setting an arbitrary small value for r. According to the original Ludwigson treatment [14], $r = 0.02$ is selected in the present study. The values of ε_L calculated by Equation (2) for the present steels subjected to warm-to-hot rolling are shown in Figure 5 as functions of the rolling temperature. Formally, this strain (ε_L) limiting the transient deformation duration can be considered as a critical point below which, the plastic flow cannot be adequately described by Hollomon-type relation, i.e., the first term in Equation (1). The flow stresses during the transient deformation can be calculated taking into account the second term in Equation (1). The strain of ε_L can be roughly related to a strain when cross slip and dislocation rearrangements, which are closely connected with dynamic recovery, impair the strain hardening [14]. Therefore, an increase in ε_L should promote plasticity, including both uniform and total elongations.

It is clearly seen in Figure 5 that ε_L increases from about 0.08 to 0.24 with an increase in the rolling temperature from 773 K to 1373 K; this suggests an improvement in plasticity with increase in rolling temperature. A decrease in SFE promotes planar slip and, thus, should increase ε_L. Indeed, the 0.4%C steel is characterized by a larger ε_L than the 0.6%C steel after hot rolling at temperatures above 1300 K (Figure 5), although the hot rolled steel with higher carbon content exhibits larger total elongation. This relatively low plasticity of the Fe-18%Mn-0.4%C steel is associated with an ε-martensitic transformation [10]. Commonly, transformation-induced plasticity (TRIP) steels demonstrate lower plasticity than TWIP steels [20,21]. After rolling at temperatures below 1300 K, the values of ε_L for the Fe-18%Mn-0.4%C steel are smaller than those for the Fe-18%Mn-0.6%C steel processed under the same conditions (Figure 5). This can be attributed to the effect of warm-to-hot rolling at the evolved dislocation density. The latter has been shown to increase as rolling temperature decreases [10]. Therefore, the transient deformation stage upon the tensile tests is shortened because of the previous plastic deformation during warm-to-hot rolling, which partially consumed the dislocation ability to planar glide.

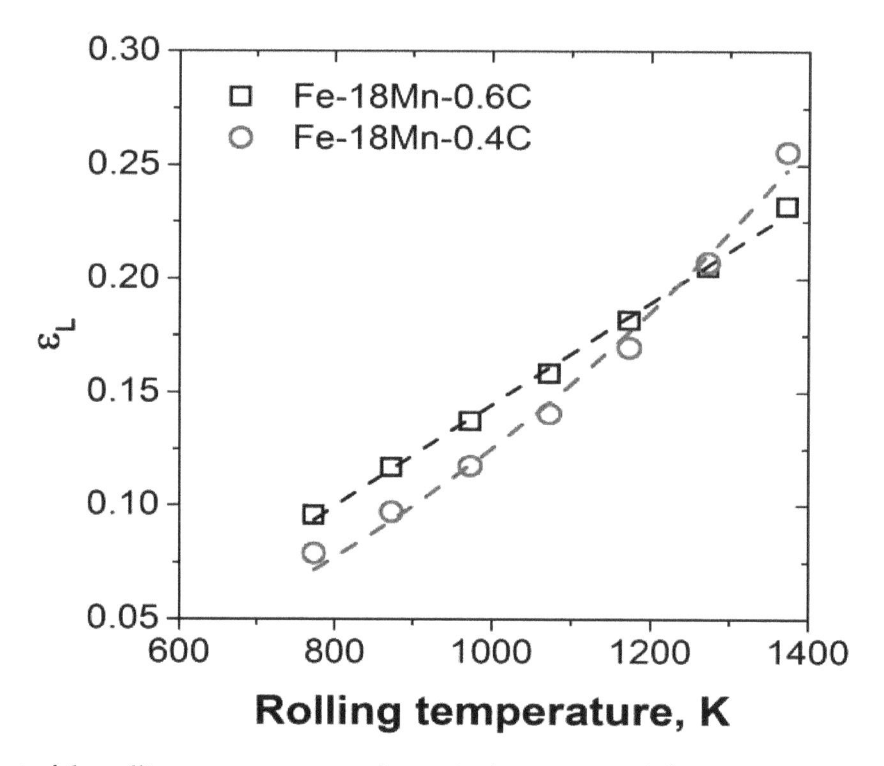

Figure 5. Effect of the rolling temperature on the strain for transient deformation during tensile tests of the Fe-18%Mn-0.4%C and Fe-18%-0.6%C steels.

4. Conclusions

The deformation behavior during tensile tests of Fe-18%Mn-0.4%C and Fe-18%-0.6%C steels subjected to warm to hot rolling was studied. The main results can be summarized as follows.

1. The hot rolling at temperatures above 1073 K was accompanied by the development of discontinuous dynamic recrystallization, leading to a decrease in the transverse grain size with a decrease in the rolling temperature. On the other hand, microstructure evolution during warm rolling at temperatures below 1073 K was controlled by the rate of dynamic recovery, which slowed down with a decrease in the rolling temperature.
2. The true stress-strain curves obtained by tensile tests at ambient temperature can be correctly represented by the Ludwigson-type relationship—$\sigma = K_1 \cdot \varepsilon^{n1} + \exp(K_2 - n_2 \cdot \varepsilon)$—where parameters of K_i and n_i depended on material and processing conditions and can be expressed by unique functions of inverse temperature of previous warm-to-hot rolling. A decrease in the rolling temperature from 1373 K to 773 K resulted in a decrease in K_1 (from approx. 2600 MPa to 2300 MPa) concurrently with n_1 (from approx. 0.7 to 0.25). Correspondingly, the true stress approached a level of about 1750 MPa during tensile tests, irrespective of the previous warm-to-hot rolling conditions. On the other hand, an increase in both K_2 and n_2 with decrease in the rolling temperature corresponded to an almost threefold increase in the yield strength and analogous degradation of plasticity.
3. The stage of transient plastic flow providing initial strain hardening and, therefore, controlling the total plasticity increases from about 0.08 to 0.24 with an increase in the rolling temperature from 773 K to 1373 K. The Fe-18%Mn-0.4%C steel is characterized by smaller values of ε_L than the Fe-18%Mn-0.6%C steel subjected to warm-to-hot rolling at the same temperatures below 1300 K, although the former should possess lower stacking fault energy. The shortening of the transient deformation stage upon tensile tests of the steels subjected to warm-to-hot rolling can be attributed to previous deformation, which partially consumed the dislocation ability to planar glide.

Author Contributions: Conceptualization, A.M.G. and R.K.; Methodology, V.T. and A.B.; Investigation, V.T.; Writing-Original Draft Preparation, A.B.; Writing-Review & Editing, A.M.G. and R.K.; Visualization, V.T.; Supervision, R.K.

Acknowledgments: The authors are grateful to the personnel of the Joint Research Centre of "Technology and Materials", Belgorod State University, for their assistance with instrumental analysis.

References

1. Bouaziz, O.; Allain, S.; Scott, C.P.; Cugy, P.; Barbier, D. High manganese austenitic twinning induced plasticity steels: A review of the microstructure properties relationships. *Curr. Opin. Solid State Mater. Sci.* **2011**, *15*, 141–168. [CrossRef]
2. De Cooman, B.C.; Estrin, Y.; Kim, S.K. Twinning-induced plasticity (TWIP) steels. *Acta Mater.* **2018**, *142*, 283–362. [CrossRef]
3. Kusakin, P.S.; Kaibyshev, R.O. High-Mn twinning-induced plasticity steels: Microstructure and mechanical properties. *Rev. Adv. Mater. Sci.* **2016**, *44*, 326–360.
4. Saeed-Akbari, A.; Mosecker, L.; Schwedt, A.; Bleck, W. Characterization and prediction of flow behavior in high-manganese twinning induced plasticity steels: Part I. mechanism maps and work-hardening behavior. *Metall. Mater. Trans. A* **2012**, *43*, 1688–1704. [CrossRef]
5. Kusakin, P.; Belyakov, A.; Haase, C.; Kaibyshev, R.; Molodov, D.A. Microstructure evolution and strengthening mechanisms of Fe–23Mn–0.3C–1.5Al TWIP steel during cold rolling. *Mater. Sci. Eng. A* **2014**, *617*, 52–60. [CrossRef]
6. Yanushkevich, Z.; Belyakov, A.; Haase, C.; Molodov, D.A.; Kaibyshev, R. Structural/textural changes and strengthening of an advanced high-Mn steel subjected to cold rolling. *Mater. Sci. Eng. A* **2016**, *651*, 763–773. [CrossRef]

7. Sakai, T.; Belyakov, A.; Kaibyshev, R.; Miura, H.; Jonas, J.J. Dynamic and post-dynamic recrystallization under hot, cold and severe plastic deformation conditions. *Prog. Mater. Sci.* **2014**, *60*, 130–207. [CrossRef]
8. Tikhonova, M.; Belyakov, A.; Kaibyshev, R. Strain-induced grain evolution in an austenitic stainless steel under warm multiple forging. *Mater. Sci. Eng. A* **2013**, *564*, 413–422. [CrossRef]
9. Tikhonova, M.; Enikeev, N.; Valiev, R.Z.; Belyakov, A.; Kaibyshev, R. Submicrocrystalline austenitic stainless steel processed by cold or warm high pressure torsion. *Mater. Sci. Forum* **2016**, *838*, 398–403. [CrossRef]
10. Torganchuk, V.; Belyakov, A.; Kaibyshev, R. Effect of rolling temperature on microstructure and mechanical properties of 18%Mn TWIP/TRIP steels. *Mater. Sci. Eng. A* **2017**, *708*, 110–117. [CrossRef]
11. Morozova, A.; Kaibyshev, R. Grain refinement and strengthening of a Cu–0.1Cr–0.06Zr alloy subjected to equal channel angular pressing. *Philos. Mag.* **2017**, *97*, 2053–2076. [CrossRef]
12. Yanushkevich, Z.; Dobatkin, S.V.; Belyakov, A.; Kaibyshev, R. Hall-Petch relationship for austenitic stainless steels processed by large strain warm rolling. *Acta Mater.* **2017**, *136*, 39–48. [CrossRef]
13. Choudhary, B.K.; Isaac Samuel, E.; Bhanu Sankara Rao, K.; Mannan, S.L. Tensile stress-strain and work hardening behaviour of 316LN austenitic stainless steel. *Mater. Sci. Technol.* **2001**, *17*, 223–231. [CrossRef]
14. Ludwigson, D.C. Modified stress-strain relation for fcc metals and alloys. *Metal. Trans.* **1972**, *2*, 2825–2828. [CrossRef]
15. Kusakin, P.; Belyakov, A.; Molodov, D.A.; Kaibyshev, R. On the effect of chemical composition on yield strength of TWIP steels. *Mater. Sci. Eng. A* **2017**, *687*, 82–84. [CrossRef]
16. Mannan, S.L.; Samuel, K.G.; Rodriguez, P. Stress-strain relation for 316 stainless steel at 300 K. *Scr. Metall.* **1982**, *16*, 255–257. [CrossRef]
17. Satyanarayana, D.V.V.; Malakondaiah, G.; Sarma, D.S. Analysis of flow behaviour of an aluminium containing austenitic steel. *Mater. Sci. Eng. A* **2007**, *452*, 244–253. [CrossRef]
18. Milititsky, M.; De Wispelaere, N.; Petrov, R.; Ramos, J.E.; Reguly, A.; Hanninen, H. Characterization of the mechanical properties of low-nickel austenitic stainless steels. *Mater. Sci. Eng. A* **2008**, *498*, 289–295. [CrossRef]
19. Liang, Z.Y.; Li, Y.Z.; Huang, M.X. The respective hardening contributions of dislocations and twins to the flow stress of a twinning-induced plasticity steel. *Scr. Mater.* **2016**, *112*, 28–31. [CrossRef]
20. Song, W.; Ingendahl, T.; Bleck, W. Control of strain hardening behavior in high-Mn austenitic steels. *Acta Metall. Sin. (Engl. Lett.)* **2014**, *27*, 546–555. [CrossRef]
21. Lee, Y.-K.; Han, J. Current opinion in medium manganese steel. *Mater. Sci. Technol.* **2015**, *31*, 843–856. [CrossRef]

18

Structural Defects in TiNi-Based Alloys after Warm ECAP

Aleksandr Lotkov [1,*], Anatoly Baturin [1], Vladimir Kopylov [2], Victor Grishkov [1] and Roman Laptev [3]

1 Institute of Strength Physics and Materials Science of the Siberian Branch of the Russian Academy of Sciences, 634055 Tomsk, Russia; abat@ispms.tsc.ru (A.B.); grish@ispms.tsc.ru (V.G.)
2 Physical-Technical Institute of the National Academy of Sciences of Belarus State Scientific Institution, 220141 Minsk, Belarus; kopylov.ecap@gmail.com
3 National Research Tomsk Polytechnic University, 634050 Tomsk, Russia; laptev.roman@gmail.com
* Correspondence: lotkov@ispms.ru

Abstract: The microstructure, martensitic transformations and crystal structure defects in the $Ti_{50}Ni_{47.3}Fe_{2.7}$ (at%) alloy after equal-channel angular pressing (ECAP, angle 90°, route B_C, 1–3 passes at T = 723 K) have been investigated. A homogeneous submicrocrystalline (SMC) structure (grains/subgrains about 300 nm) is observed after 3 ECAP passes. Crystal structure defects in the $Ti_{49.4}Ni_{50.6}$ (at%) alloy (8 ECAP passes, angle 120°, B_C route, T = 723 K, grains/subgrains about 300 nm) and $Ti_{50}Ni_{47.3}Fe_{2.7}$ (at%) alloy with SMC B2 structures after ECAP were studied by positron lifetime spectroscopy at the room temperature. The single component with the positron lifetime τ_1 = 132 ps and τ_1 = 140 ps were observed for positron lifetime spectra (PLS) obtained from ternary and binary, correspondingly, annealed alloys with coarse-grained structures. This τ_1 values correspond to the lifetime of delocalized positrons in defect-free B2 phase. The two component PLS were found for all samples exposed by ECAP. The component with τ_2 = 160 ps (annihilation of positrons trapped by dislocations) is observed for all samples after 1–8 ECAP passes. The component with τ_3 = 305 ps (annihilation of positrons trapped by vacancy nanoclusters) was detected only after the first ECAP pass. The component with τ_3 = 200 ps (annihilation of positrons trapped by vacancies in the Ti sublattice of B2 structure) is observed for all samples after 3–8 ECAP passes.

Keywords: TiNi-based alloys; ECAP; microstructure; positron lifetime spectroscopy; nanoclusters; vacancies; dislocations

---

1. Introduction

Severe plastic deformation (SPD) allows one to produce an ultrafine-grained (UFG) structure with a high strength and sufficient ductility in metal materials [1,2]. Among such materials are TiNi-based alloys distinguished for their superelasticity and shape memory effect (SME) and widely used in engineering and medicine [3]. Among the SPD methods is equal channel angular pressing (ECAP), which provides UFG TiNi-based alloys with greatly improved functional properties. It opens the most promising way of producing bulk billets from this type of materials [1,2].

By now, research data are available to judge the microstructure evolution of binary and ternary TiNi-based alloys during ECAP at 623–773 K, and the grain size effect on their mechanical and inelastic properties [4–17], and also the influence of ECAP on the temperatures of martensite transformations (MT) in TiNi-based alloys [9,10,12,17,18]. As has been shown [1,4,12], the grain structure of TiNi-based alloys can be refined to 100–500 nm in six to eight ECAP passes depending on the pressing temperature. Such alloys with an average grain-subgrain size of 300 nm show an unusual behavior with an ultimate strength of up 1200 MPa and plasticity of up to 50–60%. After ECAP at 623–773 K, the temperature of martensite transformations in TiNi-based alloys normally decreases by 20 K and more [17,18]. However, very scant data are available on the evolution of lattice defects in TiNi-based alloys exposed to ECAP. These data are necessary for the understanding both the mechanisms of grain refinement and physical factors affected on MT temperatures in alloys after ECAP. But the sole paper report that the dislocation density in such alloys after SPD increases greatly (to ~10^{15}–10^{16} m^{-2}) and that its critical value (~10^{18} m^{-2}) exists at which they assume an amorphous structure [19].

However, in addition to dislocations, SPD produces numerous excess vacancies as against thermodynamic equilibrium [20,21]. On the one hand, vacancies greatly accelerate the mass transfer, dissolution or precipitation of secondary phases [22,23], and generation of new grain boundaries in a material [24]. On the other hand, vacancies with a high concentration and low mobility can form clusters [25], and eventually decrease the long-term durability of materials, as it happens, e.g., in Al- and Ti-based alloys after ECAP [26,27]. In this context, studying the formation of vacancy-like free volumes in SPD materials can provide a better understanding of the mechanisms responsible for their properties in UFG states and of the conditions for their more efficient use. Most of the studies of SPD-induced vacancies have been performed on pure metals and their alloys [20,28–32], and almost all of them consider SPD processes at room temperature when the vacancy mobility is low. Among the methods of research in deformation-induced vacancies, their clusters, and dislocation density in metals and alloys are resistometry [21], dilatometry [29], differential scanning calorimetry [21,29], X-ray line profile analysis [33], perturbed γ-γ angular correlation (PAC) [34], and positron annihilation spectroscopy (PAS) [28–32], all of which give comparable vacancy concentrations for pure metals (10^{-2}–10^{-4}). However, little is known about SPD-induced vacancies in intermetallic compounds where their types are more diverse, compared to metals [35]. For example, vacancy-like free volumes classifiable as interface defects are detected by PAS in ball-milled nanocrystalline Fe3Si [36]. In B2 FeAl under high pressure torsion, the vacancy density can reach 10^{-2} as it follows from comparative data of differential scanning calorimetry and calculations [37]. From PAC data [38], the types of defects that dominate after ball milling are Schottky pairs in PdIn, triple defects in NiAl and FeRh, and antisite defects in FeAl. One of the studies shows that after ultrasonic shock treatment at room temperature, the relative concentration of single vacancies in equiatomic TiNi surface layers increases to ~10^{-5} [39,40].

This paper presents the results of complex research into the ECAP effect at 723 K on the microstructure, martensite transformations, and lattice defect evolution in $Ti_{50}Ni_{47.3}Fe_{2.7}$ (at%) alloy combined with experimental results about lattice defects in $Ti_{49.4}Ni_{50.6}$ alloy (at%) alloy after eight ECAP passes at the same temperature.

The choice of experimental alloys is due to the following considerations based on the results of studies of the microstructure and martensitic transformations. These alloys have the B2 phase structure at room temperature both in the initial coarse-grained state and in the submicrocrystalline state. In addition, the grain-subgrain structures of these alloys are similar after ECAP. This makes it possible to study only those defects of the crystal structure that are formed during ECAP at 723 K and to avoid the influence of defects that may appear as a result of MT.

2. Materials and Methods

The test $Ti_{50}Ni_{47.3}Fe_{2.7}$ (at%) alloy were supplied as rotary forged (1220 K) round bars of diameter 25 mm and length 140 mm (MATEK-SMA Ltd., Moscow, Russia). The round bars were forged to a square of 16 × 16 mm^2 at 1073 ± 100 K with further annealing at 773 K for 3 h. Then, the square bars were cut to 14 × 14 mm^2 and were exposed to ECAP (B_C route) with a channel angle 90°. The number of cycles was N_i, where i = 1, 2, 3. The test specimens of $Ti_{50}Ni_{47.3}Fe_{2.7}$ (at%) were pressed at the Physical-Technical Institute NASB (Minsk, Belarus). The initial coarse-grained alloy had a B2 grain size of 20–40 μm. At temperatures above 275 K, its structure was represented by a single B2 phase (the high temperature cubic phase). The sequence and the temperatures of martensite transformations were studied by temperature resistometry (four-point scheme) in cooling–heating cycles.

The test alloy $Ti_{49.4}Ni_{50.6}$ (at%) was supplied by Intrinsic Devices Inc. (San Francisco, CA, USA). In the alloy quenched from 1073 K, the start temperature of direct B2 → B19′ martensite transformation was M_S = 288 K (B19′-the monoclinic martensite phase). At room temperature, its initial coarse-grained state was represented by a B2 structure with an average grain size of 50 μm. Its submicrocrystalline state with an average grain-subgrain size of 300 nm was formed by ECAP at T = 723 K at the Ufa State Aviation Technical University (Ufa, Russia). The channel angle was 110° (B_C route), and the number of passes was $n = 8$ [6].

The samples of dimensions $10 \times 10 \times 1$ mm^3 for the positron annihilation spectroscopy and bars of dimensions $1 \times 1 \times 20$ mm^3 for the electrical resistivity measurements were prepared from bullets of initial alloy and alloy after ECAP. The foils for the transmission electron microscopy (TEM) were prepared by the mechanical polishing with subsequent electrochemical thinning of plates with initial thickness about 0.3 mm.

The microstructure of the test specimens was examined by transmission electron microscopy (JEM-2100, JEOL Ltd., Tokyo, Japan) in Nanotech Shared Use Center, ISPMS SB RAS).

The crystal defects appearing after ECAP were analyzed by positron annihilation spectroscopy (PAS) based on analysis of the positron lifetime spectra. A time resolution of spectrometer is 240 ps [41]. The positron source was ^{44}Ti with an activity of 1 MBq placed between two identical specimens. The positron beam diameter was about 6 mm. The total number of annihilation events recorded per each spectrum was no less than 5×10^6. Each specimen was analyzed from three independent the positron lifetime spectra. Processing of the spectra was carried out using LT software [42]. After subtraction of the component related to annihilation in the positron source and background, the spectra could be reliably decomposed into one lifetime component (initial state) or into two (deformed state),

$$S(t) = (I_1/\tau_1)\exp(-t/\tau_1) + (I_2/\tau_2)\exp(-t/\tau_2), \quad (1)$$

where τ_1, I_1 and τ_2, I_2 are the respective component lifetimes and intensities to judge the type of a defect and its concentration.

The Vickers microhardness was measured on a Duramin-5 microhardness tester (Struers, Ballerup, Denmark) at room temperature under a load of 100 g.

3. Results and Discussion

3.1. *Effect of Warm ECAP on Martensite Transformation Temperatures in $Ti_{50}Ni_{47.3}Fe_{2.7}$*

Figure 1 shows typical temperature dependences of resistivity on heating and cooling for $Ti_{50}Ni_{47.3}Fe_{2.7}$ (at%) alloy before and after ECAP passes. These $\rho(T)$ dependences are qualitatively similar for both initial alloy state and after ECAP.

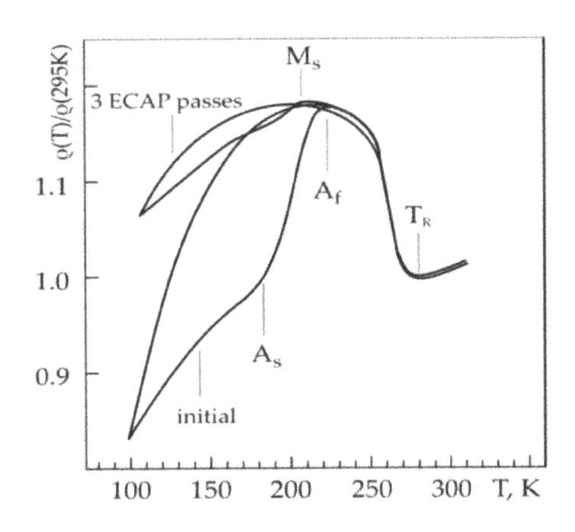

Figure 1. Typical temperature dependences of resistivity for $Ti_{50}Ni_{47.3}Fe_{2.7}$ before (initial), and after, three ECAP passes.

On cooling and heating in the temperature range from 275 K to 77 K (liquid nitrogen), the alloy is involved in the sequence of martensite transformations B2 ↔ R ↔ R + B19′ (R and B19′ are rhombohedral and monoclinic martensite phases, respectively). During B2 ↔ R transformations on cooling and heating, the resistivity of samples varies almost without hysteresis (its value is less than 3 K). The transformation R → B19′ on cooling to 77 K is incomplete throughout the volumes of all samples. The MT temperatures are denoted as T_R for B2 ↔ R MT, M_S for the start of R → B19′ MT (cooling) and A_S and A_f for B19′ → R MT (heating) The variations of MT temperatures versus ECAP passes are presented in Table 1.

Table 1. Martensite transformation temperatures in $Ti_{50}Ni_{47.3}Fe_{2.7}$.

State	T_R, K	M_s, K	A_s, K	A_f, K
Initial	275	213	184	220
1st ECAP pass	276	215	185	208
3rd ECAP pass	275	195	195	205

The main ECAP effect on MT consists in the following: it decreases M_S by 18 K, compared to the initial state, and narrows the temperature interval of reverse B19′ → R transformation by 10 K. Another effect, which is more pronounced, is a steep decrease in the value by which the resistivity changes during on cooling to 77 K. Therefore, the B19′ volume fraction appearing at T < Ms decreases greatly (2–3 times) with increasing the number of ECAP passes, i.e., with increasing the true strain. This is likely due to considerable grain-subgrain refinement in the specimens, which causes their hardening and markedly decreases the martensite transformation temperatures of the R → B19′ MT.

Thus, the B2 structure is observed at room temperature in $Ti_{50}Ni_{47.3}Fe_{2.7}$ (at%) alloy after 1–3 ECAP passes. These data correspond to the results obtained by TEM.

3.2. *Microstructure of $Ti_{50}Ni_{47.3}Fe_{2.7}$ after Warm ECAP*

Figure 2 shows typical micrographs of $Ti_{50}Ni_{47.3}Fe_{2.7}$ (at%) after one ECAP pass. As can be seen, its grain-subgrain microstructure is rather inhomogeneous. In the bulk of the deformed alloy, a clearly defined multiband structure is observed as evidence of plastic strain localization in the specimens during ECAP. The orientation angles of bands relative to each other are 45° (135°) and 90°. Inside the bands, non-equiaxial fragments of a finer grain-subgrain structure are localized. Their aspect ratio (ratio of minimum to maximum sizes) ranges from 1:5 to 1:10.

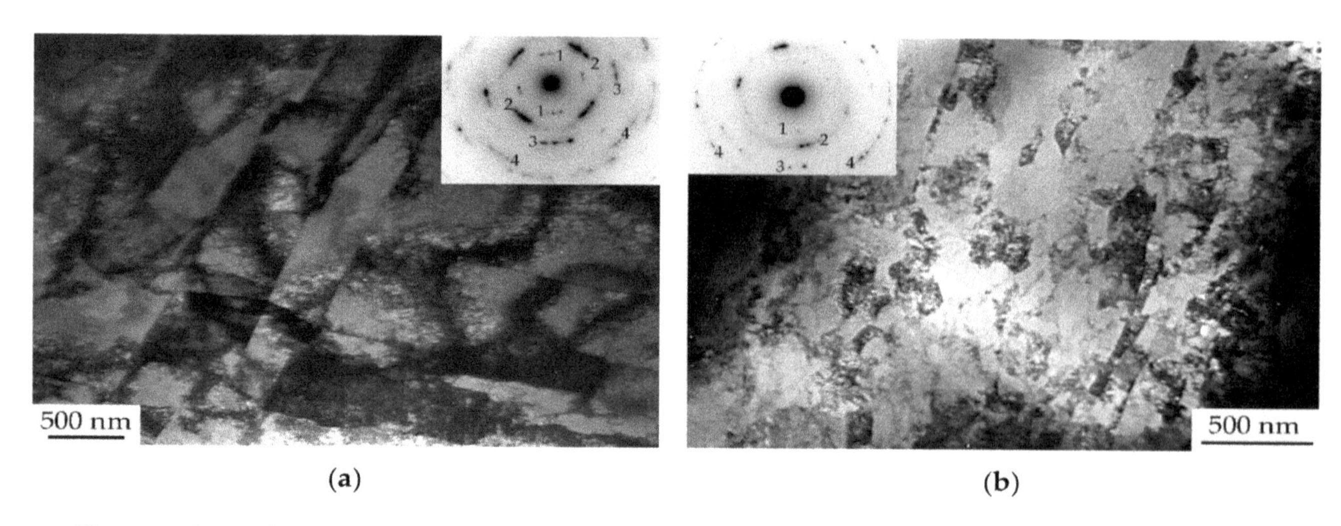

Figure 2. Typical TEM bright-field images of microvolumes with most large-scale subgrains (**a**) and fine-grained structure (**b**) in $Ti_{50}Ni_{47.3}Fe_{2.7}$ after one ECAP pass at 723 K. Details are presented in text. Quasi-rings in microdiffraction patterns include the next types of B2-reflections: 1–(100), 2–(110), 3–(200), 4–(220), 5–(310) in (**a**); 1–(100), 2–(110), 3–(200), 4–(211) in (**b**).

The misorientation of adjacent band fragments reaches 10–12°, and that of adjacent fragments inside each band measures 2–5°. The band width in different microvolumes varies widely, from a microscale size of 400–500 nm (Figure 2a) to a mesoscale one of up to ~5 μm (Figure 2b). In mesoscale bands, one can clearly see a secondary microband structure with a minimum size of grain-subgrain fragments of 100–300 nm (Figure 2b).

The fragments of the grain-subgrain structure are mostly in the state of a B2 phase, as evidenced by electron diffraction patterns in Figure 2. It is observed that, along with quasi-ring reflections, the microdiffraction patterns reveal bright peaks from fragments misoriented to less than 15° with substantial radial broadening for microband volumes (Figure 2a), and with weak radial broadening and rather uniform distribution along the Debye rings for mesoband ones (Figure 2b). From comparison of the electron diffraction patterns in Figure 2a,b, it follows that the microstructure of the alloy after one ECAP pass is spatially inhomogeneous, not only in fragment sizes, but also in internal stresses. Moreover, the stress level in microband volumes (Figure 2a) is much higher than its level in mesoband ones (Figure 2b). Another feature of the electron diffraction patterns is a clear doublet structure of B2-phase reflections typical of deformation twinning. Besides the above microstructure types, individual grains of up to 1 μm occur, but rarely, in the alloy after one ECAP pass.

Therefore, the specimens of $Ti_{50}Ni_{47.3}Fe_{2.7}$ (at%) pressed in one ECAP pass at a channel angle of 90° assumes a grain-subgrain structure with a nonuniform fragment size distribution (100 nm to 1–1.5 μm) in which the main fraction belongs to submicrocrystalline fragments (100–500 nm). Such a grain-subgrain structure is formed through fragmentation on different scales and through B2-phase twinning.

Figures 3 and 4 shows the microstructure of the alloy after three ECAP passes. As can be seen, it is qualitatively similar to, but much finer than, the microstructure after one ECAP pass. The alloy preserves the system of microbands but their width decreases to less than 1 μm. The grain-subgrain structure contains a substantial fraction of nanosized fragments (50–100 nm). The main phase in the alloy is B2, as evidenced by its electron diffraction patterns.

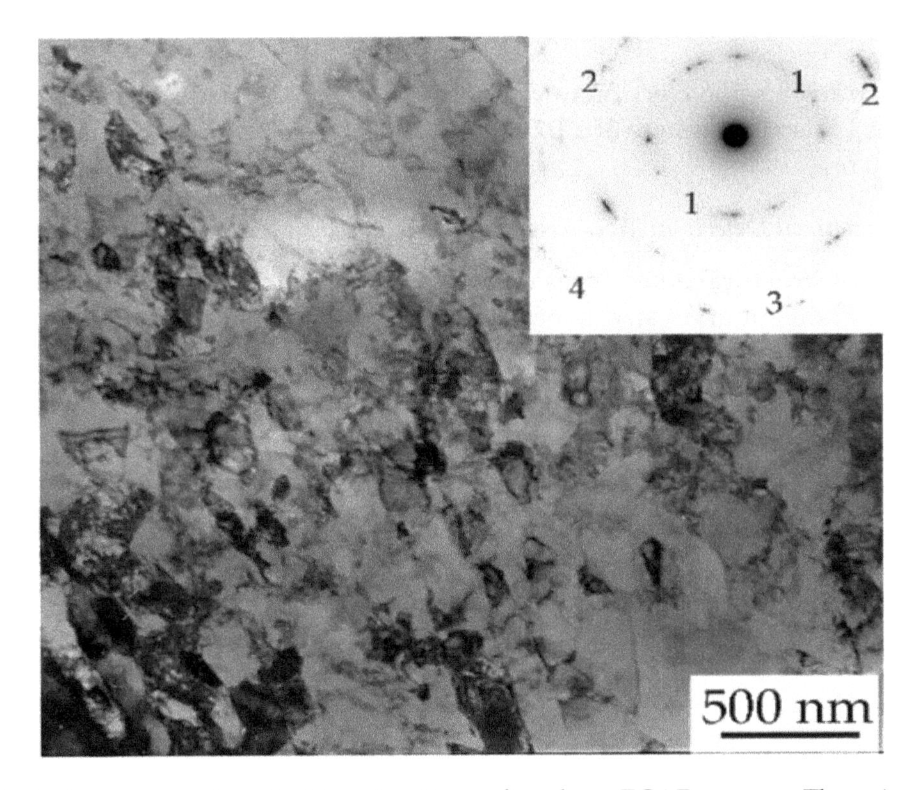

Figure 3. Typical microstructure in $Ti_{50}Ni_{47.3}Fe_{2.7}$ after three ECAP passes. The microdiffraction pattern is presented in the insert. Quasi-rings in microdiffraction pattern include the next types of B2 phase reflections: 1–(110), 2–(211), 3–(310), 4–(312).

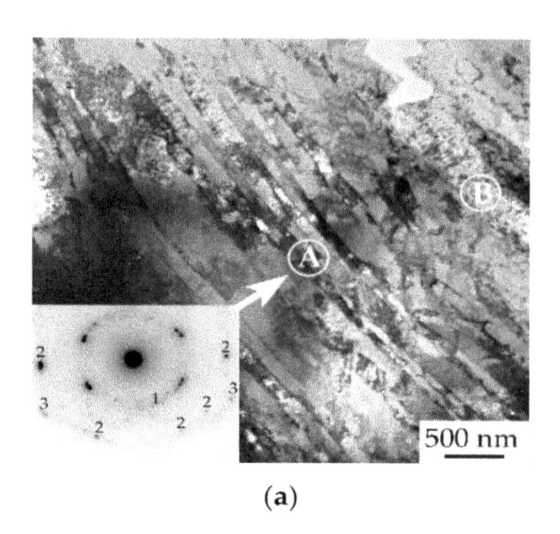

(a)

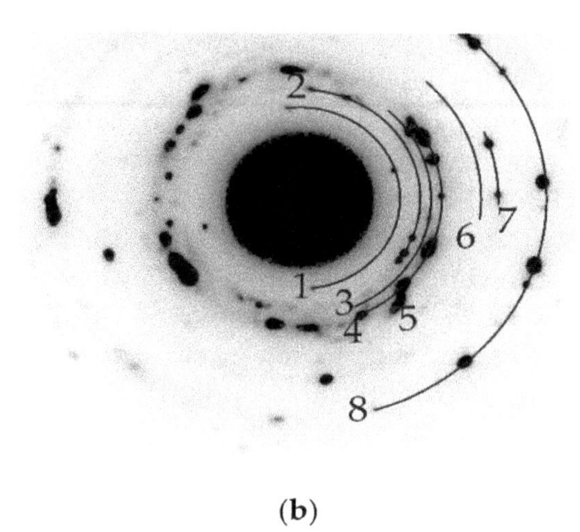

(b)

Figure 4. Microstructure in $Ti_{50}Ni_{47.3}Fe_{2.7}$ after three ECAP passes with the main B2 phase structure (region A) and small volume fraction of R phase (region B). The microdiffraction patterns are presented in the insert of (**a**,**b**), correspondingly. (**a**) The quasi-rings include the next types of B2 phase reflections: 1–(110), 2–(211), 3–(220); (**b**) The quasi-rings include the next types of B2 phase reflections: 1–(002), 2–(112), 3–(202), 4–(220), 5–(103), 6–(320), 7–(004), 8–(413).

However, several microvolumes with an R phase structure were found in the samples after three ECAP passes (e.g., region B in Figure 4a). It is most likely that the local appearance of the R phase at room temperature is due to the high level of residual stresses in these regions. According to electron microscopy data, the total proportion of regions with the R phase structure is less than 1 vol%. Therefore, the presence of these regions will not be taken into account in the subsequent discussion of the results.

Therefore, a mixed grain-subgrain structure based on a submicrocrystalline (100–500 nm) and a nanostructural fraction (50–100 nm) is formed in $Ti_{50}Ni_{47.3}Fe_{2.7}$ (at%) after ECAP at 723 K (channel angle 90°, strain rate 1 s^{-1}).

According to transmission electron microscopy [6], the microstructure of $Ti_{49.4}Ni_{50.6}$ (at%) after eight ECAP passes is also in the state of a B2 phase. The average grain-subgrain size in the specimen cross section is 300 nm.

3.3. *Evolution of Structural Defects in TiNi-based Alloys Exposed to Warm ECAP*

The evolution of structural defects was analyzed by positron annihilation spectroscopy. Figure 5 shows the average grain-subgrain size, average positron lifetime, and Vickers microhardness versus the numbers of ECAP passes for the test TiNi-based alloys.

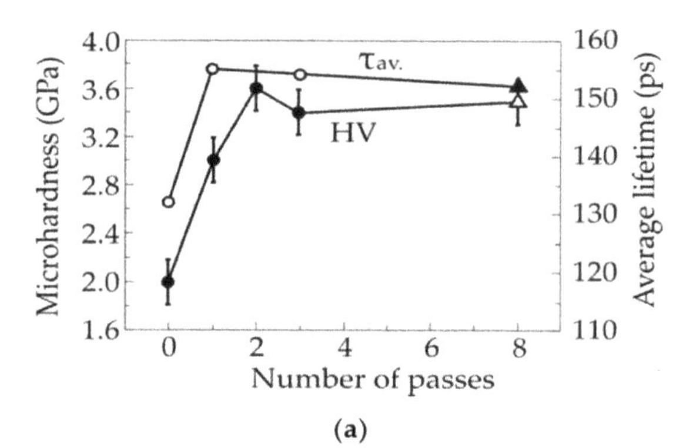

(a)

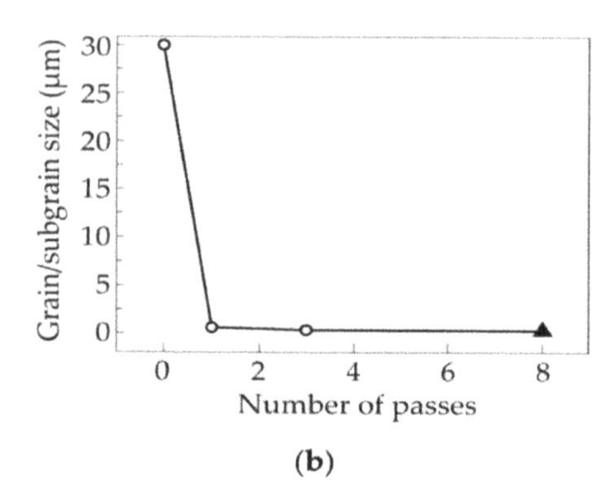

(b)

Figure 5. Average positron lifetimes, $\tau_{av.}$(error bars ± 1 ps) and microhardness, HV, (**a**) and average grain/subgrain sizes (**b**) versus number of ECAP passes: $Ti_{49.4}Ni_{50.6}$ (at%) alloy (▲, △) and $Ti_{50}Ni_{47.3}Fe_{2.7}$ (at%) alloy (○, ●).

In the specimens exposed to ECAP, the average positron lifetime τ_{av} increases greatly even after the first ECAP pass and decreases slightly with increasing the number of passes irrespective of the alloy composition (Figure 5a). The Vickers microhardness, HV, behaves in the same way (Figure 5a). The increase in the microhardness can be explained by grain refinement in ECAP (Figure 5b), while the increase in τ_{av} definitely suggests that increasing the strain increases the lattice defect density. The type of defects resulting from ECAP can be identified by decomposing the positron lifetime spectra into components (Figure 6). The initial annealed specimens have a single-component spectrum with a positron lifetime τ_1 = 132 ± 1 ps for $Ti_{50}Ni_{47.3}Fe_{2.7}$ (at%) and τ_1 = 140 ± 1 ps for $Ti_{49.4}Ni_{50.6}$ (at%) (Figure 6a). The τ_1 value corresponds to the experimental free positron lifetime 132 ps in TiNi [43] and the theoretical lifetimes of delocalized positrons in defect-free TiNi B2 structure: 120 ps [44] and 129 ps [45]. Such a decomposition shows that, even after the first ECAP pass, the spectra show no evidence of delocalized positrons: All positrons annihilate only from localized states associated with defects (so-called positron saturated trapping).

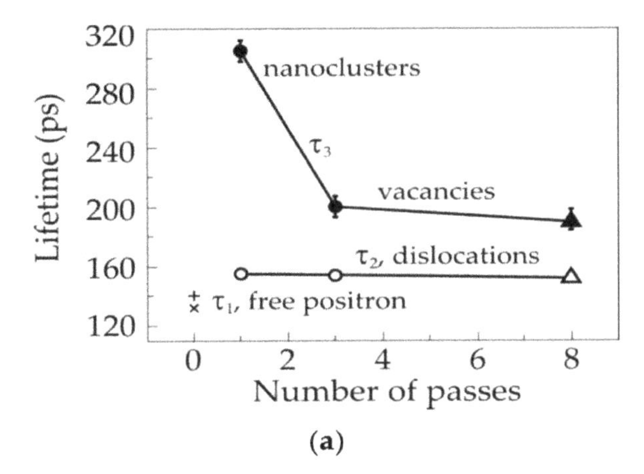

(a)

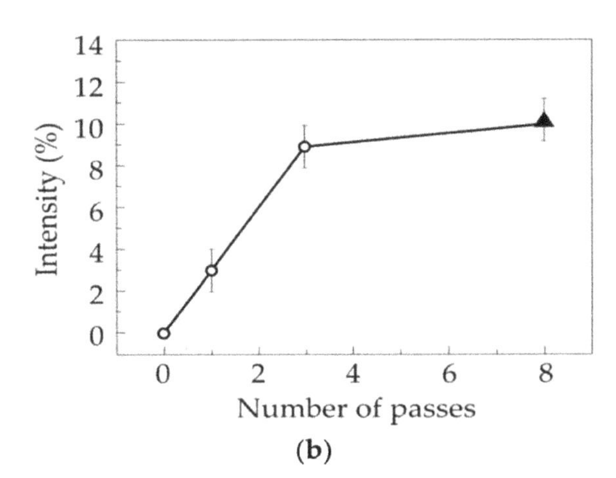

(b)

Figure 6. The positron lifetimes, corresponding to the different components of spectra; (**a**) and the intensities of I_3 component; (**b**) versus number of ECAP passes: $Ti_{49.4}Ni_{50.6}$ (at%) alloy (▲, △, +) and $Ti_{50}Ni_{47.3}Fe_{2.7}$ (at %) alloy (○, ●, ×).

Figure 6 demonstrates the evolution of positron lifetime components for $Ti_{50}Ni_{47.3}Fe_{2.7}$ (at%) and $Ti_{49.4}Ni_{50.6}$ (at%) alloys.

Noteworthy is that the positron lifetime component τ_2 is invariant with the number of ECAP passes within experimental error (Figure 6a) and hence, the defect type identified with τ_2 remains intact during ECAP strain accumulation. The same is observed in many experiments [46], and the τ_2 values correspond to the lifetime of positrons trapping by dislocations.

The lifetime of τ_3 component (Figure 6a) is far longer than the lifetime of delocalized positrons and the lifetime of positrons captured by dislocations. According to experimental data [47], the average positron lifetime in TiNi monovacancies is 195 ps. However, as has been shown by experiments [48,49] and theoretical studies [50], two types of vacancies differing in formation energy arise in B2 TiNi: vacancies V_{Ti} and V_{Ni} on its Ti and Ni sublattices, respectively. From first-principles calculations [50], it follows that the positron lifetimes in V_{Ti} and V_{Ni} vacancies differ greatly. The theoretical lifetime of positrons trapped by V_{Ti} and V_{Ni}, vacancies equal 205 ps and 134 ps, correspondingly. This situation is distinct from what is observed in pure metals with only one vacancy type and in B2-structured transition metal aluminides with very close positron lifetimes in two types of sublattice vacancies [47]. The foregoing suggests that the component τ_3 after three and eight ECAP passes corresponds to single V_{Ti} vacancies. After the first ECAP pass, the positron lifetime τ_3 = 305 ps (Figure 6a) being much longer than its value in monovacancies, can be related to nanoclusters of about five to seven vacancies, if judged from its theoretical dependence on the number of cluster vacancies in bcc metals [25]. Therefore, we have a nontrivial result since research data demonstrate that increasing the number of

ECAP passes increases both, the concentration and size of vacancy clusters in pure metals (copper, nickel, titanium) [30,51,52]. However, this conclusion follows from ECAP at room temperature when the diffusion mobility of vacancies is low and they likely fail to escape from the volume of grains to grain boundaries. The increase in vacancy concentration provides the conditions for vacancy clustering. Such an interpretation makes our results clear. After the first pass of warm ECAP, the grain-subgrain sizes remain rather large for vacancy clustering in the grain volume, but after the next passes, its level allows part of the vacancies to reach the boundaries of grains or dislocations, while the rest part falls short of the amount needed for the formation of noticeable concentrations of their complexes.

Figure 6b shows that the vacancy defect concentration associated with the intensity of the component τ_3 increases steeply in three ECAP passes and then varies slightly. The intensities of the positron lifetime components I_2 and I_3 are proportional to the dislocation density and vacancy defect concentration ($I_2 + I_3 = 1$). As the saturated positron trapping, we cannot accurately estimate the dislocation density and vacancy defect concentration from the available positron trapping model [53]. However, we can estimate the vacancy concentration from the ratio $I_2/I_3 = \nu_d\rho_d/\nu_v c_v$, considering that the specific positron trapping rates by dislocations, and vacancies are equal, respectively to $\nu_d \sim 10^{-4}$ m^{2}s^{-1} and $\nu_v \sim 10^{14}$ s^{-1} in metallic materials [54]. The dislocation density for $Ti_{50}Ni_{50}$ (at%) after eight warm ECAP passes is $\rho_d = 1.1 \times 10^{15}$ m^{-2} according to X-ray diffraction estimates [55]. The ratio I_2/I_3 equal to 10.1 and 9.1 after three, and eight ECAP passes, respectively, suggests that dislocations are the main type of defects for positron capture in TiNi-based alloys exposed to ECAP. At $I_2/I_3 \approx 10$, the estimated relative concentration of V_{Ti} vacancies after ECAP is ~10^{-4}, being many orders of magnitude higher than their thermodynamic equilibrium concentration. Surely, the estimate is rough as the exact values of ν_d and ν_v are unknown for TiN-based alloys.

It stands to reason that V_{Ni} vacancies are also produced in SPD processes, the more so their formation energy is 0.78 eV [48,49] against 0.97 eV for V_{Ti} [48,49]. However, their concentration is unassessable by positron annihilation spectroscopy because of almost the same theoretical lifetimes of delocalized positrons and positrons in V_{Ni} (134 ps). We expect that defect annealing experiments will help separate the contributions of dislocations and V_{Ni} vacancies as they differ in annealing temperature.

4. Conclusions

Even one warm ECAP pass can provide submicrocrystalline fragmentation in initially coarse-grained TiNi-based alloys, and after three passes, their structure represents a mix of grains-subgrains sized to 100–500 nm and 50–100 nm. The dislocation defects and twins contribute to such refinement. The formation of UFG structures in $Ti_{50}Ni_{47.3}Fe_{2.7}$ (at%) does not change the martensite transformation sequence B2 $\leftrightarrow$ R $\leftrightarrow$ B19′ but it decreases the R $\leftrightarrow$ B19′ start temperature by 18 K and narrows the B19′ $\rightarrow$ R temperature interval by 10 K. The B2 $\rightarrow$ R start temperature in this alloy after ECAP remains unchanged.

The refinement of the alloys to UFG structures under ECAP involves a substantial increase in the density of dislocations and vacancy-like defects. Their nanoclusters are detected only after the first ECAP pass, and only free volumes close in size to monovacancies are found after the next passes. The relative vacancy concentration estimated for the TiNi-based alloys after ECAP is about 10^{-4} being many orders of magnitude higher than their thermodynamic equilibrium concentration.

Author Contributions: Conceptualization, A.L. and A.B.; writing—original draft preparation, A.B.; writing—review and editing, A.L. and V.G.; software, R.L.; investigations, A.B., V.K., V.G. and R.L.; project administration, A.L.; funding acquisition, A.L. and R.L. All authors have read and agreed to the published version of the manuscript.

References

1. Valiev, R.Z.; Zhilyaev, A.P.; Langdon, T.G. *Bulk Nanostructured Materials: Fundamentals and Applications*; TMS-Wiley: Hoboken, NJ, USA, 2014.
2. Segal, V.M.; Beyerlein, I.J.; Tome, C.N.; Chuvil'deev, V.N.; Kopylov, V.I. *Fundamentals and Engineering of Severe Plastic Deformation*; Nova: Amityville, NY, USA, 2010.

3. Mohd Jani, J.; Leary, M.; Subic, A.; Gibson, M.A. A review of shape memory alloy research, applications and opportunities. *Mater. Des.* **2014**, *56*, 1078–1113. [CrossRef]
4. Prokoshkin, S.D.; Belousov, M.N.; Abramov, V.Y.; Korotitskii, A.V.; Makushev, S.Y.; Khmelevskaya, I.Y.; Dobatkin, S.V.; Stolyarov, V.V.; Prokof'ev, E.A.; Zharikov, A.I.; et al. Creation of submicrocrystalline structure and improvement of functional properties of shape memory alloys of the Ti–Ni–Fe system with the help of ECAP. *Met. Sci. Heat Treat.* **2007**, *49*, 51–56. [CrossRef]
5. Shahmir, H.; Nili-Ahmadabadi, M.; Langdon, T.G. Shape memory effect of NiTi alloy processed by equal-channel angular pressing followed by post deformation annealing. *IOP Conf. Series: Mater. Sci. Eng.* **2014**, *63*, 012111. [CrossRef]
6. Lukyanov, A.; Gunderov, D.; Prokofiev, E.; Churakova, A.; Pushin, V. Peculiarities of the mechanical behavior of ultrafinegrained and nanocrystalline $Ti_{49.4}Ni_{50.6}$ alloy produced by severe plastic deformation. In Proceedings of the 21st International Conference on Metallurgy and Materials, METAL 2012, Brno, Czech Republic, 23–25 May 2012; pp. 1335–1341.
7. Lotkov, A.I.; Baturin, A.A.; Grishkov, V.N.; Kopylov, V.I.; Timkin, V.N. Influence of equal-channel angular pressing on grain refinement and inelastic properties of TiNi-based alloys. *Izvestiya Visshikh Uchebnikh Zavedenii, Chernaya Metallurgiya* **2014**, *57*, 50–55. [CrossRef]
8. Pushin, V.G.; Valiev, R.Z.; Zhu, U.T.; Gunderov, D.V.; Kourov, N.I.; Kuntsevich, T.E.; Uksusnikov, A.N.; Yurchenko, L.I. Effect of Severe Plastic Deformation on the Behavior of Ti-Ni Shape Memory Alloys. *Mater. Trans.* **2006**, *47*, 694–697. [CrossRef]
9. Fan, Z.; Song, J.; Zhang, X.; Xie, C. Phase Transformations and Super-Elasticity of a Ni-rich TiNi Alloy with Ultrafine-Grained Structure. *Mater. Sci. Forum* **2010**, *667–669*, 1137–1142. [CrossRef]
10. Zhang, D.; Guo, B.; Tong, Y.; Tian, B.; Li, L.; Zheng, Y.; Gunderov, D.V.; Valiev, R.Z. Effect of annealing temperature on martensitic transformation of $Ti_{49.2}Ni_{50.8}$ alloy processed by equal channel angular pressing. *Trans. Nonferrous Met. Soc. Chin.* **2016**, *26*, 448–455. [CrossRef]
11. Karaman, I.; Kulkarni, A.V.; Luo, Z.P. Transformation behaviour and unusual twinning in a NiTi shape memory alloy ausformed using equal channel angular extrusion. *Phil. Mag.* **2005**, *85*, 1729–1745. [CrossRef]
12. Khmelevskaya, I.Y.; Prokoshkin, S.D.; Trubitsyna, I.B.; Belousov, M.N.; Dobatkin, S.V.; Tatyanin, E.V.; Korotitskiy, A.V. Structure and properties of Ti-Ni-based alloys after equal-channel angular pressing and high pressure torsion. *Mater. Sci. Eng. A* **2008**, *481–482*, 119–122. [CrossRef]
13. Prokofyev, E.; Gunderov, D.; Prokoshkin, S.; Valiev, R. Microstructure, mechanical and functional properties of NiTi alloys processed by ECAP technique. In Proceedings of the 8th European Symposium on Martensitic Transformations, ESOMAT 2009, Prague, Czech Republic, 7–11 September 2009. [CrossRef]
14. Zhang, Y.; Jiang, S. The Mechanism of Inhomogeneous Grain Refifinement in a NiTiFe Shape Memory Alloy Subjected to Single-Pass Equal-Channel Angular Extrusion. *Metals* **2017**, *7*, 400. [CrossRef]
15. Churakova, A.; Yudahina, A.; Kayumova, E.; Tolstov, N. Mechanical behavior and fractographic analysis of a TiNi alloy with various thermomechanical treatment. In Proceedings of the International Conference on Modern Trends in Manufacturing Technologies and Eguipment: Mechanical Engineering and Materials Science, ICMTMTE 2019, Sevastopol, Russia, 9–13 September 2019. [CrossRef]
16. Lucas, F.L.C.; Guido, V.; Käfer, K.A.; Bernardi, H.H.; Otubo, J. ECAE Processed NiTi Shape Memory Alloy. *Mater. Res.* **2014**, *17* (Suppl. 1), 186–190. [CrossRef]
17. Valiev, R.Z.; Langdon, T.G. Principles of equal-channel angular pressing as a processing tool for grain refifinement. *Progr. Mater. Sci.* **2006**, *51*, 881–981. [CrossRef]
18. Li, Z.; Cheng, X.; ShangGuan, Q. Effects of heat treatment and ECAE process on transformation behaviors of TiNi shape memory alloy. *Mater. Lett.* **2005**, *59*, 705–709. [CrossRef]
19. Koike, J.; Parkin, D.M.; Nastasi, M. Crystal-to-amorphous transformation of NiTi induced by cold rolling. *J. Mater. Res.* **1990**, *5*, 1414–1418. [CrossRef]
20. Čížek, J.; Janeček, M.; Vlasák, T.; Smola, B.; Melikhova, O.; Islamgaliev, R.K.; Dobatkin, S.V. The Development of Vacancies during Severe Plastic Deformation. *Mater. Trans.* **2019**, *60*, 1533–1542. [CrossRef]
21. Zehetbauer, M.J.; Steiner, G.; Schafler, G.; Korznikov, A.; Korznikova, E. Deformation Induced Vacancies with Severe Plastic Deformation: Measurements and Modeling. *Mater. Sci. Forum* **2006**, *503–504*, 57–65. [CrossRef]
22. Xue, K.-M.; Wang, B.-X.-T.; Yan, S.-L.; Bo, D.-Q.; Li, P. Strain-induced dissolution and precipitation of secondary phases and synergetic stengthening mechanisms of Al-Zn-Mg-Cu alloy during ECAP. *Adv. Eng. Mater.* **2019**, 1801182. [CrossRef]

23. Straumal, B.B.; Pontikis, V.; Kilmametov, A.R.; Mazilkin, A.A.; Dobatkin, S.V.; Baretzky, B. Competition between precipitation and dissolution in Cu–Ag alloys under high pressure torsion. *Acta Mater.* **2017**, *122*, 60–71. [CrossRef]
24. Farber, V.M. Contribution of diffusion processes to structure formation in intense cold plastic deformation of metals. *Metal Sci. Heat Treat.* **2002**, *44*, 317–323. [CrossRef]
25. Čížek, J.; Melikhova, O.; Barnovská, Z.; Procházka, I.; Islamgaliev, R.K. Vacancy clusters in ultra-fine grained metals prepared by severe plastic deformation. *J. Phys. Conf. Ser.* **2013**, *443*, 012008. [CrossRef]
26. Betekhtin, V.I.; Kadomtsev, A.G.; Kral, P.; Dvorak, J.; Svoboda, M.; Saxl, I.; Sklenička, V. Significance of Microdefects Induced by ECAP in Aluminium, Al 0.2%Sc Alloy and Copper. *Mater. Sci. Forum* **2008**, *567–568*, 93–96. [CrossRef]
27. Betekhtin, V.I.; Kadomtsev, A.G.; Narykova, M.V.; Amosova, O.V.; Sklenicka, V. Defect Structure and Mechanical Stability of Microcrystalline Titanium Produced by Equal Channel Angular Pressing. *Technol. Phys. Lett.* **2017**, *43*, 61–63. [CrossRef]
28. Kuznetsov, P.V.; Mironov, Y.P.; Tolmachev, A.I.; Bordulev, Y.S.; Laptev, R.S.; Lider, A.M.; Korznikov, A.V. Positron spectroscopy of defects in submicrocrystalline nickel after low-temperature annealing. *Phys. Sol. State* **2015**, *57*, 219–228. [CrossRef]
29. Reglitz, G.; Oberdorfer, B.; Fleischmann, N.; Kotzurek, J.A.; Divinski, S.V.; Sprengel, W.; Wilde, G.; Würschum, R. Combined volumetric, energetic and microstructural defect analysis of ECAP-processed nickel. *Acta Mater.* **2016**, *103*, 396–406. [CrossRef]
30. Lukáč, F.; Čížek, J.; Knapp, J.; Procházka, I.; Zháňal, P.; Islamgaliev, R.K. Ultra-fine grained Ti prepared by severe plastic deformation. *J. Phys. Conf. Ser.* **2016**, *674*, 012007. [CrossRef]
31. Bartha, K.; Zháňal, P.; Stráský, J.; Čížek, J.; Dopita, M.; Lukáč, F.; Harcuba, P.; Hájek, M.; Polyakova, V.; Semenova, I.; et al. Lattice defects in severely deformed biomedical Ti-6Al-7Nb alloy and thermal stability of its ultra-fine grained microstructure. *J. Alloys Compd.* **2019**, *788*, 881–890. [CrossRef]
32. Domınguez-Reyes, R.; Savoini, B.; Monge, M.Á.; Muñoz, Á.; Ballesteros, C. Thermal Stability Study of Vacancy Type Defects in Commercial Pure Titanium Using Positron Annihilation Spectroscopy. *Adv. Eng. Mater.* **2017**, *19*, 1500649. [CrossRef]
33. Gubicza, J.; Ungár, T. Characterization of defect structures in nanocrystalline materials by X-ray line profile analysis. *Z. Kristallogr.* **2007**, *222*, 567–579. [CrossRef]
34. Petry, W.; Brussler, M.; Gröger, V.; Müller, H.G.; Vogl, G. The nature of point defects produced by cold working of metals studied with Mössbauer spectroscopy and perturbed γ-γ angular correlation. *Hyperfine Interact.* **1983**, *15–16*, 371–374. [CrossRef]
35. Schaefer, H.-E.; Baier, F.; Müller, M.A.; Reichle, K.J.; Reimann, K.; Rempel, A.A.; Sato, K.; Ye, F.; Zhang, X.; Sprengel, W. Vacancies and atomic processes in intermetallics–From crystals to quasicrystals and bulk metallic glasses. *Phys. Stat. Sol. B* **2011**, *48*, 2290–2299. [CrossRef]
36. Würschum, R.; Greiner, W.; Valiev, R.Z.; Rapp, M.; Sigle, W.; Schneeweiss, O.; Schaefer, H.-E. Interfacial free volumes in ultra-fine grained metals prepared by severe plastic deformation, by spark erosion, or by crystallization of amorphous alloys. *Scr. Metal. Mater.* **1991**, *251*, 2451–2456. [CrossRef]
37. Gammer, C.; Karnthaler, H.P.; Rentenberger, C. Reordering a deformation disordered intermetallic compound by antiphase boundary movement. *J. Alloys Compd.* **2017**, *713*, 148–155. [CrossRef]
38. Collins, G.S.; Sinha, P. Structural, thermal and deformation-induced point defects in PdIn. *Hyperfine Interactions* **2000**, *130*, 151–179. [CrossRef]
39. Lotkov, A.I.; Baturin, A.A.; Grishkov, V.N.; Kopylov, V.I. Possible role of crystal structure defects in grain structure nanofragmentation under severe cold plastic deformation of metals and alloys. *Phys. Mesomech.* **2007**, *10*, 179–189. [CrossRef]
40. Lotkov, A.I.; Baturin, A.A.; Grishkov, V.N.; Kuznetsov, P.V.; Klimenov, V.A.; Panin, V.E. Structural defects and mesorelief of the titanium nickelide surface after severe plastic deformation by an ultrasonic method. *Fizicheskaya Mesomekhanika* **2005**, *8*, 109–112.
41. Laptev, R.S.; Lider, A.M.; Bordulev, Y.S.; Kudiiarov, V.N.; Garanin, G.V.; Wang, W.; Kuznetsov, P.V. Investigation of defects in hydrogen-saturated titanium by means of positron annihilation techniques. *Defect Diffus. Forum* **2015**, *365*, 232–236. [CrossRef]
42. Giebel, D.; Kansy, J. LT10 Program for Solving Basic Problems Connected with Defect Detection. *Phys. Procedia* **2012**, *35*, 122–127. [CrossRef]

43. Würschum, R.; Donner, P.; Hornbogen, E.; Schaefer, H.-E. Vacancy Studies in Melt-Spun Shape Memory Alloys. *Mater. Sci. Forum* **1992**, *105–110*, 1333–1336. [CrossRef]
44. Mizuno, M.; Araki, H.; Shirai, Y. Theoretical calculation of positron lifetimes for defects in solids. *Adv. Quant. Chem.* **2003**, *42*, 109–126. [CrossRef]
45. Kalchikhin, V.V.; Kulkova, S.E. Kinetic properties and characteristics of electron-positron annihilation in NiMn and NiTi. *Russ. Phys. J.* **1992**, *35*, 922–926. [CrossRef]
46. Janeček, M.; Stráský, J.; Cizek, J.; Harcuba, P.; Václavová, K.; Polyakova, V.V.; Semenova, I.P. Mechanical Properties and Dislocation Structure Evolution in Ti6Al7Nb Alloy Processed by High Pressure Torsion. *Met. Mat. Trans. A* **2014**, *45*, 7–15. [CrossRef]
47. Würschum, R.; Badura-Gergen, K.; Kümmerle, E.; Grupp, C.; Schaefer, H.-E. Characterization of radiation-induced lattice vacancies in intermetallic compounds by means of positron-lifetime studies. *Phys. Rev. B* **1996**, *54*, 849–856. [CrossRef]
48. Baturin, A.A.; Lotkov, A.I. Determination of vacancy formation energy in TiNi compound with B2 structure by positron annihilation. *Fiz. Met. Metalloved.* **1993**, *76*, 168–170.
49. Baturin, A.; Lotkov, A.; Grishkov, V.; Lider, A. Formation of vacancy-type defects in titanium nickelide. In Proceedings of the 10th European Symposium on Martensitic Transformations, ESOMAT 2015, Antverp, Belgium, 14–18 September 2015. [CrossRef]
50. Lu, J.M.; Hu, Q.M.; Wang, L.; Li, Y.J.; Xu, D.S.; Yang, R. Point defects and their interaction in TiNi from first-principles calculations. *Phys. Rev. B* **2007**, *75*, 094108. [CrossRef]
51. Janeček, M.; Čížek, J.; Dopita, M.; Král, R.; Srba, O. Mechanical properties and microstructure development of ultrafine-grained Cu processed by ECAP. *Mater. Sci. Forum* **2008**, *584–586*, 440–445. [CrossRef]
52. Setman, D.; Schafler, E.; Korznikova, E.; Zehetbauer, M.J. The presence and nature of vacancy type defects in nanometals detained by severe plastic deformation. *Mater. Sci. Eng. A* **2008**, *493*, 116–122. [CrossRef]
53. Hautojärvi, P.; Corbel, C. Positron Spectroscopy in Metals and Superconductors. In *Positron Spectroscopy of Solids, Proceedings of the Internation School of Physics "Enrico Fermi", Course CXXV, Varenna, Italy, 1993*; Dupasquier, A., Mills, A.P., Eds.; IOS: Amsterdam, The Nétherlands, 1995; pp. 491–532. [CrossRef]
54. Čížek, J.; Janeček, M.; Srba, O.; Kužel, R.; Barnovská, Z.; Procházka, I.; Dobatkin, S. Evolution of defects in copper deformed by high-pressure torsion. *Acta Mater.* **2011**, *59*, 2322–2329. [CrossRef]
55. Churakova, A.; Gunderov, D. Increase in the dislocation density and yield stress of the $Ti_{50}Ni_{50}$ alloy caused by thermal cycling. *Mater. Today: Proc.* **2017**, *4*, 4732–4736. [CrossRef]

Permissions

All chapters in this book were first published by MDPI; hereby published with permission under the Creative Commons Attribution License or equivalent. Every chapter published in this book has been scrutinized by our experts. Their significance has been extensively debated. The topics covered herein carry significant findings which will fuel the growth of the discipline. They may even be implemented as practical applications or may be referred to as a beginning point for another development.

The contributors of this book come from diverse backgrounds, making this book a truly international effort. This book will bring forth new frontiers with its revolutionizing research information and detailed analysis of the nascent developments around the world.

We would like to thank all the contributing authors for lending their expertise to make the book truly unique. They have played a crucial role in the development of this book. Without their invaluable contributions this book wouldn't have been possible. They have made vital efforts to compile up to date information on the varied aspects of this subject to make this book a valuable addition to the collection of many professionals and students.

This book was conceptualized with the vision of imparting up-to-date information and advanced data in this field. To ensure the same, a matchless editorial board was set up. Every individual on the board went through rigorous rounds of assessment to prove their worth. After which they invested a large part of their time researching and compiling the most relevant data for our readers.

The editorial board has been involved in producing this book since its inception. They have spent rigorous hours researching and exploring the diverse topics which have resulted in the successful publishing of this book. They have passed on their knowledge of decades through this book. To expedite this challenging task, the publisher supported the team at every step. A small team of assistant editors was also appointed to further simplify the editing procedure and attain best results for the readers.

Apart from the editorial board, the designing team has also invested a significant amount of their time in understanding the subject and creating the most relevant covers. They scrutinized every image to scout for the most suitable representation of the subject and create an appropriate cover for the book.

The publishing team has been an ardent support to the editorial, designing and production team. Their endless efforts to recruit the best for this project, has resulted in the accomplishment of this book. They are a veteran in the field of academics and their pool of knowledge is as vast as their experience in printing. Their expertise and guidance has proved useful at every step. Their uncompromising quality standards have made this book an exceptional effort. Their encouragement from time to time has been an inspiration for everyone.

The publisher and the editorial board hope that this book will prove to be a valuable piece of knowledge for researchers, students, practitioners and scholars across the globe.

List of Contributors

Andrea Školáková, Jana Körberová, Jan Pinc, Pavel Salvetr and Pavel Novák
Department of Metals and Corrosion Engineering, University of Chemistry and Technology Prague, Technická 5, 166 28 Prague 6, Czech Republic

Jaroslav Málek
UJP Praha a.s., Nad Kamínkou 1345, 156 10 Prague 16, Czech Republic

Dana Rohanová and Eva Gregorová
Department of Glass and Ceramics, University of Chemistry and Technology Prague, Technická 5, 166 28 Prague 6, Czech Republic

Eva Jablonská
Department of Biochemistry and Microbiology, University of Chemistry and Technology Prague, Technická 5, 166 28 Prague 6, Czech Republic

Hui Wang, Haiyou Huang and Jianxin Xie
Key Laboratory for Advanced Materials Processing of the Ministry of Education, Institute for Advanced Materials and Technology, University of Science and Technology Beijing, Beijing 100083, China

Mien-Chung Chen, Yang-Chun Chiu, Tse-An Pan and Sheng-Long Lee
Institute of Material Science and Engineering, National Central University, Taoyuan 320, Taiwan

Ming-Che Wen
Department of Mechanical Engineering, National Central University, Taoyuan 320, Taiwan

Yu-Chih Tzeng
Department of Power Vehicle and Systems Engineering, Chung-Cheng Institute of Technology, National Defense University, Taoyuan 334, Taiwan

Sergey Zherebtsov, Nikita Stepanov, Dmitry Shaysultanov, Nikita Yurchenko, Margarita Klimova and Gennady Salishchev
Laboratory of Bulk Nanostructured Materials, Belgorod State University, 308015 Belgorod, Russia

Yulia Ivanisenko
Karlsruhe Institute of Technology, Institute of Nanotechnology, 76021 Karlsruhe, Germany

Gaetano Palumbo
Department of Chemistry and Corrosion of Metals, Faculty of Foundry Engineering, AGH University of Science and Technology, 30-059 Krakow, Poland

Kamila Kollbek
Academic Centre for Materials and Nanotechnology, AGH University of Science and Technology, Mickiewicza St. 30, 30-059 Kraków, Poland

Andrzej Bernasik
Department of Condensed Matter Physics, Faculty of Physics and Applied Computer Science, AGH University of Science and Technology, Mickiewicza St. 30, 30-059 Krakow, Poland

Roma Wirecka
Academic Centre for Materials and Nanotechnology, AGH University of Science and Technology, Mickiewicza St. 30, 30-059 Kraków, Poland
Department of Condensed Matter Physics, Faculty of Physics and Applied Computer Science, AGH University of Science and Technology, Mickiewicza St. 30, 30-059 Krakow, Poland

Marcin Górny
Department of Cast Alloys and Composites Engineering, Faculty of Foundry Engineering, AGH University of Science and Technology, 30-059 Krakow, Poland

Siqian Zhang, Haoyu Zhang, Junhong Hao, Jing Liu and Lijia Chen
School of Materials Science and Engineering, Shenyang University of Technology, Shenyang 110870, China

Jie Sun
State Key Laboratory of Rolling and Automation, Northeastern University, Shenyang 110004, China

Alexandra Fedoseeva, Nadezhda Dudova, Rustam Kaibyshev and Andrey Belyakov
Laboratory of Mechanical Properties of Nanostructured Materials and Superalloys, Belgorod National Research University, Pobeda 85, Belgorod 308015, Russia

Paixian Fu, Hongwei Liu, Hanghang Liu and Yanfei Cao
Shenyang National Laboratory for Materials Science, Institute of Metal Research, Chinese Academy of Sciences, Shenyang 110016, China

Chen Sun and Ningyu Du
Shenyang National Laboratory for Materials Science, Institute of Metal Research, Chinese Academy of Sciences, Shenyang 110016, China
School of Materials Science and Engineering, University of Science and Technology of China, Hefei 230026, China

Marina Fedorischeva and Olga Perevalova
Institute of Strength Physics and Materials Science SB RAS, 634055 Tomsk, Russia

Mark Kalashnikov, Irina Bozhko and Victor Sergeev
Institute of Strength Physics and Materials Science SB RAS, 634055 Tomsk, Russia
Department of Materials Science, National Research Tomsk Polytechnic University, 634050 Tomsk, Russia

Hong He, Long Zhang, Shikang Li and Luoxing Li
State Key Laboratory of Advanced Design and Manufacturing for Vehicle Body, College of Mechanical and Vehicle Engineering, Hunan University, Changsha 410082, China

Hui Zhang
College of Materials Science and Engineering, Hunan University, Changsha 410082, China

Chaoyang Li and Guangjie Huang
International Joint Laboratory for Light Alloys (Ministry of Education), College of Materials Science and Engineering, Chongqing University, Chongqing 400044, China

Lingfei Cao
International Joint Laboratory for Light Alloys (Ministry of Education), College of Materials Science and Engineering, Chongqing University, Chongqing 400044, China
Shenyang National Laboratory for Materials Science, Chongqing University, Chongqing 400044, China

Xiaodong Wu
State Key Laboratory of Advanced Design and Manufacturing for Vehicle Body, College of Mechanical and Vehicle Engineering, Hunan University, Changsha 410082, China
International Joint Laboratory for Light Alloys (Ministry of Education), College of Materials Science and Engineering, Chongqing University, Chongqing 400044, China
Shenyang National Laboratory for Materials Science, Chongqing University, Chongqing 400044, China

Bin Liao
Alnan Aluminium Co., Ltd., Nanning Guangxi 530031, China

Nanpu Cheng
School of Materials and Energy, Southwest University, Chongqing 400715, China

Runar Larsen Broks
Department of Materials Science and Engineering, Norwegian University of Science and Technology (NTNU), NO-7491 Trondheim, Norway

Christian Oen Paulsen and Ida Westermann
Department of Materials Science and Engineering, Norwegian University of Science and Technology (NTNU), NO-7491 Trondheim, Norway
Centre for Advanced Structural Analysis (CASA), Norwegian University of Science and Technology (NTNU), NO-7491 Trondheim, Norway

Morten Karlsen
Department of Materials Science and Engineering, Norwegian University of Science and Technology (NTNU), NO-7491 Trondheim, Norway
Equinor ASA, NO-7053 Trondheim, Norway

Jarle Hjelen
Department of Materials Science and Engineering, Norwegian University of Science and Technology (NTNU), NO-7491 Trondheim, Norway
Department of Geoscience and Petroleum, Norwegian University of Science and Technology (NTNU), NO-7491 Trondheim, Norway

Changsheng Tan, Yiduo Fan and Guojun Zhang
School of Materials Science and Engineering, Xi'an University of Technology, Xi'an 710048, China

Qiaoyan Sun
State Key Laboratory for Mechanical Behavior of Materials, Xi'an Jiaotong University, Xi'an 710049, China

Xiaofang Shi, Weiwei Hu and Yun Tan
College of Material and Metallurgy, Guizhou University, Guiyang 550025, China

Zhenglai Zhang
Zhejiang Huashuo Technology Co., Ltd., Ningbo 315000, China

Wei Li
College of Material and Metallurgy, Guizhou University, Guiyang 550025, China
Zhejiang Huashuo Technology Co., Ltd., Ningbo 315000, China

Liang Tian
Guizhou Province Technology Innovation Service Center, Guiyang 550004, China

Olivia D. Underwood
Connector & LAC Technology, Sandia National Laboratories, Albuquerque, NM 87185, USA

Jonathan D. Madison
Materials Mechanics, Sandia National Laboratories, Albuquerque, NM 87185, USA

Gregory B. Thompson
Metallurgical and Materials Engineering, The University of Alabama, Tuscaloosa, AL 35487, USA

Marcin Stawarz
Department of Foundry Engineering, Silesian University of Technology, 7 Towarowa Street, 44-100 Gliwice, Poland

Paweł M. Nuckowski
Department of Engineering Materials and Biomaterials, Silesian University of Technology, 18A Konarskiego Street, 44-100 Gliwice, Poland

Vladimir Torganchuk
Belgorod State University, Belgorod 308015, Russia

Aleksandr M. Glezer
National University of Science & Technology (MISIS), Moscow 119049, Russia

Aleksandr Lotkov, Anatoly Baturin and Victor Grishkov
Institute of Strength Physics and Materials Science of the Siberian Branch of the Russian Academy of Sciences, 634055 Tomsk, Russia

Vladimir Kopylov
Physical-Technical Institute of the National Academy of Sciences of Belarus State Scientific Institution, 220141 Minsk, Belarus

Roman Laptev
National Research Tomsk Polytechnic University, 634050 Tomsk, Russia

Index

P

R

S

T

U

Y